1D and 2D Nanomaterials for Sensor Applications

1D and 2D Nanomaterials for Sensor Applications

Editors

Andrew F. Zhou
Peter X. Feng

Basel • Beijing • Wuhan • Barcelona • Belgrade • Novi Sad • Cluj • Manchester

Editors

Andrew F. Zhou
Department of Chemistry,
Biochemistry, and Physics
Indiana University
of Pennsylvania
Indiana
United States

Peter X. Feng
Department of Physics
University of Puerto Rico
San Juan
United States

Editorial Office
MDPI AG
Grosspeteranlage 5
4052 Basel, Switzerland

This is a reprint of articles from the Special Issue published online in the open access journal *Crystals* (ISSN 2073-4352) (available at: https://www.mdpi.com/journal/crystals/special_issues/1D_2D_Nanomaterials_Sensor_Applications).

For citation purposes, cite each article independently as indicated on the article page online and as indicated below:

Lastname, A.A.; Lastname, B.B. Article Title. *Journal Name* **Year**, *Volume Number*, Page Range.

ISBN 978-3-7258-1708-5 (Hbk)
ISBN 978-3-7258-1707-8 (PDF)
doi.org/10.3390/books978-3-7258-1707-8

© 2024 by the authors. Articles in this book are Open Access and distributed under the Creative Commons Attribution (CC BY) license. The book as a whole is distributed by MDPI under the terms and conditions of the Creative Commons Attribution-NonCommercial-NoDerivs (CC BY-NC-ND) license.

Contents

About the Editors vii

Preface ix

Andrew F. Zhou and Peter X. Feng
One-Dimensional and Two-Dimensional Nanomaterials for Sensor Applications
Reprinted from: *Crystals* 2024, *14*, 622, doi:10.3390/cryst14070622 1

Sedigheh Barzegar, Mahmood Karimi Abdolmaleki, William B. Connick and Ghodratollah Absalan
Enhancing Vapochromic Properties of Platinum(II) Terpyridine Chloride Hexaflouro Phosphate in Terms of Sensitivity through Nanocrystalization for Fluorometric Detection of Acetonitrile Vapors
Reprinted from: *Crystals* 2024, *14*, 347, doi:10.3390/cryst14040347 4

Vera M. Kalygina, Alexander V. Tsymbalov, Petr M. Korusenko, Aleksandra V. Koroleva and Evgeniy V. Zhizhin
Effect of Traps on the UV Sensitivity of Gallium Oxide-Based Structures
Reprinted from: *Crystals* 2024, *14*, 268, doi:10.3390/cryst14030268 23

Shuo Li and Hai Yang
Strain-Modulated Electronic Transport Properties in Two-Dimensional Green Phosphorene with Different Edge Morphologies
Reprinted from: *Crystals* 2024, *14*, 239, doi:10.3390/cryst14030239 38

Pham Hong Thach and Tran Van Khai
Thermal Evaporation Synthesis, Optical and Gas-Sensing Properties of ZnO Nanowires
Reprinted from: *Crystals* 2023, *13*, 1380, doi:10.3390/cryst13091380 47

Vera Kalygina, Sergey Podzyvalov, Nikolay Yudin, Elena Slyunko, Mikhail Zinoviev, Vladimir Kuznetsov, et al.
Effect of UV and IR Radiation on the Electrical Characteristics of $Ga_2O_3/ZnGeP_2$ Hetero-Structures
Reprinted from: *Crystals* 2023, *13*, 1203, doi:10.3390/cryst13081203 68

Yuanyuan Xiong, Mengxiao Chen, Zhen Mao, Yiqing Deng, Jing He, Huaixuan Mu, et al
Synthesis of Up-Conversion Fluorescence N-Doped Carbon Dots with High Selectivity and Sensitivity for Detection of Cu^{2+} Ions
Reprinted from: *Crystals* 2023, *13*, 812, doi:10.3390/cryst13050812 80

Eric Y. Li, Andrew F. Zhou and Peter X. Feng
High-Performance Nanoplasmonic Enhanced Indium Oxide—UV Photodetectors
Reprinted from: *Crystals* 2023, *13*, 689, doi:10.3390/cryst13040689 98

Majid Zarei, Seyedeh M. Hamidi and K. -W. -A. Chee
Colorimetric Plasmonic Hydrogen Gas Sensor Based on One-Dimensional Nano-Gratings
Reprinted from: *Crystals* 2024, *14*, 363, doi:10.3390/cryst13020363 108

Sujit Kumar Saini and Suneet Kumar Awasthi
Sensing and Detection Capabilities of One-Dimensional Defective Photonic Crystal Suitable for Malaria Infection Diagnosis from Preliminary to Advanced Stage: Theoretical Study
Reprinted from: *Crystals* 2023, *13*, 128, doi:10.3390/cryst13010128 119

Cristina Dumitriu, Alexandra Constantinescu, Alina Dumitru and Cristian Pîrvu
Modified Electrode with ZnO Nanostructures Obtained from Silk Fibroin for Amoxicillin Detection
Reprinted from: *Crystals* **2022**, *12*, 1511, doi:10.3390/cryst12111511 . . . . **136**

Vanga Ganesh, Mai S. A. Hussien, Ummar Pasha Shaik, Ramesh Ade, Mervat I. Mohammed, Thekrayat H. AlAbdulaal, et al.
Impact of Mo-Doping on the Structural, Optical, and Electrocatalytic Degradation of ZnO Nanoparticles: Novel Approach
Reprinted from: *Crystals* **2022**, *12*, 1239, doi:10.3390/cryst12091239 . . . . **150**

About the Editors

Andrew F. Zhou

Dr. Andrew F. Zhou is a professor of Physics at Indiana University of Pennsylvania, specializing in optics and solid-state lasers. His current research focuses on nanophotonics and the application of 1D and 2D nanomaterials in sensor development. He has published extensively in his field, aiming to advance sensor technology through innovative material synthesis and device fabrication.

Peter X. Feng

Dr. Peter X. Feng is a professor of Physics at the University of Puerto Rico, Rio Piedras campus. Over the past decade, his research has concentrated on the plasma processing of nanomaterials and their applications in gas sensors and photodetectors. Currently, his primary interest lies in studying the super-wide bands and super-broad bandwidths of materials.

Preface

Rapid advancements in nanotechnology have opened up new frontiers in sensor technologies, particularly with one-dimensional (1D) and two-dimensional (2D) nanomaterials. This reprint, "1D and 2D Nanomaterials for Sensor Applications", compiles 12 selected papers from the Special Issue of the MDPI *Journal of Crystals*, showcasing the latest research in this field.

This reprint explores advancements in the unique synthesis and preparation techniques of 1D and 2D nanomaterials, novel integration routes for new and enhanced functionalities, and advanced device simulation and design strategies for sensing applications. It highlights the diverse applications and vast potential of 1D and 2D nanomaterials in sensing, emphasizing the importance of continued research. Applications in various sensing fields, such as gas sensing, biosensing, and environmental monitoring, are demonstrated.

We extend our sincere gratitude to the reviewers for their invaluable comments and suggestions, which significantly improved the quality of the papers. Special thanks to Dr. Supattra Panthai, the section managing editor, whose dedicated efforts were instrumental in the success of this Special Issue and reprint.

We hope this compilation serves as a valuable resource for researchers and scientists in academia, industry, and government labs, providing a deeper understanding of nanomaterial research and its applications in sensor technology. We trust that readers will find this reprint both informative and inspiring, paving the way for future innovations in nanomaterials and sensors.

Andrew F. Zhou and Peter X. Feng
Editors

Editorial

One-Dimensional and Two-Dimensional Nanomaterials for Sensor Applications

Andrew F. Zhou [1],* and Peter X. Feng [2]

1 Department of Chemistry, Biochemistry, and Physics, Indiana University of Pennsylvania, Indiana, PA 15705, USA
2 Department of Physics, University of Puerto Rico, San Juan, PR 00936, USA; peter.feng@upr.edu
* Correspondence: fzhou@iup.edu

Citation: Zhou, A.F.; Feng, P.X. One-Dimensional and Two-Dimensional Nanomaterials for Sensor Applications. *Crystals* 2024, 14, 622. https://doi.org/10.3390/cryst14070622

Received: 4 July 2024
Accepted: 5 July 2024
Published: 6 July 2024

Copyright: © 2024 by the authors. Licensee MDPI, Basel, Switzerland. This article is an open access article distributed under the terms and conditions of the Creative Commons Attribution (CC BY) license (https://creativecommons.org/licenses/by/4.0/).

The significance of 1D and 2D nanomaterials in sensor technology lies in their unique properties and the potential for high-performance sensing [1,2]. These materials offer several advantages over traditional sensor materials, including a high surface-to-volume ratio, quantum confinement effects, mechanical flexibility, versatile synthesis methods, and multifunctional capabilities [3–5]. This Special Issue entitled "One-Dimensional and Two-Dimensional Nanomaterials for Sensor Applications" presents a collection of 11 papers that delve into various aspects of nanomaterials in sensing applications. These studies cover the synthesis, characterization, and application of different nanomaterials, highlighting their potential to enhance sensor performance, showcasing the latest research in the field, and providing insights into future directions.

This issue focuses on the synthesis and fabrication of 1D and 2D materials, such as nanotubes, nanowires, nanorods, graphene, and other 2D materials. These studies explore different synthesis methods and their impact on the material's properties and sensing capabilities. For example, Ganesh et al. present a study on the synthesis and characterization of pure and Mo-doped ZnO nanoparticles, which have potential applications in photocatalysis, water purification, and sensor technology (contribution 1). Dumitriu et al. report on the fabrication of ZnO nanostructures derived from silk fibroin for amoxicillin sensing, demonstrating good performance for electrochemical detection (contribution 2). Xiong et al. demonstrated the synthsis of N-doped carbon dots using a one-step hydrothermal method for Cu^{2+} ion detection with better selectivity and sensitivity (contribution 6).

The published papers highlight the research on surface functionalization and defect engineering of nanomaterials for enhanced sensor performance. This includes modifying the surface of nanomaterials to improve their selectivity, sensitivity, and stability towards specific analytes. For instance, Saini et al. explore the sensing and detection capabilities of a one-dimensional defective photonic crystal for malaria infection diagnosis, based on the minute sensing of the refractive index of different RBC samples (contribution 3). Zarei et al. present a colorimetric plasmonic hydrogen gas sensor based on a one-dimensional nano-grating and thin-film Pd, which shows high sensitivity and selectivity for hydrogen gas detection (contribution 4).

The applications of 1D and 2D nanomaterials in various sensing fields have been explored, such as gas sensing, biosensing, and environmental monitoring. The papers in this Special Issue demonstrate the potential of these materials in real-world applications. Li et al. report a high-performance In_2O_3 UV photodetector with Pt nanoparticle surface functionalization, which shows excellent responsivity and repeatability to a wide range of UV lights (contribution 5). Kalygina et al. study the effect of UV and IR radiation on the electrical characteristics of $Ga_2O_3/ZnGeP_2$ heterostructures, which can be used as radiation detectors in the IR range (contribution 7). Thach and Khai report on the synthesis and characterization of ZnO nanowires and their application as gas sensors for NO_2 detection (contribution 8).

This Special Issue also highlights the need for collaboration between different disciplines to address the challenges and opportunities in this area. Li and Yang's study on the strain-modulated electronic transport properties of two-dimensional green phosphorene with different edge morphologies provides valuable insights into the design of electronic nano-devices (contribution 9). Kalygina et al.'s investigation of resistive metal/β-Ga_2O_3/metal structures with different interelectrode distances and electrode topologies for UV sensing contributes to the development of efficient UV detectors (contribution 10). Barzegar et al.'s work on the synthesis and characterization of Pt(tpy)Cl nanocrystals and their application as a vapochromic sensor for the detection of acetonitrile vapors showcases the potential of nanomaterials in chemical sensing (contribution 11).

Overall, the studies presented in this Special Issue highlight the diverse range of applications and the potential of 1D and 2D nanomaterials in sensing, and emphasize the importance of continued research in this field to further advance the development of nanomaterial-based sensors. However, there are still challenges to be addressed, such as improving the sensitivity, selectivity, and stability of sensors, as well as developing scalable and cost-effective fabrication methods [6,7]. Future research should focus on these areas to realize the full potential of nanomaterials in sensor technology. Additionally, interdisciplinary collaboration between materials science, chemistry, physics, and engineering will be crucial for the successful translation of these research findings into practical sensor applications [8–10].

Conflicts of Interest: The editors declare no conflicts of interest.

List of Contributions

1. Ganesh, V.; Hussien, M.S.; Shaik, U.P.; Ade, R.; Mohammed, M.I.; AlAbdulaal, T.H.; Zahran, H.Y.; Yahia, I.S.; Abdel-wahab, M.S. Impact of Mo-Doping on the Structural, Optical, and Electrocatalytic Degradation of ZnO Nanoparticles: Novel Approach. *Crystals* **2022**, *12*, 1239.
2. Dumitriu, C.; Constantinescu, A.; Dumitru, A.; Pîrvu, C. ZnO Nanostructures Derived from Silk Fibroin for Amoxicillin Sensing. *Crystals* **2022**, *12*, 1511.
3. Saini, S.K.; Awasthi, S.K. Sensing and Detection Capabilities of One-Dimensional Defective Photonic Crystal Suitable for Malaria Infection Diagnosis from Preliminary to Advanced Stage: Theoretical Study. *Crystals* **2023**, *13*, 128.
4. Zarei, M.; Hamidi, S.M.; Chee, K.W.A. Colorimetric Plasmonic Hydrogen Gas Sensor Based on One-Dimensional Nano-Gratings. *Crystals* **2023**, *13*, 363.
5. Li, E.Y.; Zhou, A.F.; Feng, P.X. High-Performance Nanoplasmonic Enhanced Indium Oxide-UV Photodetectors. *Crystals* **2023**, *13*, 689.
6. Xiong, Y.; Chen, M.; Mao, Z.; Deng, Y.; He, J.; Mu, H.; Li, P.; Zou, W.; Zhao, Q. Synthesis of Up-Conversion Fluorescence N-Doped Carbon Dots with High Selectivity and Sensitivity for Detection of Cu^{2+} Ions. *Crystals* **2023**, *13*, 812.
7. Kalygina, V.; Podzyvalov, S.; Yudin, N.; Slyunko, E.; Zinoviev, M.; Kuznetsov, V.; Lysenko, A.; Kalsin, A.; Kopiev, V.; Kushnarev, B.; et al. Effect of UV and IR Radiation on the Electrical Characteristics of Ga_2O_3/ZnGeP2 Hetero-Structures. *Crystals* **2023**, *13*, 1203.
8. Thach, P.H.; Khai, T.V. Thermal Evaporation Synthesis, Optical and Gas-Sensing Properties of ZnO Nanowires. *Crystals* **2023**, *13*, 1380.
9. Li, S.; Yang, H. Strain-Modulated Electronic Transport Properties in Two-Dimensional Green Phosphorene with Different Edge Morphologies. *Crystals* **2024**, *14*, 239.
10. Kalygina, V.; Tsymbalov, A.V.; Korusenko, P.M.; Koroleva, A.V.; Zhizhin, E.V. Effect of Traps on the UV Sensitivity of Gallium Oxide-Based Structures. *Crystals* **2024**, *14*, 268.
11. Barzegar, S.; Karimi Abdolmaleki, M.; Connick, W.B.; Absalan, G. Enhancing Vapochromic Properties of Platinum(II) Terpyridine Chloride Hexaflouro Phosphate in Terms of Sensitivity through Nanocrystalization for Fluorometric Detection of Acetonitrile Vapors. *Crystals* **2024**, *14*, 347.

References

1. Korotcenkov, G. Current trends in nanomaterials for metal oxide-based conductometric gas sensors: Advantages and limitations. part 1: 1D and 2D nanostructures. *Nanomaterials* **2020**, *10*, 1392. [CrossRef] [PubMed]
2. Yadav, V.K.; Malik, P.; Khan, A.H.; Pandit, P.R.; Hasan, M.A.; Cabral-Pinto, M.M.; Islam, S.; Suriyaprabha, R.; Yadav, K.K.; Dinis, P.A.; et al. Recent advances on properties and utility of nanomaterials generated from industrial and biological activities. *Crystals* **2021**, *11*, 634. [CrossRef]
3. Zhou, T.; Zhang, T. Recent progress of nanostructured sensing materials from 0D to 3D: Overview of structure–property-application relationship for gas sensors. *Small Methods* **2021**, *5*, 2100515. [CrossRef] [PubMed]
4. Baig, N. Two-dimensional nanomaterials: A critical review of recent progress, properties, applications, and future directions. *Compos. Part A Appl. Sci. Manuf.* **2023**, *165*, 107362. [CrossRef]
5. Zhou, A.F.; Wang, X.; Pacheco, E.; Feng, P.X. Ultrananocrystalline diamond nanowires: Fabrication, characterization, and sensor applications. *Materials* **2021**, *14*, 661. [CrossRef] [PubMed]
6. Hunter, G.W.; Akbar, S.; Bhansali, S.; Daniele, M.; Erb, P.D.; Johnson, K.; Liu, C.C.; Miller, D.; Oralkan, O.; Hesketh, P.J.; et al. Editors' choice—Critical review—A critical review of solid state gas sensors. *J. Electrochem. Soc.* **2020**, *167*, 037570. [CrossRef]
7. Luo, Y.; Abidian, M.R.; Ahn, J.H.; Akinwande, D.; Andrews, A.M.; Antonietti, M.; Bao, Z.; Berggren, M.; Berkey, C.A.; Bettinger, C.J.; et al. Technology roadmap for flexible sensors. *ACS Nano* **2023**, *17*, 5211–5295. [CrossRef] [PubMed]
8. Tyagi, D.; Wang, H.; Huang, W.; Hu, L.; Tang, Y.; Guo, Z.; Ouyang, Z.; Zhang, H. Recent advances in two-dimensional-material-based sensing technology toward health and environmental monitoring applications. *Nanoscale* **2020**, *12*, 3535–3559. [CrossRef] [PubMed]
9. DeCost, B.L.; Hattrick-Simpers, J.R.; Trautt, Z.; Kusne, A.G.; Campo, E.; Green, M.L. Scientific AI in materials science: A path to a sustainable and scalable paradigm. *Mach. Learn. Sci. Technol.* **2020**, *1*, 033001. [CrossRef] [PubMed]
10. Sobczyk, M.; Wiesenhütter, S.; Noennig, J.R.; Wallmersperger, T. Smart materials in architecture for actuator and sensor applications: A review. *J. Intell. Mater. Syst. Struct.* **2022**, *33*, 379–399. [CrossRef]

Disclaimer/Publisher's Note: The statements, opinions and data contained in all publications are solely those of the individual author(s) and contributor(s) and not of MDPI and/or the editor(s). MDPI and/or the editor(s) disclaim responsibility for any injury to people or property resulting from any ideas, methods, instructions or products referred to in the content.

Article

Enhancing Vapochromic Properties of Platinum(II) Terpyridine Chloride Hexaflouro Phosphate in Terms of Sensitivity through Nanocrystalization for Fluorometric Detection of Acetonitrile Vapors

Sedigheh Barzegar [1,2], Mahmood Karimi Abdolmaleki [3,*], William B. Connick [2] and Ghodratollah Absalan [1]

1. Massoumi Laboratory, Department of Chemistry, College of Sciences, Shiraz University, Shiraz 71454, Iran; sedighechem62@yahoo.com (S.B.)
2. Department of Chemistry, University of Cincinnati, Cincinnati, OH 45221, USA
3. Department of Physical and Environmental Sciences, Texas A&M University-Corpus Christi, 6300 Ocean Drive, Corpus Christi, TX 78412, USA
* Correspondence: ben.karimi@tamucc.edu

Abstract: The vapochromic properties of [Pt(tpy)Cl](PF$_6$) crystals in the presence of acetonitrile and its effect on the crystal structure as well as the fluorescence spectrum of this complex have already been studied in the past. We synthesized nanocrystals of this compound for the first time, and discussed different parameters and methods that affect nanocrystal structure modulation. The study demonstrates the vapochromic properties of the nanocrystals toward acetonitrile vapor by investigating the morphology and fluorescence spectra of the nanocrystals. Vapochromic studies were conducted on [Pt(tpy)Cl](PF$_6$) nanocrystals for five cycles of absorption and desorption of acetonitrile, demonstrating shorter response times compared to regular bulk crystals.

Keywords: Pt(II) complex; [Pt(tpy)Cl](PF$_6$) crystals; [Pt(tpy)Cl](PF$_6$) nanocrystals; vapochromic crystals; vapochromic nanocrystals

1. Introduction

Vapochromic and vapoluminescent materials have garnered significant scientific interest due to their potential applications as chemical sensors and modules for detecting various environmental factors such as temperature, pressure, and the concentration of chemical vapors, including volatile organic compounds (VOCs) [1–9]. Investigations in this area have reported the color, fluorescence, and/or phosphorescence change in vapochromic materials through adsorption and/or desorption of vapors [10–18]. For example, a study examined how [Pt(tpy)Cl](PF$_6$) crystals change with the exposure to acetonitrile (MeCN), a VOC, and how this affects their structure [10]. Another study explored how exposure to water vapor causes a transition between hydrated and dehydrated forms of Pt(tpy)ClClO$_4$, a square-planar Pt(II) complex, with reversible spectroscopic changes, indicating recyclability [19]. Additionally, research on three specific square-planar Pt(II) complexes found that exposure to VOC vapors induces luminescent switching properties in all three complexes [20].

This vapochromic/vapoluminescent phenomenon occurs due to the splitting of the complex ligand field and the alteration of distances between the Pt(II) ions in square-planar Pt(II) complexes. However, designing a vapor-sensitive and selective sensor with these properties has remained challenging [10,21]. The spectroscopic characteristics of square planar Pt(II) complexes stem from interactions between Pt atoms. These interactions, which are noncovalent, form quasi-one-dimensional stacked systems depending on the ligands incorporated into the complex [22,23]. Initially, the Pt(II) units in the parent solid complex have normal distances, but they rearrange into a continuous chain with short

Pt···Pt overlaps in the new lattice formed after exposure to vapor [10]. Any variation in these Pt···Pt interactions, caused by substitutions in the ligand and/or counterions as well as various intermolecular interactions (e.g., hydrogen bonds, coordination bonds, hydrophobic interactions), result in modulation of spectroscopic properties [24–27] due to the changes in the complex's lattice parameters. These changes in spectroscopic properties make these complexes suitable for use as sensors in analyte detection.

When VOCs fill the gaps in the lattice of the Pt(II) complex, they rearrange the Pt···Pt interaction, creating a new lattice. This rearrangement causes a change in the spectroscopic properties of the complex. VOCs can be detected sensitively and selectively because they match the size and shape of the voids [10]. Notably, VOCs are chemicals with high vapor pressures at room temperature, posing environmental risks and health hazards. Therefore, detecting them is crucially important.

Using fluorometry improves the vapochromic properties of the complex because fluorescence spectroscopy is approximately a thousand times more sensitive and selective compared to the UV/Vis absorption spectrophotometry. This increased sensitivity leads to better detection limits. Therefore, the main purpose of our study is to achieve more precise and accurate measurements. We synthesized nanocrystals of [Pt(tpy)Cl](PF$_6$), to the best of our knowledge for the first time. We discuss various methods and parameters that influence the modulation of the nanocrystal structure. Then, we investigate the vapochromic properties of these nanocrystals in response to MeCN vapor by analyzing their morphology and fluorescence spectra after exposure to MeCN. This allows us to examine the differences in vapochromism between crystals and nanocrystals and determine their suitability for use in vapor sensors.

2. Experimental

2.1. Reagents

We purchased K$_2$PtCl$_4$ from Pressure Chemical, while both COD (1,5-cyclooctadiene) and tpy (2,2′:6′2″-terpyridine) were obtained from Aldrich. Additionally, NH$_4$PF$_6$ was purchased from Alfa Aesar, and acetonitrile (MeCN), acetone, acetic acid, and hexane were bought from Merck chemical company. Poly-alanine and sodium iodide were bought from Sigma and Fisher Scientific, respectively. Deionized water was utilized in all experiments.

2.2. Apparatus

We used an Acton SpectraPro 300i spectrophotometer equipped with a photon counter photomultiplier tube (PMT) detector to record the fluorescence spectra. Mass spectra were obtained using a Micromass Q-time of flight (TOF) 2 instrument from Waters, Milford, MA, USA. A Bruker AC 400 MHz NMR was used to characterize the synthesized compounds. SEM images of both crystals and nanocrystals were collected using an FEI/Phillip XL 30 ESEM-FEG from F.E.I. Company (Hillsboro, OR, USA). X-ray powder diffraction (XRD) patterns were acquired using a Philips X'Pert Powder Diffractometer from Malvern Panalytical company (Malvern, UK), employing 0.413620 Å photons in transmission setup. We used platinum striped mirrors to select incident photons, along with a double Si(111) monochromator featuring an adjustable sagittal focus. Scattered beams were filtered using (12) perfect Si(111) analyzer crystals. Discrete detectors spanning an angular range from −6 to 16° 2θ were scanned across a 34° 2θ range. Data points were collected at intervals of 0.001° 2θ, with a scanning speed set at 0.01°/s.

2.3. Synthesis of [Pt(tpy)Cl](PF$_6$)

A total of 2.5 g of K$_2$PtCl$_4$ was weighed and added to a round bottom flask. Then, 60.0 mL of water and 80.0 mL of acetic acid were added, and the mixture was left under an Ar bubble for 30 min. Afterward, 2.44 mL of COD was added using a syringe, and the mixture was refluxed for one hour. To obtain a precipitate, the solution was filtered under vacuum, and the obtained precipitate was washed with deionized water several times. The precipitate was then placed under low vacuum in the hood overnight to obtain a powder of Pt(COD)Cl$_2$. In the next step, 1.5 g of Pt(COD)Cl$_2$ and 0.94 g of 2,2':6'2''-terpyridine (tpy) were added to a 1000 mL round-bottom volumetric flask, followed by the addition of 800 mL water and 50 mL acetone. The mixture was refluxed with gentle stirring overnight to dissolve [Pt(tpy)Cl]Cl. The solution was then transferred to a beaker, and a small amount of NH$_4$PF$_6$ was added while the solution was still warm from the refluxing step. An emulsion of [Pt(tpy)Cl](PF$_6$) was formed (Figure 1) as a yellow precipitate via an anion metathesis reaction between [Pt(tpy)Cl]Cl and NH$_4$PF$_6$ in water. The emulsion was filtered under vacuum to obtain a powdered form of [Pt(tpy)Cl](PF$_6$) [28,29]. If the filtrate remains yellow, some additional NH$_4$PF$_6$ should be added to use the remaining Pt(tpy)Cl$_2$.

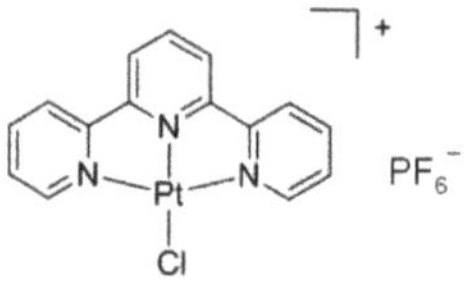

Figure 1. The chemical structure of the [Pt(tpy)Cl](PF$_6$) [11].

Long, yellow crystalline needles of [Pt(tpy)Cl](PF$_6$) are readily obtained through the evaporation of a 1:1 acetone/water mixture at room temperature. Subsequently, the crystals in the remaining water should be filtered using filter paper and a gravity funnel.

2.4. Vapochromic Studies of [Pt(tpy)Cl](PF$_6$) Crystals

To record fluorescence spectra, the crystals of [Pt(tpy)Cl](PF$_6$) were placed on a microscope slide with a layer of grease. Fluorescence detection was achieved by placing the crystals at a right angle in front of the photon beam. Samples were exposed to air saturated with MeCN at room temperature using a sealed plastic chamber (~1.0 L) with a MeCN solvent reservoir (Scheme 1) to initiate the MeCN vapor absorption process. Subsequently, the MeCN reservoir was removed, and the desorption process was monitored by recording the fluorescence spectra. The absorption and desorption of the MeCN onto and from [Pt(tpy)Cl](PF$_6$) crystals were tracked by recording the fluorescence spectra at each step. The excitation wavelength (λ_{ex}) was 436 nm. The experiments were conducted at room temperature (25 °C) and atmospheric pressure, where the vapor pressure of MeCN was 0.117 atm [29].

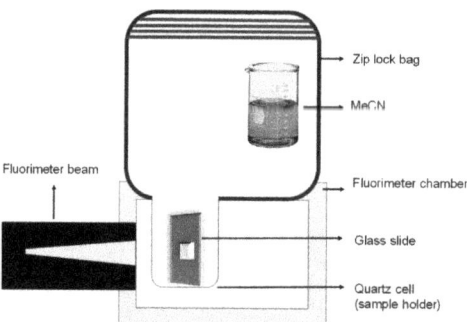

Scheme 1. Instrumental set up for the fluorescence study of absorption and desorption of acetonitrile onto and from [Pt(tpy)Cl](PF$_6$) crystals.

2.5. Synthesis of Pt[(tpy)Cl]PF$_6$ Nanocrystals

A total of 0.1 g of [Pt(tpy)Cl](PF$_6$) powder was dissolved in a mixture of acetone and water (95:5 v/v) in a 10 mL volumetric flask. A silicon surface, previously cleaned in acetone and distilled water using ultrasound, was employed as a substrate. Using a micro syringe, 1.0 µL of complex solution was dispensed onto the silicon surface.

To investigate the effect of dispersion and heat on the size of [Pt(tpy)Cl](PF$_6$) nanocrystals, three different methods were employed, and scanning electron microscopic (SEM) images were subsequently examined:

In Methods 1 and 2: a single drop of a solution containing [Pt(tpy)Cl](PF$_6$) at a concentration of 1.6×10^{-3} mol L^{-1} in a mixture of acetone and water (95:5, v/v) was added to three drops of hexane (solution A).

Method #1. Subsequently, 1.0 µL of solution A was applied onto the silicon substrate using a syringe (Scheme 2A).

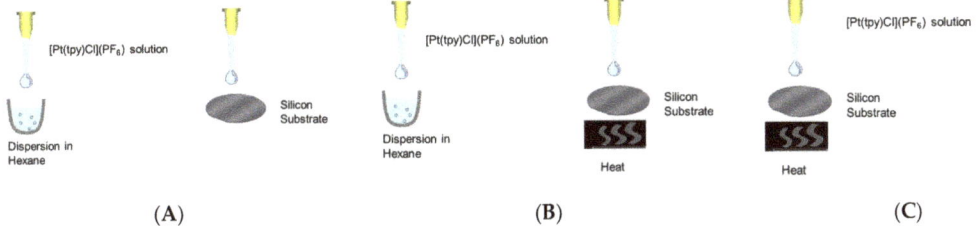

Scheme 2. Synthesis of Pt[(tpy)Cl]PF$_6$ nanocrystals using [Pt(tpy)Cl](PF$_6$) at a concentration of 1.6×10^{-3} mol L^{-1} in a mixture of acetone and water (95:5, v/v). (**A**) Method #1: dispersion of the Pt[(tpy)Cl]PF$_6$ solution in hexane, followed by deposition onto a silicon substrate. (**B**) Method #2: dispersion of the Pt[(tpy)Cl]PF$_6$ solution in hexane, followed by deposition onto a preheated silicon substrate. (**C**) Method #3: deposition of the Pt[(tpy)Cl]PF$_6$ solution onto a preheated silicon substrate.

Method #2. Using a syringe, 1.0 µL of solution A was dispensed onto a silicon substrate preheated for 5 min in an oven at 105 °C (Scheme 2B).

Method #3. A total of 1.0 µL of 1.6×10^{-3} mol L^{-1} [Pt(tpy)Cl](PF$_6$) solution in an acetone and water mixture (95:5, v/v) was poured onto the preheated silicon substrate using a syringe directly (Scheme 2C). The silicon substrate was preheated for 5 min in an oven at 105 °C to facilitate rapid solvent evaporation.

To investigate the type of substrate on [Pt(tpy)Cl](PF$_6$) nanocrystal structures, the following procedure was implemented: First, 0.1 g of [Pt(tpy)Cl](PF$_6$) was dissolved in a mixture of acetone and water (95:5, v/v) in a 10.0 mL volumetric flask. Then, 1.0 µL of the solution was dropped onto either a silicon or glass slide substrates, preheated in the oven (105 °C) for 5 min. Then, SEM images were examined.

To study the effect of Pt[(tpy)Cl]PF$_6$ stock solution concentration on nanocrystal structures, the following procedure was performed: First, 0.02 g, 0.01 g, and 0.005 g of [Pt(tpy)Cl](PF$_6$) were individually dissolved in 10 mL of acetone/water mixture (95:5, v/v) to produce 3.2×10^{-3}, 1.6×10^{-3}, and 0.8×10^{-3} mol L^{-1} stock solutions, respectively. Then, 1.0 µL of each solution was dispensed onto the preheated silicon substrate (using Method 3) to fabricate nanocrystals. Subsequently, SEM images were studied.

2.6. Vapochromic Studies of [Pt(tpy)Cl](PF$_6$) Nanocrystals

A 3.2×10^{-3} mol L^{-1} [Pt(tpy)Cl](PF$_6$) solution (saturated solution) was prepared using a mixture of acetone/water (95:5 v/v) in a 10.0 mL volumetric flask. A total of 1.0 µL of the solution was dispensed onto a preheated silicon substrate at 105 °C. The silicon, held on a glass slide, was positioned in front of the fluorimeter light source. The exposure to MeCN vapor was conducted at room temperature using a sealed plastic chamber with a MeCN solvent reservoir (Scheme 1). The absorption and desorption processes of MeCN onto and from the [Pt(tpy)Cl](PF$_6$) nanocrystals were monitored by recording the fluorescence spectra for five cycles. The excitation wavelength (λ_{ex}) was 436 nm. The experiments were conducted at room temperature (25 °C) and atmospheric pressure, where the vapor pressure of MeCN was to be 0.117 atm [29].

3. Results and Discussion

3.1. Characterization of [Pt(tpy)Cl](PF$_6$) Crystals by ^{1}HNMR

The ^{1}HNMR spectra of [Pt(tpy)Cl](PF$_6$) were taken at room temperature to confirm the synthesis of the compound. Chemical shifts were compared with the solvent (Figure 2). Each peak in the NMR spectrum represented a hydrogen in the structure, as shown in the figure. Therefore, the spectra could be analyzed separately as a four-spin system for the protons of the terminal rings (protons 1, 2, 3, and 4, along with their symmetry-related protons) and as an independent three-spin system for the protons of the central ring (protons 5, 6, and the symmetry-related proton of 5). The chemical shifts corresponding to the protons of terpyridine appeared at approximately 7.9 ppm for protons 2 and 2′, 8.5 ppm for protons 3 and 3′, 8.6 ppm for protons 5, 6, 5′, 4, and 4′, and 8.9 ppm for protons 1 and 1′. The peak at 3.3 ppm was attributed to water, and the peak at 2.5 ppm was attributed to dimethyl sulfoxide (DMSO), which was used to dissolve the crystals of [Pt(tpy)Cl](PF$_6$).

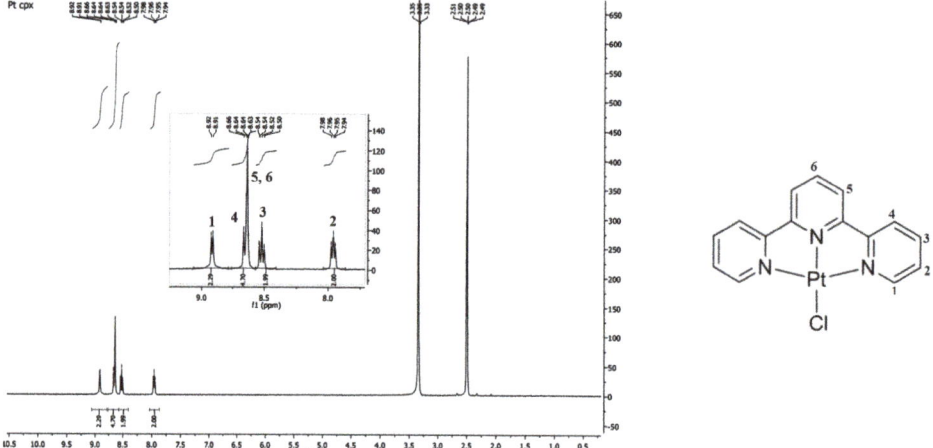

Figure 2. Characterization of [Pt(tpy)Cl](PF$_6$) crystals by ^{1}HNMR.

3.2. Characterization of [Pt(tpy)Cl](PF$_6$) Crystals by X-ray Powder Diffraction

The crystallinity of [Pt(tpy)Cl](PF$_6$) was investigated by powder XRD in comparison with the simulated powder pattern from Mercury software (version 3.7) [30] by using the following parameters including temperature: 298.0 K; Start 2θ: 0 degree; End 2θ: 10 degree; 2θ step: 0.002 degree; calibration wavelength: 0.41362; FWHM: 0.05; and h, k, l: 0, 0, 1. The X-ray powder diffractograms are presented in Figure 3. Both diffractograms have the same pattern and characteristic diffraction peaks at 2θ degrees, confirming the formation and purity of [Pt(tpy)Cl](PF$_6$) crystals [30].

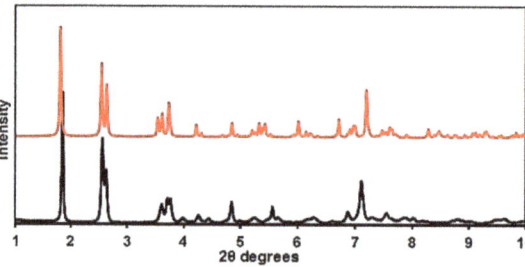

Figure 3. The X-ray powder diffractogram of the [Pt(tpy)Cl](PF$_6$) crystals (—) and the simulated pattern from the mercury software (—) [30].

3.3. Vapochromic Studies of the [Pt(tpy)Cl](PF$_6$) Crystals

3.3.1. Studying Absorption/Desorption Cycle of MeCN Vapor in the Structure of [Pt(tpy)Cl](PF$_6$) Crystals

Figure 4A displays the fluorescence spectra of [Pt(tpy)Cl](PF$_6$) (λ_{ex} = 436 nm) crystals before and after exposure to MeCN vapors. When excited at 436 nm, the crystals show a consistent broad fluorescence band centered at 550 nm, which indicates high-quality crystals of [Pt(tpy)Cl](PF$_6$). After exposure to MeCN vapor in a sealed chamber (Scheme 1), the fluorescence spectrum broadens further, with a peak at 680 nm and a shoulder around 600–610 nm (Figure 4A), indicating the formation of an intermediate structure. This red shift suggests an increase in Pt...Pt interactions. To assess the fluorescence changes during the desorption, spectra were collected after exposing [Pt(tpy)Cl](PF$_6$) crystals to MeCN vapors and then removing the MeCN source. The peak at 680 nm decreased while the peak at 600 nm increased (Figure 4B). The final fluorescence spectrum post-desorption differed from the initial one, suggesting that the original high-quality crystals could not be restored after MeCN desorption. The increased fluorescence intensity at 600 nm suggests the formation of solvated [Pt(tpy)Cl](PF$_6$), consistent with the absorption fluorescence spectra. When was exposed to MeCN vapors, the room-temperature fluorescence spectrum (λ_{ex} = 436 nm) showed a broad band at 685 nm, attributed to the primary spin-forbidden metal–metal-to-ligand-charge transfer (MMLCT) transition [10,31]. The blue shift to 600 nm indicated that MeCN vapor desorption led to an increased Pt...Pt distance, weakening these interactions and destabilizing MMLCT states [10,31]. The 600 nm fluorescence band might be linked to MMLCT from Pt...Pt dimers, while the 685 nm fluorescence band in MeCN-vapor-exposed crystals was associated with MMLCT from a linear chain of closely interacting Pt centers [10,31]. In more detail, in the yellow crystals, the dz^2 orbitals are further apart, but in the red form, they are closer, as shown in Scheme 3. This closer distance boosts orbital interaction, forming dσ and dσ* orbitals. The higher energy of the dσ* orbital allows easier electron transfer to ligand π* orbitals, causing a red shift in fluorescence. Thus, yellow crystals exhibit metal-to-ligand charge transfer (MLCT), while MeCN-exposed red crystals display MMLCT [10,31]. These interaction changes, combined with alterations in crystal morphology due to MeCN exposure, result in asymmetrical behavior in absorption/desorption fluorescence spectra. Previous studies have examined single-crystal XRD measurements of [Pt(tpy)Cl](PF$_6$) crystals and MeCN-exposed ones [10].

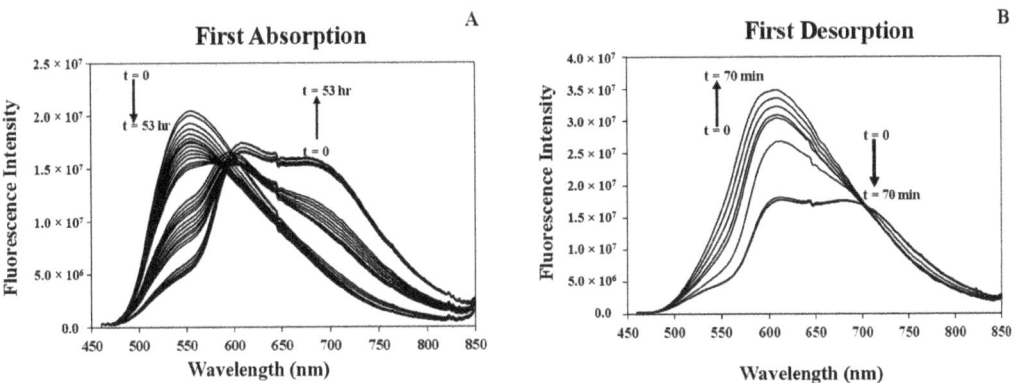

Figure 4. Fluorescence spectra of the crystals of Pt(II) complexes (by time) during the first absorption (**A**) and desorption (**B**) of acetonitrile vapor (λ_{ex} = 436 nm).

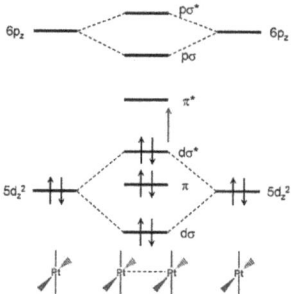

Scheme 3. The interaction of Pt-Pt orbitals to form dσ and dσ* orbitals and the possibility of MMLCT [10,31]. In detail, the closer distance of the dz^2 orbitals in MeCN-vapor-exposed crystals produces dσ and dσ* orbitals. The higher energy of the dσ* orbital allows easier electron transfer to ligand π* orbitals.

The process of absorption/desorption cycling of MeCN vapors in [Pt(tpy)Cl](PF$_6$) crystals was repeated multiple times. In the second cycle (Figure 5A,B), the maximum wavelength of the fluorescence shifted to 700 nm during MeCN absorption and then to 600 nm during desorption. Repeating the cycles five times showed that the maximum absorption peak remained at 700 nm (Figure 5C,E,G), while the desorption peak stayed at 600 nm (Figure 5D,F,H), consistent with the peaks in the second absorption/desorption cycle. It is worth noting that the peaks' intensity during the absorption and desorption cycles is significant enough to confidently suggest this crystal as a sensor for MeCN vapors.

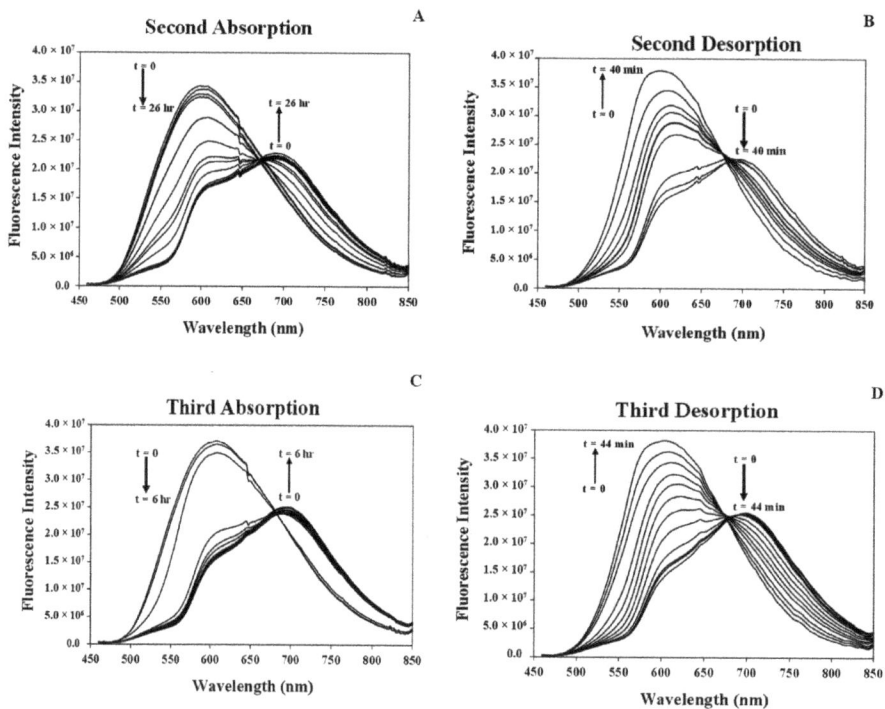

Figure 5. *Cont.*

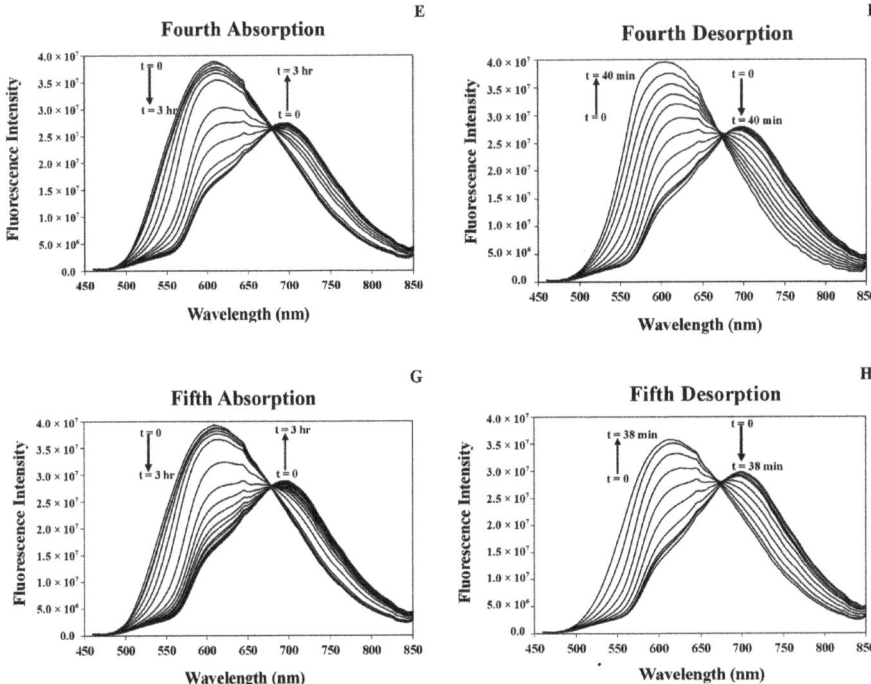

Figure 5. Fluorescence spectra of the crystals of Pt(II) complexes (by time) during the second (**A,B**), third (**C,D**), fourth (**E,F**), and fifth (**G,H**) absorption and desorption of acetonitrile vapor (λ_{ex} = 436 nm).

3.3.2. Vapochromic Studies on the Microscopic and SEM Images of the [Pt(tpy)Cl](PF$_6$) Crystals before and after Being Exposed to MeCN Vapor

Long, yellow crystalline needles of [Pt(tpy)Cl](PF$_6$) were grown by evaporating a solution of [Pt(tpy)Cl](PF$_6$) in a 1:1 acetone/water mixture at room temperature. The crystals were then filtered using a gravity funnel and filter paper to obtain high quality crystals. It is crucial to avoid using a vacuum funnel, as it can create defects in the crystal structure. SEM images of the crystals are shown in Figure 6A–D, with different magnifications at (A) 100 μm, (B) 20 μm, (C) 20 μm, and (D) 10 μm scales, indicating that the crystals are defect-free. Microscopic images of these high-quality yellow crystals are presented in Figure 7A. When exposed to MeCN vapor, the color changes from yellow to red (Figure 7B) due to alterations in Pt...Pt interactions. Figure 8 shows SEM images of the crystals after the first (A), second (B), and fifth (C) absorption/desorption cycle of MeCN vapor. After the fifth cycle, changes in the crystal structure are evident. Cracks and defects begin to appear along the edges of the crystal (Figure 8A). These eventually cover most of the crystal surface and penetrate deep into the lattice structure (Figure 8B). During desorption, noticeable color changes from red to yellow spread from the crystal's long edge, resulting in a yellow crystal with visible defects. The accumulated observations suggest that stress from vapor absorption and desorption leads to fracturing and the formation of defects, resulting in a loss of long-range order, as seen in Figure 8C. Previous XRD and single-crystal XRD data for [Pt(tpy)Cl](PF$_6$) and MeCN-exposed [Pt(tpy)Cl](PF$_6$) were collected to confirm changes in Pt...Pt distances upon vapor absorption/desorption [10].

Figure 6. The SEM images of the high-quality [Pt(tpy)Cl](PF$_6$) crystals with different magnifications: (**A**) 100 μm, (**B**) 20 μm, (**C**) 20 μm, and (**D**) 10 μm.

Figure 7. The microscopic images of [Pt(tpy)Cl](PF$_6$) crystals' color change (**A**) before and (**B**) after exposure to MeCN vapor.

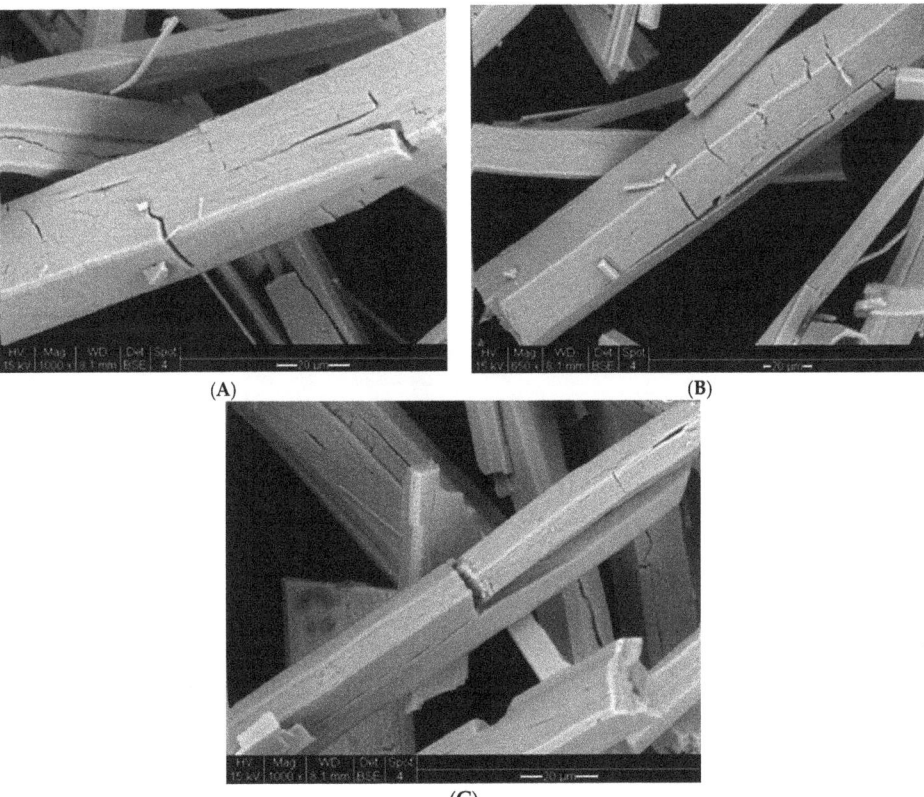

Figure 8. The SEM images of the [Pt(tpy)Cl](PF$_6$) crystals after (**A**) first, (**B**) second, and (**C**) fifth absorption/desorption of MeCN vapor.

3.4. Characterization of [Pt(tpy)Cl](PF$_6$) Nanocrystals by Mass Spectroscopy

The nanocrystals formed on the silicon substrate were rinsed with MeCN and used for mass spectroscopic analysis. The [Pt(tpy)Cl](PF$_6$) complex weighed 12.0 µg, and the MeCN volume was 50.0 µL. Using mass spectrometry–electrospray ionization (MS-ESI) in positive ion mode, MeCN (CH$_3$CN) showed an m/z of 463.0 for [Pt(tpy)Cl$^+$]. In the negative ion mode, MS-ESI showed an m/z of 144.9 for PF$_6^-$. Notably, the instrument was calibrated in positive ion mode using poly-alanine and in negative ion mode using sodium iodide. The observed isotope patterns closely matched those predicted based on natural isotopic abundances.

3.5. Investigation of the Effect of Dispersion and Heat on [Pt(tpy)Cl](PF$_6$) Nanocrystal Sizes

The SEM images of the samples prepared using three methods (described in Section 2.5) are displayed in Figure 9, with the corresponding width, length, and aspect ratio provided in Table 1. In Figure 9A, nanocrystals created by dispersing the [Pt(tpy)Cl](PF$_6$) solution in hexane exhibited a large width of 170 nm, length of 3960 nm, and an aspect ratio of 26.1 owing to nanocrystal aggregation. This aggregation is attributed to the slow evaporation of the solvent in the presence of hexane on a cold silicon substrate. Figure 9B shows structures formed by first dispersing the solution in hexane and then pouring it onto a preheated silicon substrate. These nanocrystals had the smallest widths of 22 nm, lengths of 101.3 nm, and the lowest aspect ratio of 4.5. This likely occurred due to acetone's higher volatility compared to hexane, resulting in the formation of partially needle-shaped nanocrystals and partial evaporation over extended periods, hindering nanocrystal formation. Evaporating the solution on a silicon substrate preheated in the oven, without initial dispersion in hexane, resulted in nanocrystals with widths of 89 nm, lengths of 2519 nm, and a high aspect ratio of 26.4, as shown in Figure 9C. This method generated more nanocrystals with reduced size and less aggregation compared to those produced by Methods 1 and 2, without requiring dispersion in hexane. This phenomenon is likely due to the miscibility of acetone and water, as well as the volatility of acetone compared to hexane.

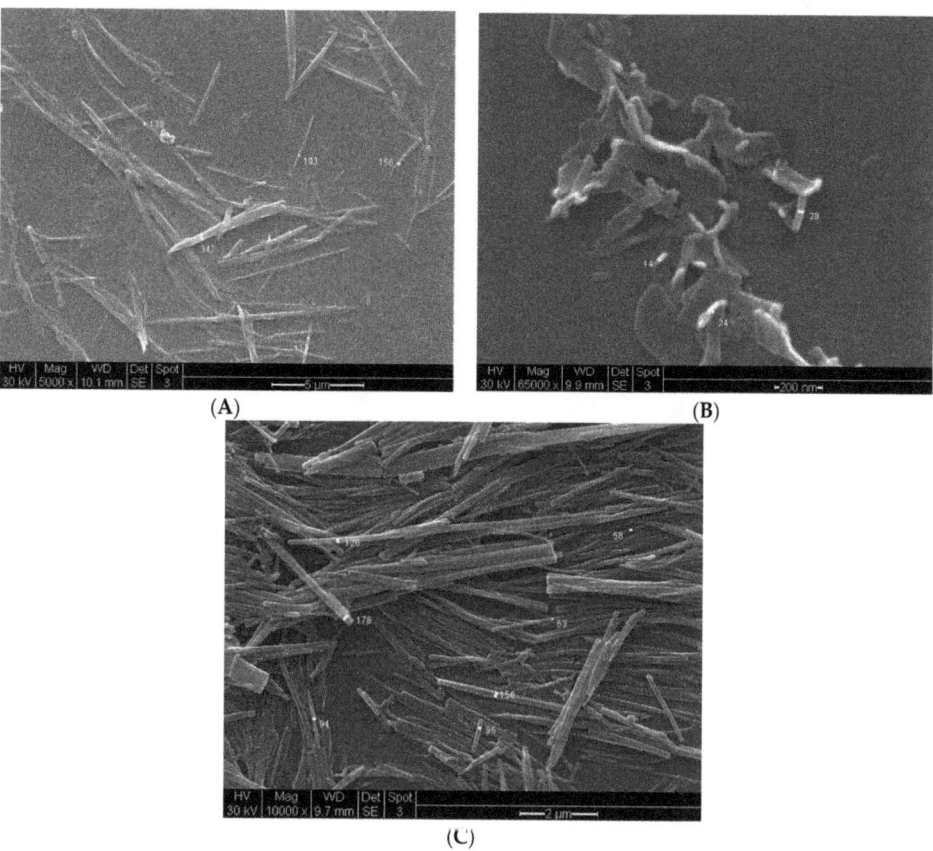

Figure 9. The SEM images of the [Pt(tpy)Cl](PF$_6$) nanocrystals synthesized under different preparation conditions: (**A**) Method 1: Deposition of dispersed [Pt(tpy)Cl](PF$_6$) onto a silicon substrate, (**B**) Method 2: Deposition of dispersed [Pt(tpy)Cl](PF$_6$) onto a preheated silicon substrate, (**C**) Method 3: Direct deposition of [Pt(tpy)Cl](PF$_6$) onto a preheated silicon substrate.

Table 1. The width, length, and aspect ratio of [Pt(tpy)Cl](PF$_6$) nanocrystals generated using 1.6×10^{-3} mol L^{-1} [Pt(tpy)Cl](PF$_6$) solution by various methods.

Method	Mean Width (nm) (n = 10)	Mean Length (nm) (n = 10)	Mean Aspect Ratio (n = 10)
1 [a]	170.0	3960.0	26.1
2 [b]	22.0	101.3	4.5
3 [c]	89.0	2519.0	26.4

[a] Method 1: Deposition of dispersed [Pt(tpy)Cl](PF$_6$) onto a silicon substrate. [b] Method 2: Deposition of dispersed [Pt(tpy)Cl](PF$_6$) onto a preheated silicon substrate. [c] Method 3: Direct deposition of [Pt(tpy)Cl](PF$_6$) onto a preheated silicon substrate.

The SEM images and the energy-dispersive X-ray spectroscopy (EDX) analysis, presented in Figures 9 and 10, respectively, confirm the formation of nanocrystals of [Pt(tpy)Cl](PF$_6$) and reveal the presence of the elements from the compound, including Pt, C, N, Cl, P, and F. The SEM images suggest that the size of [Pt(tpy)Cl](PF$_6$) nanocrystals can be modulated by changing the preparation conditions. Generating nanocrystals by depositing 1.0 µL of 1.6×10^{-3} mol L^{-1} [Pt(tpy)Cl](PF$_6$) solution onto a preheated silicon substrate (for 5 min in an oven at 105 °C) using a syringe (Method 3) resulted in satisfactory nanometer-sized nanocrystal widths with a high aspect ratio, which were selected for further experiments.

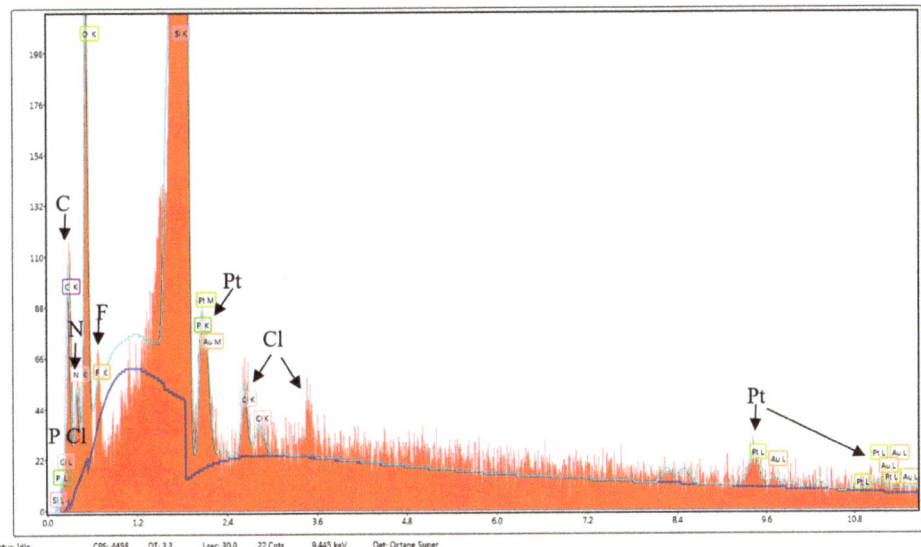

Figure 10. The EDX obtained from nanocrystals of [Pt(tpy)Cl](PF$_6$), which confirms the presence of the elements of the compound.

3.6. Effect of Pt[(tpy)Cl]PF$_6$ Stock Solution Concentration on Nanocrystal Structures

The SEM images of the nanocrystals, produced by using various concentrations of [Pt(tpy)Cl](PF$_6$) stock solutions, are depicted in Figure 11, illustrating the sizes and shapes of nanocrystals modulated by different solution concentrations. As shown in Figure 11A, nanocrystals formed from an 8.0×10^{-4} mol L^{-1} [Pt(tpy)Cl](PF$_6$) solution had a mean width of 83 nm, a mean length of 233 nm, and an aspect ratio of 3.6. In Figure 11B, nanocrystals from a 1.6×10^{-3} mol L^{-1} stock solution exhibited a mean width of 88 nm, a mean length of 209 nm, and an aspect ratio of 2.5. However, these structures appeared less uniform compared to the nanocrystals shown in Figure 11C, which originated from a 3.2×10^{-3} mol L^{-1} solution. The nanocrystals from this higher concentration displayed a mean width of 53 nm, a mean length of 951 nm, and an aspect ratio of 17.9. Notably, these nanocrystals not only exhibited uniformity but also narrower width and a satisfactory aspect ratio.

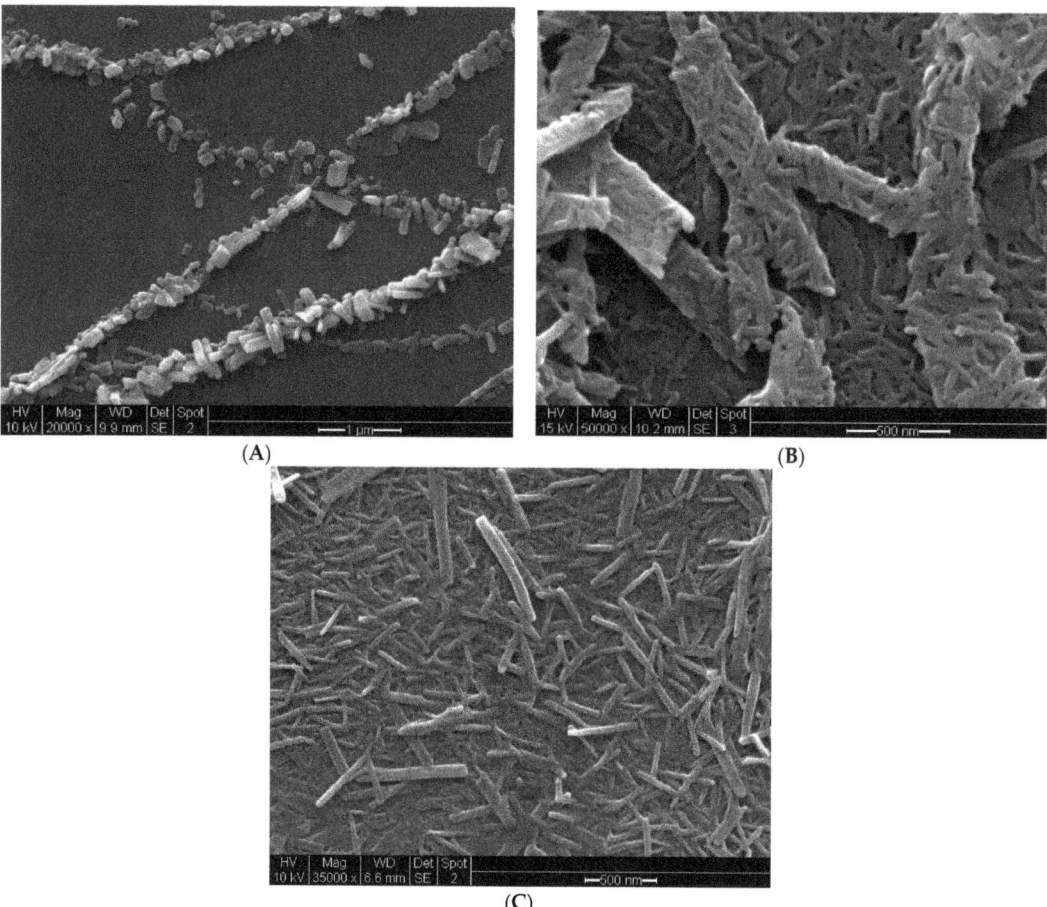

Figure 11. Effect of Pt[(tpy)Cl]PF$_6$ stock solution concentration ((**A**) 8.0×10^{-4} mol L^{-1}, (**B**) 1.6×10^{-3} mol L^{-1}, (**C**) 3.2×10^{-3} mol L^{-1}) on nanocrystal structures.

3.7. Vapochromic Studies of [Pt(tpy)Cl](PF$_6$) Nanocrystals

3.7.1. Studying Absorption/Desorption of MeCN Vapor in the Structure of [Pt(tpy)Cl](PF$_6$) Nanocrystals

Figure 12 shows the fluorescence spectra of [Pt(tpy)Cl](PF$_6$) nanocrystals (λ_{ex} = 436 nm) before and after exposure to MeCN vapor. Exciting the nanocrystals at 436 nm results in a consistent broad fluorescence band centered at 565 nm. Exposing the nanocrystal to MeCN vapor in a sealed chamber results in a slight decrease in intensity at 565 nm during initial MeCN absorption, but a red shift was observed from 565 to 585 nm. This exposure led to a broad fluorescence band with a maximum fluorescence at 715 nm and a shoulder near 580–585 nm (Figure 12A) due to the formation of an intermediate structure of solvated [Pt(tpy)Cl](PF$_6$) nanocrystals. The red shift in the fluorescence band is consistent with an increase in Pt...Pt interactions [31]. During desorption, upon removal of the MeCN vapor source, a decrease in intensity at 715 nm and an increase in intensity at 585 nm were observed (Figure 12B). The intensity of the 585 nm fluorescence band rapidly increases over a period of 16 min, but the final fluorescence spectrum differed from that of the initial [Pt(tpy)Cl](PF$_6$) nanocrystals, indicating that the initial nanocrystals could not be restored after desorption. The increased intensity in the 585 nm shoulder with the loss of MeCN suggests that the short wavelength shoulder results from deformation of [Pt(tpy)Cl](PF$_6$) nanocrystals.

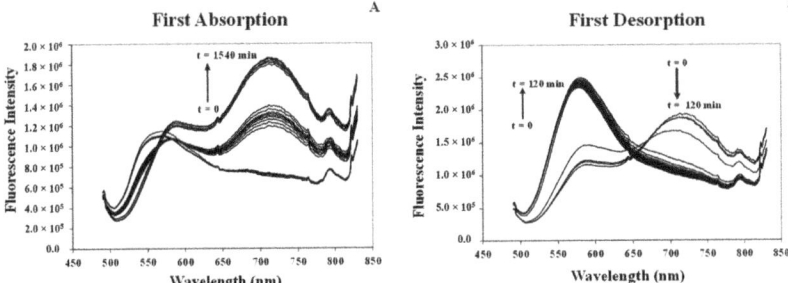

Figure 12. Fluorescence spectra of the nanocrystals of [Pt(tpy)Cl](PF$_6$) (by time) during the first (**A**) absorption and (**B**) desorption of MeCN vapor (λ_{ex} = 436 nm).

At room temperature, the fluorescence spectrum (λ_{ex} = 436 nm) of [Pt(tpy)Cl](PF$_6$) nanocrystals exhibited a broad band at 715 nm when exposed to MeCN vapor, preliminarily attributed to the primary spin-forbidden MMLCT transition of the lowest energy level [31]. The blue shift to 585 nm is consistent with MeCN desorption, causing an increase in the Pt...Pt distances, weakening these interactions and destabilizing MMLCT states [31]. The 585 nm fluorescence band was indecisively attributed to a MMLCT transition resulting from Pt...Pt dimers, whereas the 715 nm fluorescence band of the MeCN-exposed film was assigned to an MMLCT transition resulting from a linear chain of closely interacting Pt centers [31].

The process of absorbing and desorbing MeCN in the structure of [Pt(tpy)Cl](PF$_6$) nanocrystals was repeated multiple times. In the second cycle (Figure 13A,B), it was observed that the maximum wavelength for the MeCN vapor absorption shifted to 720 nm and then to 585 nm during desorption. Repeating the cycles five times showed that the absorption peak remained at 720 nm (Figure 13C,E,G) while the desorption peak remained at 585 nm (Figure 13D,F,H), consistent with the second desorption peak. Importantly, the intensity of the peaks during the absorption and desorption cycles is sufficiently high to propose this compound as a sensor for MeCN vapor detection.

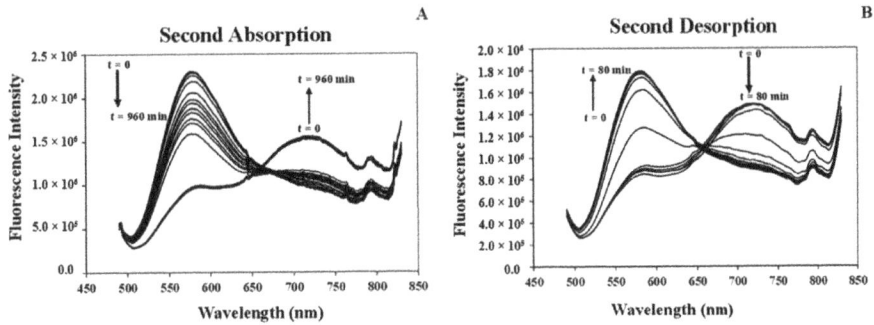

Figure 13. *Cont.*

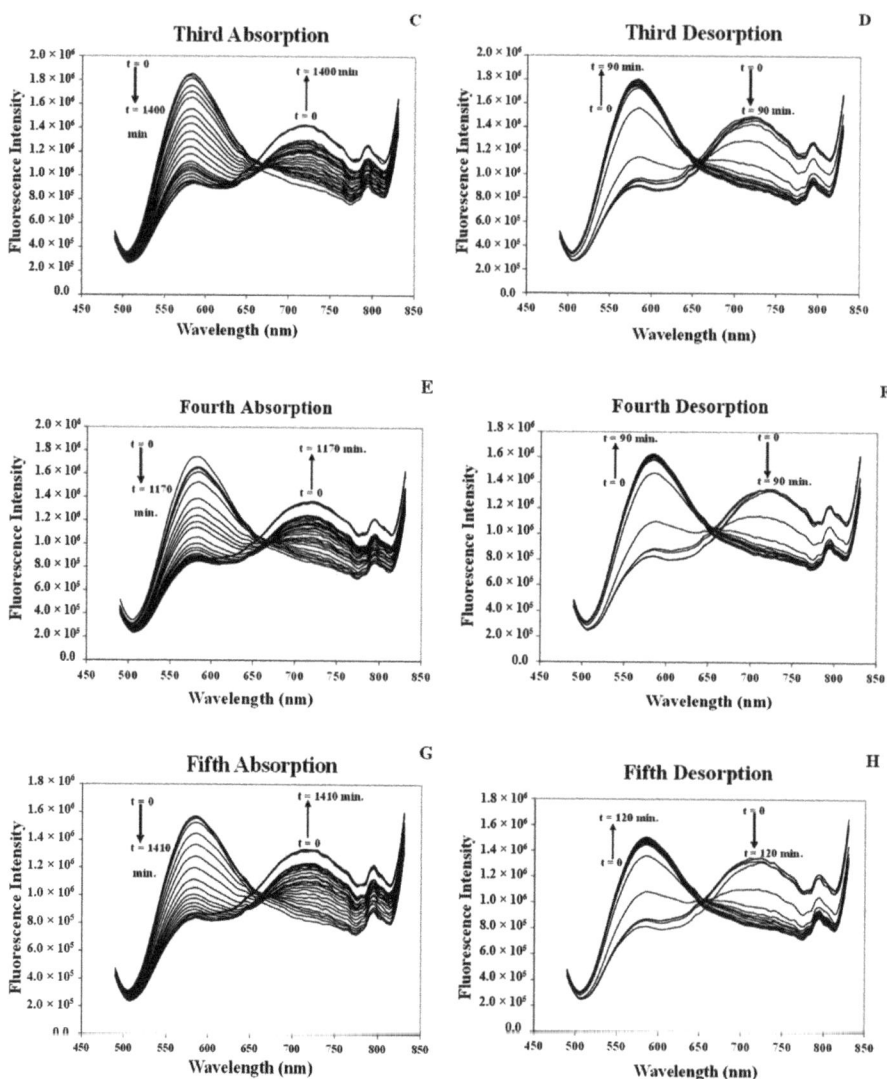

Figure 13. Fluorescence spectra of the nanocrystals of Pt(II) complexes (by time) during the second to fifth absorption (**A,C,E,G**) and desorption (**B,D,F,H**) of MeCN vapor (λ_{ex} = 436 nm).

3.7.2. Vapochromic Studies on the SEM Images of the [Pt(tpy)Cl](PF$_6$) Nanocrystals before and after Being Exposed to MeCN Vapor

Initially, the [Pt(tpy)Cl](PF$_6$) nanocrystals appear well-shaped, with each nanocrystal being individually distinguishable (See Figure 14). However, when exposed to MeCN vapor, the nanocrystals appear to be destroyed, deformed, and merged (Figure 14B). Additionally, SEM images of nanocrystals after five cycles of MeCN vapor absorption/desorption were recorded (Figure 14C), demonstrating significant changes in the nanocrystal structure. Unlike the crystals, there were no defects on the nanocrystal structures, but they had merged and lost their original shapes. These observations suggest that the stress associated with vapor absorption and desorption led to the malformation of nanocrystals, resulting in the loss of their nanosized structure.

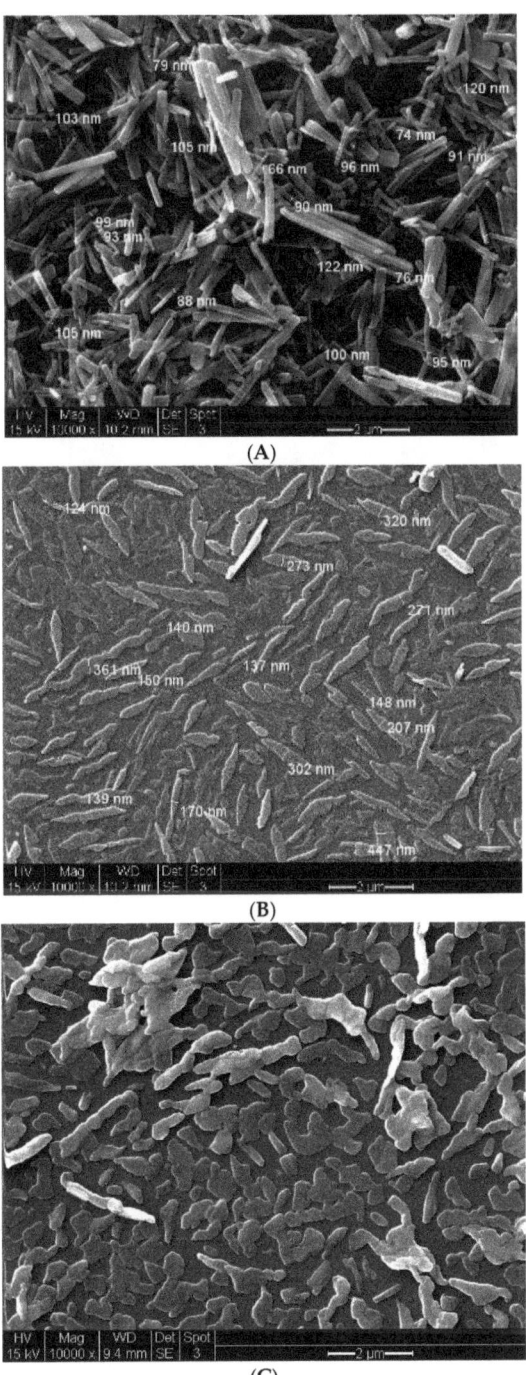

Figure 14. The SEM images of [Pt(tpy)Cl](PF$_6$) nanocrystals (**A**) before exposure to MeCN vapor (66–122 nm), (**B**) merged together (120–450 nm) after exposure to MeCN vapor, (**C**) after 5 cycles of MeCN vapor absorption/desorption.

3.7.3. Response Time of the [Pt(tpy)Cl](PF$_6$) Nanocrystals for the Absorption/Desorption of MeCN Vapor

The fluorescence signal intensity of [Pt(tpy)Cl](PF$_6$) nanocrystals during the MeCN vapor absorption over time was plotted. Figure 15 shows that during absorption, signal intensity decreased over time at λ = 585 nm and increased at λ = 720 nm. Hence, either of these wavelengths can be used to detect changes. Notably, during the first exposure cycle, there was no change in signal intensity at λ = 585 nm, but changes in the first absorption process could be tracked at λ = 720 nm. It is concluded from the response times that MeCN vapor absorption onto [Pt(tpy)Cl](PF$_6$) nanocrystals resulted in signal changes within a period of less than 100 min.

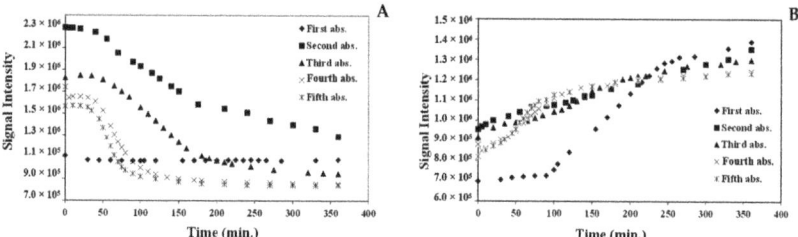

Figure 15. The [Pt(tpy)Cl](PF$_6$) nanocrystals response time during MeCN vapor absorption at (**A**) λ = 585 nm and (**B**) λ = 720 nm.

Similarly, the fluorescence signal intensity of [Pt(tpy)Cl](PF$_6$) nanocrystals during MeCN vapor desorption was recorded over time and plotted (Figure 16). It was observed that during desorption, the signal intensity decreased over time at λ = 720 nm and increased at λ = 585 nm. Hence, either of these wavelengths can be employed to detect changes. The response times confirmed MeCN vapor desorption from [Pt(tpy)Cl](PF$_6$) nanocrystals occurred within a time period of less than 26 min.

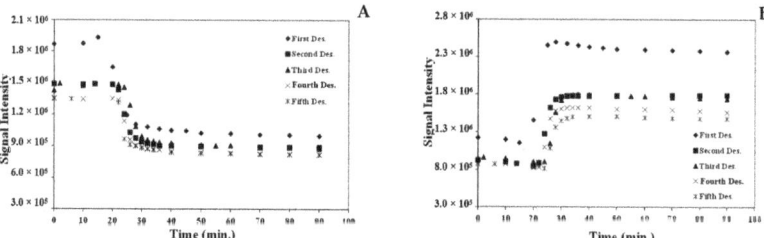

Figure 16. The [Pt(tpy)Cl](PF$_6$) nanocrystals response time during MeCN vapor desorption at (**A**) λ = 720 nm and (**B**) λ = 585 nm.

3.8. Comparison of the Response Times of MeCN Vapor Absorption/Desorption of the [Pt(tpy)Cl](PF$_6$) Crystals and Nanocrystals

The fluorescence-spectroscopy-monitored vapor absorption/desorption cycles, repeated several times, have demonstrated the reliability and effectiveness of these materials in real-world sensing applications. According to the data in Table 2, during the first cycle of MeCN vapor absorption, [Pt(tpy)Cl](PF$_6$) nanocrystals indicated a response time of 100 min, compared to 11 h for [Pt(tpy)Cl](PF$_6$) crystals. Subsequent cycles of MeCN vapor absorption showed a response time of 10 min for nanocrystals, compared to 9 h or 75 min for crystals, confirming that nanocrystals serve as superior sensors. Although crystal defects in later cycles contribute to faster responses to MeCN, the desorption time of MeCN vapor from nanocrystals and crystals remains almost the same except for the first desorption from crystals.

Table 2. Comparison of the response times to MeCN vapor absorption/desorption in the [Pt(tpy)Cl](PF$_6$) crystals and nanocrystals in 5 cycles.

Cycles	Whole Absorption Time (h)		Absorption Response (min.)		Whole Desorption Time (min.)		Desorption Response (min.)	
	Crystal	Nanocrystal	Crystal (I700/I600)	Nanocrystal (I720/I585)	Crystal	Nanocrystal	Crystal (I600/I700)	Nanocrystal (I585/I720)
1	48	11.5	11 h	100	70	36	45	20
2	18	7.5	9 h	10	40	36	25	24
3	2	9.3	75	10	44	34	28	24
4	2	6.0	75	10	40	36	24	24
5	2	6.0	75	10	38	60	26	26

4. Conclusions

This paper investigated the effects of various factors, including dispersion in hexane, temperature, and stock solution concentration, on the fabrication of [Pt(tpy)Cl](PF$_6$) nanocrystals. Consequently, the fluorescence studies and SEM images confirmed that the [Pt(tpy)Cl](PF$_6$) compound is vapochromic with nanosized dimensions as well. Finally, sensors constructed from [Pt(tpy)Cl](PF$_6$) nanocrystals are expected to exhibit faster response times compared to bulk sensors in MeCN vapor absorption/desorption detection.

Author Contributions: Conceptualization: W.B.C.; formal analysis: S.B.; investigation: S.B. and M.K.A.; methodology: S.B. and M.K.A.; supervision: M.K.A. and G.A.; roles/writing—original draft and writing: S.B.; review and editing: G.A. and M.K.A. All authors have read and agreed to the published version of the manuscript.

Funding: This research was funded by the National Science Foundation (CHE-1152853).

Data Availability Statement: The original contributions presented in the study are included in the article, further inquiries can be directed to the corresponding author.

Acknowledgments: The authors wish to acknowledge the support of this work by Shiraz University Research Council.

Conflicts of Interest: The authors declare no conflict of interest.

References

1. Kato, M.; Ito, H.; Hasegawa, M.; Ishii, K. Soft Crystals: Flexible Response Systems with High Structural Order. *Chem. A Eur. J.* **2019**, *25*, 5105–5112. [CrossRef]
2. Evariste, S.; Khalil, A.M.; Kerneis, S.; Xu, C.; Calvez, G.; Costuas, K.; Lescop, C. Luminescent Vapochromic Single Crystal to Single Crystal Transition in One-Dimensional Coordination Polymer Featuring the First Cu(I) Dimer Bridged by an Aqua Ligand. *Inorg. Chem. Front.* **2020**, *7*, 3402–3411. [CrossRef]
3. Utrera-Melero, R.; Huitorel, B.; Cordier, M.; Mevellec, J.-Y.; Massuyeau, F.; Latouche, C.; Martineau-Corcos, C.; Perruchas, S. Combining Theory and Experiment to Get Insight into the Amorphous Phase of Luminescent Mechanochromic Copper Iodide Clusters. *Inorg. Chem.* **2020**, *59*, 13607–13620. [CrossRef]
4. Sergeenko, A.S.; Ovens, J.S.; Leznoff, D.B. Copper(II) Dihalotetracyanoplatinate(IV) Coordination Polymers and Their Vapochromic Behavior. *Inorg. Chem.* **2017**, *56*, 7870–7881. [CrossRef]
5. Karabacak, S.; Qun, D.L.C.; Ammanath, G.; Yeasmin, S.; Yagmurcukardes, M.; Alagappan, P.; Liedberg, B.; Yıldız, Ü.H. Polarity Induced Vapochromism and Vapoluminescence of Polythiophene Derivatives for Volatile Organic Compounds Classification. *Sens. Actuators B Chem.* **2023**, *389*, 133884. [CrossRef]
6. Zhang, X.; Li, B.; Chen, Z.-H.; Chen, Z.-N. Luminescence Vapochromism in Solid Materials Based on Metal Complexes for Detection of Volatile Organic Compounds (VOCs). *J. Mater. Chem.* **2012**, *22*, 11427. [CrossRef]
7. Yam, V.W.-W.; Au, V.K.-M.; Leung, S.Y.-L. Light-Emitting Self-Assembled Materials Based on D8 and D10 Transition Metal Complexes. *Chem. Rev.* **2015**, *115*, 7589–7728. [CrossRef]
8. Li, E.; Jie, K.; Liu, M.; Sheng, X.; Zhu, W.; Huang, F. Vapochromic Crystals: Understanding Vapochromism from the Perspective of Crystal Engineering. *Chem. Soc. Rev.* **2020**, *49*, 1517–1544. [CrossRef]
9. Taylor, S.D.; Howard, W.; Kaval, N.; Hart, R.; Krause, J.A.; Connick, W.B. Solid-State Materials for Anion Sensing in Aqueous Solution: Highly Selective Colorimetric and Luminescence-Based Detection of Perchlorate Using a Platinum(II) Salt. *Chem. Commun.* **2010**, *46*, 1070. [CrossRef]

10. Taylor, S. *Solid-State Structures and Electronic Properties of Platinum(II) Terpyridyl Complexes: Implications for Vapor and Aqueous Anion Sensing*; University of Cincinnati: Cincinnati, OH, USA, 2011.
11. Ni, J.; Zheng, W.; Qi, W.-J.; Guo, Z.-C.; Liu, S.-Q.; Zhang, J.-J. Synthesis, Structure and Luminescent Switching Properties of Cycloplatinated(II) Complexes Bearing Phenyl β-Diketone Ligands. *J. Organomet. Chem.* **2021**, *952*, 122048. [CrossRef]
12. Ohno, K.; Shiraishi, K.; Sugaya, T.; Nagasawa, A.; Fujihara, T. Cyclometalated Platinum(II) Complexes in a *Cis-N, N* Configuration: Photophysical Properties and Isomerization to Trans Isomers. *Inorg. Chem.* **2022**, *61*, 3420–3433. [CrossRef]
13. Li, B.; Liang, Z.; Yan, H.; Li, Y. Visual Self-Assembly and Stimuli-Responsive Materials Based on Recent Phosphorescent Platinum(II) Complexes. *Mol. Syst. Des. Eng.* **2020**, *5*, 1578–1605. [CrossRef]
14. Soto, M.A.; Kandel, R.; MacLachlan, M.J. Chromic Platinum Complexes Containing Multidentate Ligands. *Eur. J. Inorg. Chem.* **2021**, *2021*, 894–906. [CrossRef]
15. Shiotsuka, M.; Ono, R.; Kurono, Y.; Asano, T.; Sakae, Y. Photoluminescence of Platinum(II) Diethynylphenanthroline Organometallic Complexes with Bis-Arylethynyl Derivatives in Solution and Solid State. *J. Organomet. Chem.* **2019**, *880*, 116–123. [CrossRef]
16. Shiotsuka, M.; Asano, T.; Kurono, Y.; Ono, R.; Kawabe, R. Synthesis and Photophysical Characterization of Phosphorescent Platinum(II) Bis-(Trimethylsilyl)Ethynyl-Phenanthroline Organometallic Complexes with Bis-Arylethynyl Derivatives. *J. Organomet. Chem.* **2017**, *851*, 1–8. [CrossRef]
17. Shiotsuka, M.; Goto, A.; Miura, S.; Uekusa, H.; Ono, R. Vapochromism and Vapoluminescence of Platinum(II) 3,8-Bis-(3-Hydroxy-3-Methylbut-1-Yn-1-Yl)-Phenanthroline Organometallic Complexes with Bis-Arylethynyl Derivatives. *J. Organomet. Chem.* **2020**, *929*, 121554. [CrossRef]
18. Shiotsuka, M.; Ogihara, M.; Hanada, T.; Kasai, K. Multicolor Detection with Vapochromism of Platinum(II) 3,8-Bis-(2-Triethylsilylethynyl)-Phenanthroline Organometallic Complexes with Bis-Arylethynyl Derivatives. *J. Organomet. Chem.* **2022**, *965–966*, 122334. [CrossRef]
19. Norton, A.E.; Karimi Abdolmaleki, M.; Zhao, D.; Taylor, S.D.; Kennedy, S.R.; Ball, T.D.; Bovee, M.O.; Connick, W.B.; Chatterjee, S. Vapoluminescence Hysteresis in a Platinum(II) Salt-Based Humidity Sensor: Mapping the Vapochromic Response to Water Vapor. *Sens. Actuators B Chem.* **2022**, *359*, 131502. [CrossRef]
20. Ni, J.; Guo, Z.; Zhu, Q.; Liu, S.; Zhang, J. The Two-Stepwise Luminescent Switching Properties of Triple-Stimuli-Responsive Platinum(II) Complexes Bearing 4,4′-Bis(2-Phenylethynyl)-2,2′-Bipyridine Ligand. *Dye. Pigment.* **2023**, *217*, 111406. [CrossRef]
21. Kobayashi, A.; Kato, M. Vapochromic Platinum(II) Complexes: Crystal Engineering toward Intelligent Sensing Devices. *Eur. J. Inorg. Chem.* **2014**, *2014*, 4469–4483. [CrossRef]
22. Lai, S.-W.; Chan, M.C.W.; Cheung, K.-K.; Che, C.-M. Spectroscopic Properties of Luminescent Platinum(II) Complexes Containing 4,4′,4″-Tri-Tert-Butyl-2,2′:6′,2″-Terpyridine (^{t}Bu$_3$Tpy). Crystal Structures of [Pt(^{t}Bu$_3$Tpy)Cl]ClO$_4$ and [Pt(^{t}Bu$_3$Tpy){CH$_2$C(O)Me}]ClO$_4$. *Inorg. Chem.* **1999**, *38*, 4262–4267. [CrossRef]
23. Büchner, R.; Cunningham, C.T.; Field, J.S.; Haines, R.J.; McMillin, D.R.; Summerton, G.C. Luminescence Properties of Salts of the [Pt(4′Ph-Terpy)Cl]+ Chromophore: Crystal Structure of the Red Form of [Pt(4′Ph-Terpy)Cl]BF4 (4′Ph-Terpy = 4′-Phenyl-2,2′:6′,2″-Terpyridine). *J. Chem. Soc. Dalton Trans.* **1999**, *999*, 711–718. [CrossRef]
24. Hill, M.G.; Bailey, J.A.; Miskowski, V.M.; Gray, H.B. Spectroelectrochemistry and Dimerization Equilibria of Chloro(Terpyridine)-Platinum(II). Nature of the Reduced Complexes. *Inorg. Chem.* **1996**, *35*, 4585–4590. [CrossRef]
25. Kui, S.C.F.; Law, Y.-C.; Tong, G.S.M.; Lu, W.; Yuen, M.-Y.; Che, C.-M. Spectacular Luminescent Behaviour of Tandem Terpyridyl Platinum(II) Acetylide Complexes Attributed to Solvent Effect on Ordering of Excited States, "Ion-Pair" Formation and Molecular Conformations. *Chem. Sci.* **2011**, *2*, 221–228. [CrossRef]
26. Yam, V.W.-W.; Chan, K.H.-Y.; Wong, K.M.-C.; Zhu, N. Luminescent Platinum(II) Terpyridyl Complexes: Effect of Counter Ions on Solvent-Induced Aggregation and Color Changes. *Chem. A Eur. J.* **2005**, *11*, 4535–4543. [CrossRef]
27. Bailey, J.A.; Hill, M.G.; Marsh, R.E.; Miskowski, V.M.; Schaefer, W.P.; Gray, H.B. Electronic Spectroscopy of Chloro(Terpyridine)-Platinum(II). *Inorg. Chem.* **1995**, *34*, 4591–4599. [CrossRef]
28. Kato, M.; Kishi, S.; Wakamatsu, Y.; Sugi, Y.; Osamura, Y.; Koshiyama, T.; Hasegawa, M. Outstanding Vapochromism and PH-Dependent Coloration of Dicyano(4,4′-Dicarboxy-2,2′-Bipyridine)Platinum(II) with a Three-Dimensional Network Structure. *Chem. Lett.* **2005**, *34*, 1368–1369. [CrossRef]
29. National Library of Medicine. Available online: https://pubchem.ncbi.nlm.nih.gov (accessed on 29 May 2023).
30. Karimi Abdolmaleki, M. *Synthesis, Characterization, Thermodynamic, and Kinetic Studies of Vapochromic Pt(II) Complexes*; University of Cincinnati: Cincinnati, OH, USA, 2018.
31. Taylor, S.D.; Norton, A.E.; Hart, R.T.; Abdolmaleki, M.K.; Krause, J.A.; Connick, W.B. Between Red and Yellow: Evidence of Intermediates in a Vapochromic Pt(II) Salt. *Chem. Commun.* **2013**, *49*, 9161–9163. [CrossRef]

Disclaimer/Publisher's Note: The statements, opinions and data contained in all publications are solely those of the individual author(s) and contributor(s) and not of MDPI and/or the editor(s). MDPI and/or the editor(s) disclaim responsibility for any injury to people or property resulting from any ideas, methods, instructions or products referred to in the content.

Article

Effect of Traps on the UV Sensitivity of Gallium Oxide-Based Structures

Vera M. Kalygina [1,*], Alexander V. Tsymbalov [1], Petr M. Korusenko [2,3], Aleksandra V. Koroleva [4] and Evgeniy V. Zhizhin [4]

[1] Research and Development Centre for Advanced Technologies in Microelectronics, National Research Tomsk State University, 634050 Tomsk, Russia; zoldmine@gmail.com
[2] Department of Solid State Electronics, Saint Petersburg State University, 199034 Saint Petersburg, Russia; korusenko_petr@mail.ru
[3] Department of Physics, Omsk State Technical University, 644050 Omsk, Russia
[4] Research Park, Saint Petersburg State University, 199034 Saint Petersburg, Russia; koroleva.alexandra.22@gmail.com (A.V.K.); evgeny_liquid@mail.ru (E.V.Z.)
* Correspondence: vmkalygina@mail.ru

Abstract: Resistive metal/β-Ga_2O_3/metal structures with different interelectrode distances and electrode topologies were investigated. The oxide films were deposited by radio-frequency magnetron sputtering of a Ga_2O_3 (99.999%) target onto an unheated sapphire c-plane substrate (0001) in an Ar/O_2 gas mixture. The films are sensitive to ultraviolet radiation with wavelength λ = 254. Structures with interdigital electrode topology have pronounced persistent conductivity. It is shown that the magnitude of responsivity, response time $τ_r$, and recovery time $τ_d$ are determined by the concentration of free holes p involved in recombination processes. For the first time, it is proposed to consider hole trapping both by surface states N_{ts} at the metal/Ga_2O_3 interface and by traps in the bulk of the film.

Keywords: gallium oxide; MSM structures; UV radiation; traps; persistent conductivity; RFMS

Citation: Kalygina, V.M.; Tsymbalov, A.V.; Korusenko, P.M.; Koroleva, A.V.; Zhizhin, E.V. Effect of Traps on the UV Sensitivity of Gallium Oxide-Based Structures. *Crystals* **2024**, *14*, 268. https://doi.org/10.3390/cryst14030268

Academic Editors: Andrew F. Zhou and Peter X. Feng

Received: 8 February 2024
Revised: 27 February 2024
Accepted: 6 March 2024
Published: 9 March 2024

Copyright: © 2024 by the authors. Licensee MDPI, Basel, Switzerland. This article is an open access article distributed under the terms and conditions of the Creative Commons Attribution (CC BY) license (https://creativecommons.org/licenses/by/4.0/).

1. Introduction

In recent years, the exploration of advanced materials for electronic devices has led to the emergence of gallium oxide (Ga_2O_3) as a promising semiconductor for various applications, particularly in the development of detectors [1,2]. Gallium oxide, a wide-bandgap semiconductor (E_g = 4.4–5.3 eV), exhibits unique properties that make it well suited for high-performance ultraviolet sensors [3,4]. This compound has attracted considerable attention in the field of optoelectronics due to its wide bandgap energy, excellent thermal stability, and robust chemical properties [5,6]. Ultraviolet detectors (UVDs) based on gallium oxide have found a wide range of applications in areas such as environmental monitoring, flame detection, communication systems, biomedical applications, etc. [7,8].

The sensitivity of UVDs to UV radiation and their speed of operation are the main parameters characterizing their ability to detect UV radiation and the time resolution of the devices, respectively [9]. Previous works have demonstrated detectors based on Ga_2O_3 with high values of responsivity (R), detectivity (D*), and external quantum efficiency (EQE) [10]. However, response times $τ_r$ and recovery times $τ_d$ exceeded several seconds. In this regard, the development of ultraviolet detectors with minimal response $τ_r$ and recovery times $τ_d$ as well as high sensitivity is an important task.

In β-Ga_2O_3, there are many deep traps in the bandgap, formed by structural defects and impurities [11]. Deep traps in β-Ga_2O_3 can be classified into two types: majority and minority traps. This classification is either based on their capability to capture an electron (majority trap) or a hole (minority trap), which is clearly manifested in the capture cross-section rate value, or depends on how close they are to the valence band maximum and conduction band minimum [12].

Recent research on deep electron traps in n-type β-Ga$_2$O$_3$ Schottky barrier diodes (SBDs) has enabled the mapping of the upper half of the bandgap and revealed the presence of four deep electron traps with activation energy values near Ec-0.6 eV (E1), Ec-0.79 eV (E2), Ec-0.75 eV (E2*), and Ec-1 eV (E3) [12].

Ga$_2$O$_3$-based UVDs exhibit persistent conductivity [13]. This is due to holes that can self-localize on defects in the gallium oxide crystal lattice [14]. As a result, the response and recovery times of gallium oxide UVDs can be several minutes. Minority traps H1, H2, and H3, with energies of about 0.2, 0.3, and 1.3 eV, respectively, above the valence band maximum are the most known defects that are related to gallium vacancy [12]. Therefore, the study of the effect of traps on photoelectric performance plays a key role in the development and optimization of gallium oxide detectors [15].

This work considers ultraviolet radiation detectors based on M/β-Ga$_2$O$_3$/M with different interelectrode distances and topologies. The responsivity values of the structures were compared with their response τ_r and recovery times τ_d. An analysis of their current–voltage characteristics (I–V characteristics) was carried out over a wide voltage range. For the first time, it was proposed to take into account surface states at the gallium oxide/metal interface which are involved in the capture of free holes.

2. Experimental Procedure

2.1. Methodology of Structure Production

In this work, two types of detectors based on β-Ga$_2$O$_3$ were studied: detectors with two parallel electrodes (the first sample type) and detectors with interdigitated electrodes (the second sample type).

Electrodes for detectors of the first type were created by depositing a layer of platinum on smooth sapphire substrates using the vacuum deposition method. In the next stage, contacts were formed using explosive photolithography. The distance between the electrodes was 250 μm (Figure 1a). A gallium oxide film 150–200 nm thick was deposited by radio-frequency magnetron sputtering (RFMS) of a Ga$_2$O$_3$ target (99.999%) onto unheated sapphire substrates using an AUTO-500 setup (Edwards, UK) in an Ar/O$_2$ gas mixture. The oxygen concentration in the mixture was (56.1 ± 0.5) vol. %. The distance between the target and the substrate was 70 mm. The pressure in the chamber during spraying was maintained at 7 × 10^{-6} bar. Then, the gallium oxide film was annealed in argon at 900 °C for 30 min. After annealing, the film transformed from an amorphous state to a polycrystalline beta phase of Ga$_2$O$_3$ [16]. The wafer was then cut into 1.4 × 1.4 mm^2 chips, as shown in Figure 1a.

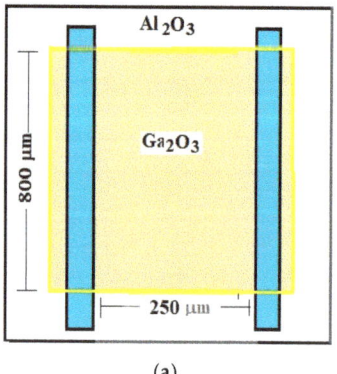 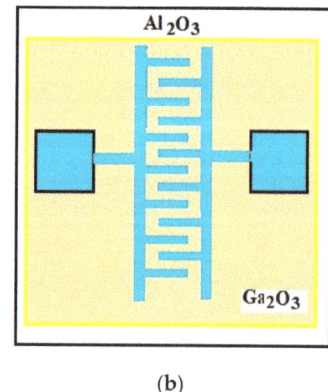

(a)   (b)

Figure 1. Schematic representation of detectors with two parallel electrodes (**a**) and detectors with interdigitated electrodes (**b**).

The second type of detectors was prepared similarly, with the exception that a gallium oxide film was first deposited onto a sapphire substrate. Following this, an interdigitated topology of Ti/V-based electrodes was formed. Samples with interelectrode distance d equal to 50, 30, 10, and 5 μm were obtained. The number of electrodes was 50, 75, 150, and 200, respectively. The chips were 3.5 × 3.5 mm² in size, regardless of interelectrode distance (Figure 1b).

2.2. Structural Property Characterization and Electrical Measurements

X-ray diffraction (XRD) spectra of Ga_2O_3 films before and after annealing were measured by a Bruker D8 Discover (Bruker AXS GmbH, Germany) diffractometer (CuK α radiation λ = 1.54 Å) with a position- sensitive linear detector LynxEye in Bragg–Brentano geometry at an ω = 0.5° angle offset from the normal position of the sapphire substrate.

The chemical composition of the samples was studied by X-ray photoelectron spectroscopy (XPS). XPS measurements were carried out using a hemispherical analyzer included in the ESCALAB 250Xi (Thermo Fisher Scientific, Waltham, MA, USA) laboratory spectrometer. The measurements were carried out using a monochromatized AlKa radiation (hν = 1486.6 eV). Survey and core-level (Ga 2p, O 1s, Ga 3d) photoemission (PE) spectra were recorded in the constant-energy mode of the analyzer with pass energies of 100 and 50 eV, respectively. The film's surface was irradiated with argon ions at an average energy of 500 eV for 300 s to remove adsorbed atoms and molecules of contaminants. The analysis of the core-level spectra was processed using Avantage Data System software.

The spectral dependencies of the photoelectric properties were measured by means of the spectrometric system based on a MonoScan 2000 monochromator (Ocean Insight, Orlando, FL, USA) and a DH-2000 Micropack lamp, described in detail in [10]. A krypton-fluorine lamp VL-6.C was used as the source of irradiation at λ = 254 nm and the light power density P = 780 μW/cm². The photoelectric and electrical properties of the samples were measured by means of Keithley 2611B.

3. Results

Figure 2 shows the XRD spectra of the gallium oxide film before and after annealing at 900 °C. Peaks at 2θ = 41.7° and 90.7° correspond to (0006) and (00012) crystals planes of sapphire substrates. The gallium oxide thin film without annealing is amorphous (Figure 2, black line). The film annealed at 900 °C has peaks of the monoclinic β-Ga_2O_3 phase (Figure 2, red line and blue bars). A partial texture with a preferred orientation of (20-1) Ga_2O_3 planes parallel to the substrate surface is formed in the annealed film.

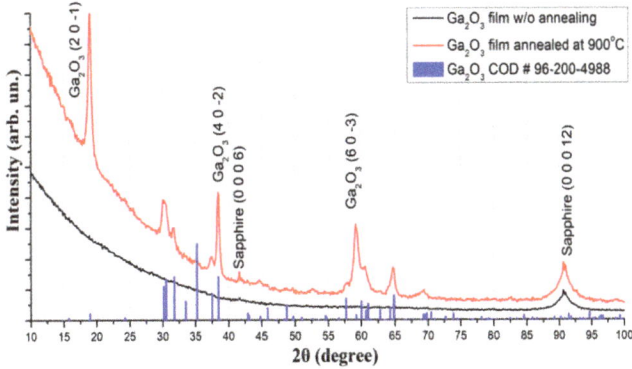

Figure 2. XRD pattern of gallium oxide film without annealing and annealed at 900 °C.

Figure 3 shows the survey spectra of the studied sample before and after 300 s etching with argon ions. In both survey PE spectra, gallium (Ga 2s~1303 eV, Ga 2p~1118.1 eV, Ga LMM~424.5 eV, Ga 3s~161.2 eV, Ga 3p~106.1 eV, and Ga 3d~20.8 eV) and oxygen lines

(O KLL~976.5 eV, O 1s~531.4 eV, O 2s~24 eV) are present [17]. At the same time, the survey spectrum measured from the surface before etching also contains carbon lines (C KLL~1224 eV, C 1s~285 eV). After etching with argon ions, the intensity of the O 1s PE line decreases significantly, and the intensity of the G 2p PE line, on the contrary, increases. This result indicates a change in the [O]/[Ga] ratio. It is important to note that for a more correct quantitative analysis and determination of the [O]/[Ga] ratio, the gallium Ga 3s and the oxygen O 1s lines were used. This is due to the fact that the escape depths of photoelectrons, which are determined by the inelastic mean free path (IMFP) of a photoelectron, for these shells are ~2.2 and ~1.7 nm, respectively, while for the Ga 2p and O 1s shells, the IMFP differs by almost 2 times (see Table 1). It is clearly seen that the [O]/[Ga] ratio differs before and after ion etching; the reasons for this are outlined in the Discussion section.

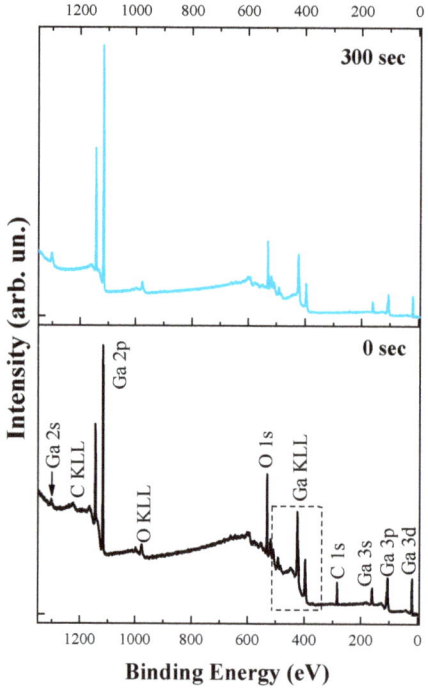

Figure 3. Survey PE spectra of the RFMS-deposited β-Ga_2O_3 film, measured before and after 300 s etching with argon ions.

Table 1. Elemental composition of the RFMS-deposited β-Ga_2O_3 film.

Ga_2O_3 Film	Concentration, at.%		IMFP *, nm					[O]/[Ga]
	[O] O 1s	[Ga] Ga 3s	Ga 2p	O 1s	Ga 3s	Ga 3p	Ga 3d	
Surface (no etching)	59.80	40.20	0.88	1.72	2.21	2.28	2.39	1.48
After 300 s Ar⁺ etching	54.4	45.6						1.19

* IMFP—inelastic mean free path.

A detailed analysis of the core-level PE spectra measured after 300 s Ar⁺ etching showed that the positions of the Ga $2p_{3/2}$, Ga 3d, and O 1s line maxima at the binding energy are 1118.1, 20.6, and 531.1 eV, respectively. This result, together with the [O]/[Ga] ratio, indicates that gallium is in the Ga_2O_3 compound (Figure 4) [18].

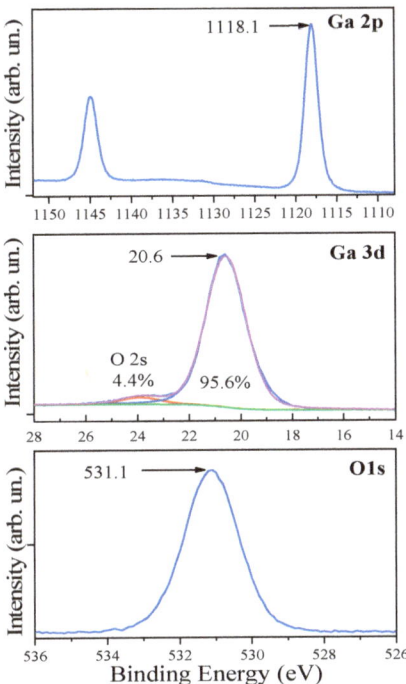

Figure 4. Ga 2p, Ga 3d, and O 1s PE spectra of the RFMS-deposited β-Ga$_2$O$_3$ film, measured after 300 s etching with argon ions.

Figure 5 shows the transmission spectrum for β-Ga$_2$O$_3$ films. Gallium oxide films are transparent in the near-ultraviolet (UVA) range. The transmittance values T decrease from 83% to 74.5% between wavelengths of 320 nm and 280 nm. The structures are solar-blind. The optical bandgap of the β-Ga$_2$O$_3$ film was determined by analyzing the dependence of the absorption coefficient α on photon energy hν, resulting in E$_g$ = 4.8 ± 0.05 eV (Figure 5, inset).

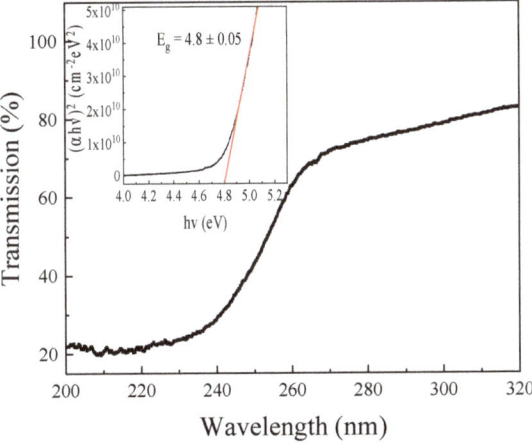

Figure 5. Transmission spectrum of the RFMS-deposited β-Ga$_2$O$_3$ films. Dependence of α^2 on photons energy (black curve) is shown in insertion (red curve—approximation of the linear section).

3.1. Structures of the First Type

The current–voltage characteristics of both sample types are symmetrical regarding voltage polarity. The dependence of the dark current (I_D) on the voltage (U) is linear for detectors of the first type. I_D values do not exceed 10-180 pA in the voltage range $-200 \leq U \leq 200$ V (Figure 6a). The current–voltage characteristic maintains a linear shape under UV exposure (Figure 6b). Ultraviolet radiation with a wavelength of 254 nm leads to an increase in current by 3–4 orders of magnitude. Total current (I_L) values increase linearly under UV exposure to 1.6–1.7 µA with increasing voltage up to ±200 V. In Figure 6, the I_{L1} and I_{L2} curves correspond to the currents measured during the first and second measurements of the structures during continuous exposure to UV radiation.

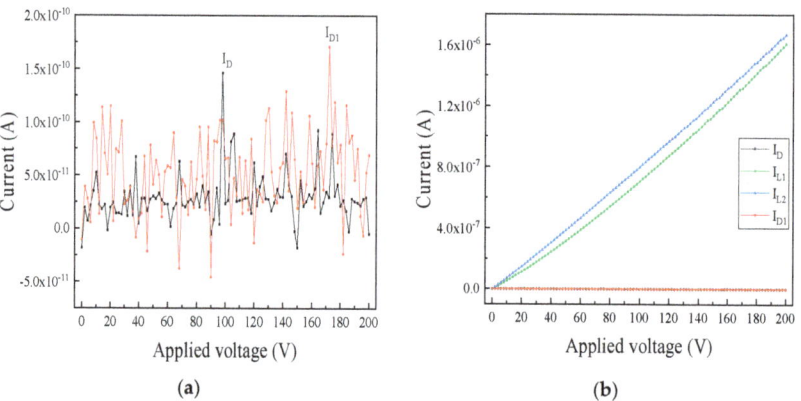

Figure 6. Dark current–voltage characteristics (**a**) and current–voltage characteristics measured under exposure to radiation with λ = 254 nm (**b**). I_D and I_{D1} are dark currents measured before (I_D) and immediately after turning off the UV radiation (I_{D1}); I_{L1} and I_{L2} are currents measured during exposure to radiation with λ = 254 nm.

Figure 6a shows that the dark current I_{D1} returns to its original values I_D almost immediately after exposure to UV radiation. The dark currents measured before and immediately after turning off the UV light are consistent within experimental error (Figure 6a). Thus, the structures do not exhibit "persistent conductivity".

Figure 7 shows the time profile of the current structure under UV exposure with λ = 254 nm. The response times τ_r and recovery times τ_d of the photocurrent are in the millisecond range, confirming the absence of "persistent conductivity" in samples of this type.

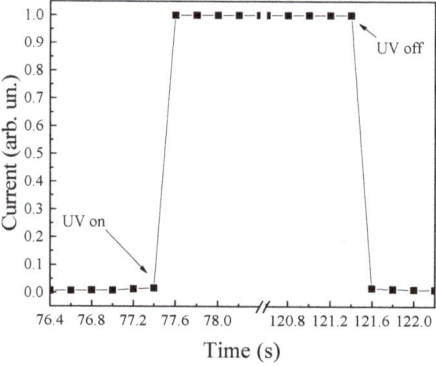

Figure 7. Time dependence of the normalized I_L at λ = 254 nm, P = 780 µW/cm^2, and U = 1 V.

3.2. Structures of the Second Type

The current–voltage characteristics of structures with interdigitated contacts (structures of the second type) and interelectrode distances of 50 and 30 µm are linear in the range of 0 to 200 V (Figure 8).

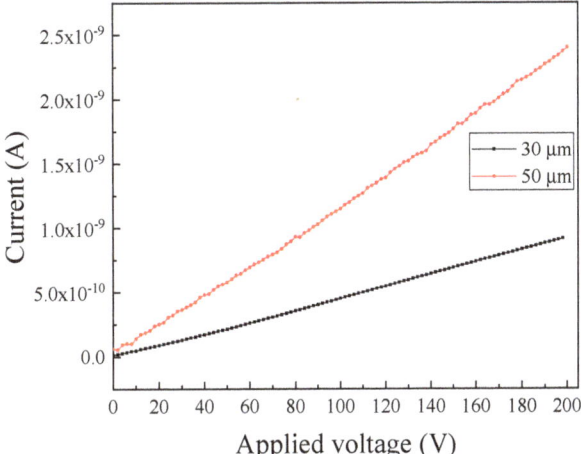

Figure 8. Dark current–voltage characteristics of samples with interdigitated electrodes and interelectrode distances of 50 µm (red curve) and 30 µm (black curve).

The dark I–V shape changes as the interelectrode distance decreases in structures with interdigitated electrodes. Thus, the dark current–voltage characteristics of the samples with an interelectrode distance of 5 µm and 10 µm cannot be represented by a linear dependence of current I on voltage U in the range of the indicated voltages (Figure 9, inset). Figure 9 shows the dark current–voltage characteristics on a double logarithmic scale for samples with an interelectrode distance of 5 µm and 10 µm in the voltage range $0 < V \leq 200$ V.

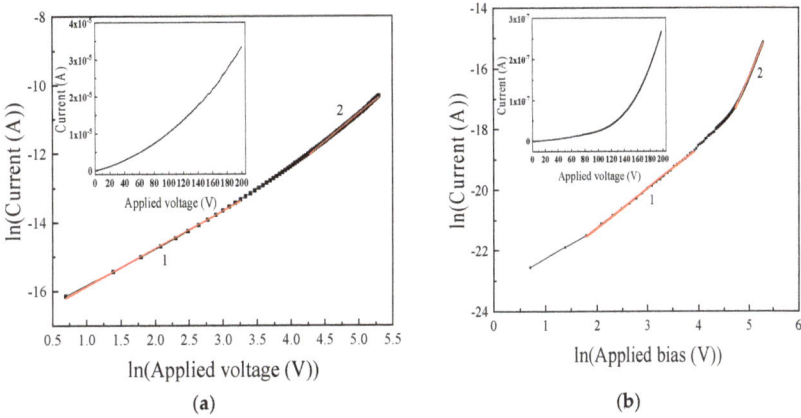

Figure 9. Dark current–voltage characteristics of samples with interdigitated electrodes and interelectrode distances of 5 µm (**a**) and 10 (**b**) µm on a double logarithmic scale (insert current–voltage curve in linear coordinates).

The coordinates ln(I) from ln(U) contain linear sections. The slope of the first section is 1.10 for the structure with an interelectrode distance of 5 µm and 1.63 for the second section. For a structure with an interelectrode distance of 10 µm, the slope of the first section is 1.31, and for the second—3.80. Thus, the dark current–voltage characteristics of samples with interelectrode distances d = 5 and 10 μm can be represented by power law dependence I~U^m, where m = 1 in the range of low voltages and m > 1 at higher electric fields. Thus, the conductivity of structures of the second type in the absence of radiation is determined by space-charge-limited currents (SCLCs) in a semiconductor with traps unevenly distributed in energy (Figure 9). The relationship between current and voltage in this case is determined by the following expression [19]:

$$J = N_c \mu e^{1-l} \left[\frac{\varepsilon \varepsilon_0 l}{N_t(l+1)} \right]^l \left(\frac{2l+1}{l+1} \right)^{l+1} \frac{U^{l+1}}{d^{2l+1}} \quad (1)$$

where N_c is the effective density of quantum states in the conduction band of gallium oxide; ε—the relative permittivity of Ga_2O_3; ε_0—the electrical constant; μ—electron mobility; N_t is the concentration of traps in the Ga_2O_3 film; e—electron charge; l = m − 1. Exponent l was determined from the slope of the current–voltage characteristic on a double logarithmic scale. The transition voltage from Ohm's law to SCLC is described by the following expression:

$$U_{\Omega \to T} = \frac{ed^2}{\varepsilon \varepsilon_0} \left(\frac{l+1}{2l+1} \right)^{(1+l)/l} N_t \left(\frac{n_0}{N_c} \right)^{1/l} \left(\frac{l+1}{l} \right). \quad (2)$$

From experimental data and Expression (2), it was found that the trap concentration N_t was 2×10^{17} cm^{-3} and 6×10^{17} cm^{-3} for structures with d = 5 and 10 μm, respectively. The value of N_t was calculated using $n_0 = 1 \times 10^{17}$ cm^{-3} and $N_c = 4 \times 10^{18}$ cm^{-3} [20]. The conclusion regarding SCLCs in these samples is consistent with the established concept of multiple traps in the bandgap of β-Ga_2O_3 [21–23].

In structures with interdigitated electrodes (structures of the second type), the dependence of current I_L on voltage cannot be represented by a linear dependence, regardless of the interelectrode distance (Figure 10a).

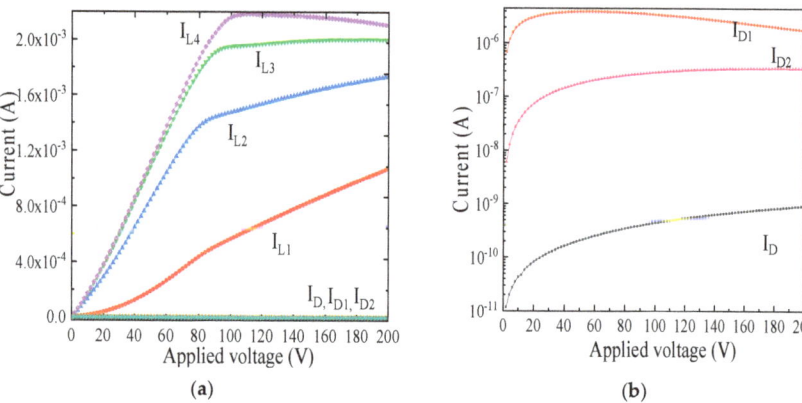

Figure 10. Current–voltage characteristics of the second sample type with d = 30 μm. (**a**): I_{L1}–I_{L4}—currents measured during exposure to radiation with λ = 254 nm; (**b**): I_D—initial dark current, I_{D1}—dark current measured immediately after UV exposure, I_{D2}—dark current measured immediately after I_{D1}.

Initially, I_L increases linearly with increasing voltage, and then a slowdown in the growth of the light current is observed, followed by a decrease, for example, in the I_{L4} curve in Figure 10a.

The light current increases with each subsequent measurement: $I_{L1} < I_{L2} < I_{L3} < I_{L4}$ during continuous exposure to radiation with λ = 254 nm (Figure 10a). Further measurements lead to the stabilization of the light current: it reaches a stationary state.

A similar behavior of the current–voltage characteristic under continuous UV exposure is typical for samples with d = 10 μm (Figure 11a). In the region of strong electric fields, a noticeable decrease in light current is observed with increasing voltage in the structure (Figure 11a, curves I_{L1}–I_{L8}). The maximum points of I_L from U were determined by differentiation (Figure 11b).

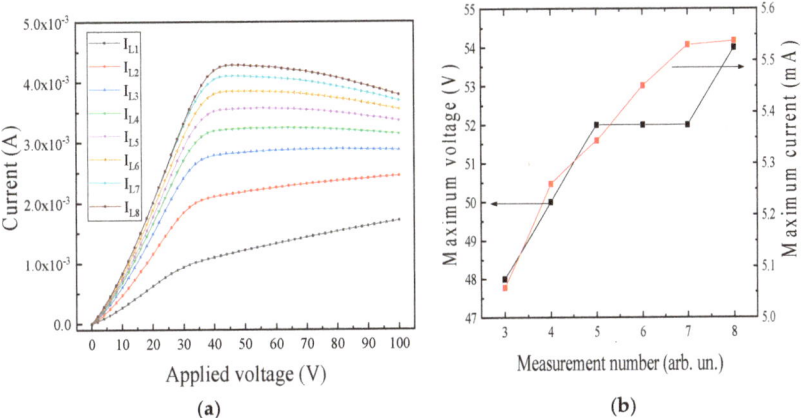

Figure 11. I–V characteristics of the second type of sample with d = 10 μm. (**a**) I_{L1}–I_{L8} currents measured during exposure to radiation with λ = 254 nm; (**b**) dependence of the maximum current value and the corresponding voltage on the measurement number.

It should be noted that the behavior of the current–voltage characteristic is also affected by radiation with energy hν < E_g. In [16], the influence of pre-exposure of the structure to radiation with quanta of energy hν less than the bandgap E_g of gallium oxide on the behavior of the current–voltage characteristic is considered. Pre-exposure to broadband radiation increases the photocurrent I_{L1} generated by short-wavelength radiation and reduces the deviation of photocurrents I_{L2}, I_{L3}, and I_{L4} from I_{L1} (first measurement) (Figure 12).

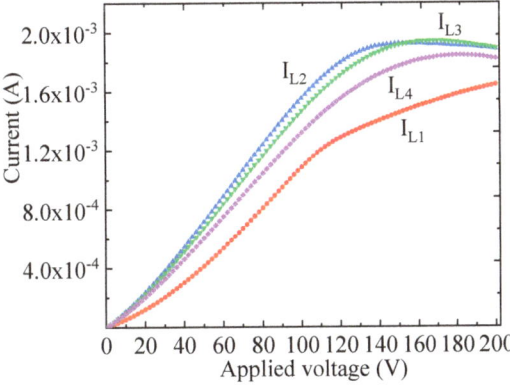

Figure 12. Current–voltage characteristics of a sample with d = 30 μm after exposure to broadband radiation with hν < E_g; the designations are the same as in Figure 10.

Exposure to radiation with hν < E_g reduces the concentration of active trap centers due to changes in their charge state. This leads to an increase in the photocurrent and stability of detectors in the UV range [16].

Dark current I_D takes some time to return to its initial value after exposure to UV radiation. In other words, the samples have persistent photoconductivity. Figure 10b shows dark current–voltage characteristics measured at different times: before UV exposure (I_D); after UV exposure (I_{D1}); and 30 s after UV exposure (I_{D2}). The dark current measured immediately after turning off the UV is 3–4 orders of magnitude higher than the initial values of dark current I_D at sample voltages $0 \leq U \leq 200$ V. The presence of a maximum in the I_{D1} curve is explained by the gradual relaxation of the residual current as the voltage across the structure increases.

Thus, characteristics of structures with interdigitated electrodes are a large responsivity value and the presence of "persistent conductivity". Table 2 shows the responsivity R and detectivity D* values for structures of the first and second types. The responsivity and detectivity values were calculated using the following expressions [24]:

$$R = \frac{I_L - I_D}{Ps}, \qquad (3)$$

$$D^* = R\sqrt{\frac{s}{2eI_D}} \qquad (4)$$

where P—light power density, s—effective area, e—electron charge [7].

Table 2. Photoelectric characteristics of structures with different interelectrode distances under UV irradiation, λ = 254 nm.

d, μm	s, cm²	I_D, A	I_L, A	R, A/W	D*, Jones	Bias, V
5	0.0125	1.96×10^{-10}	7.5×10^{-3}	770	1.1×10^{16}	30
10	0.0174	1.01×10^{-10}	4.7×10^{-3}	350	8.1×10^{15}	30
30	0.0247	1.20×10^{-9}	2.1×10^{-3}	110	8.8×10^{14}	200
50	0.0270	2.77×10^{-9}	1.0×10^{-3}	47	2.6×10^{14}	200
250	0.0020	4.30×10^{-11}	1.6×10^{-6}	1.0	1.2×10^{13}	200

4. Discussions

Based on the XRD and XPS spectra, as well as the dependence of the absorption coefficient α on energy hν, it can be concluded that the film deposited using the RFMS method is β-Ga_2O_3. However, in the XPS data, a significant decrease in the [O]/[Ga] ratio from 1.48 to 1.19 was found after ion beam etching (see Table 1). This result is not associated with the presence of an oxygen-depleted layer in the Ga_2O_3 film but is explained by the removal of oxygen atoms from the Ga_2O_3 crystal lattice under the influence of argon ions during long-term treatment of the film surface. This effect was previously observed during the etching of various metal oxides with argon ions, as presented in [25]. Thus, the film obtained in this work is represented by stoichiometric Ga_2O_3 with a low content of oxygen vacancies.

Figure 13 shows the energy diagram of the M/Ga_2O_3/M structure. The energy diagram shown takes into account the ratios of the work function of electrons from metals χ_0 used in our experiments (4.3 eV—Ti, V and 5.7 eV—Pt), and the electron affinity of Ga_2O_3 (4.0 eV) [26,27].

The conductivity of structures for both types in the absence of radiation is determined by space-charge-limited currents (SCLCs) in a trap semiconductor. The transition voltage from Ohm's law to the power law dependence of current on voltage depends on the concentration of traps in the oxide film N_t and is proportional to d^2, as follows from Expression (2). Transition voltage increases with the increasing concentration of free traps (those that have not captured electrons). It is assumed that in structures of the first type with two parallel electrodes, the concentration of free traps is high. Therefore, it is not possible to achieve a transition voltage in the voltage range $0 \leq U \leq 200$ V. The high concentration of free traps causes low responsivity values R (Table 2) and relatively short response and recovery times.

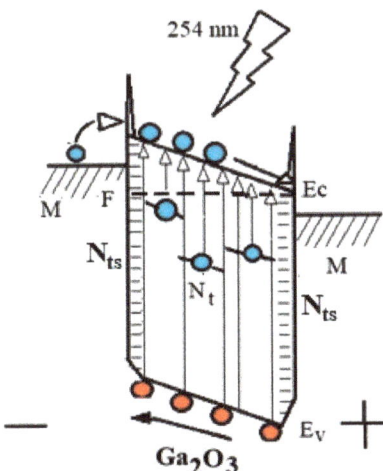

Figure 13. Schematic representation of the energy diagram of M/Ga$_2$O$_3$/M.

It is assumed that structures with interdigitated electrodes have a higher density of surface states (N$_{ts}$) at the M/Ga$_2$O$_3$ interface, which are divided into fast and slow. The average energy density of surface states should be the same for all samples obtained in one technological cycle. Then, the total number of N$_{ts}$ is determined by the contact perimeter W. The W values for the studied samples depend on the topology of the contacts and are given in Table 3. States above the Fermi level are able to trap holes at negative potentials on contact (Figure 13). Trapped holes do not contribute to recombination. As a result, a high concentration of generated electrons remains in the conduction band, which explains the large responsivity values presented in Table 3. In addition, the accumulation of holes on the negative electrode leads to the formation of an internal field.

Table 3. Time characteristics of structures under UV exposure with λ = 254 nm as a function of electrode perimeter.

d, μm	W, cm	τ_{r1}, s	τ_{r2}, s	τ_{d1}, s	τ_{d2}, s
5	39.8	0.63	15.5	0.02	2.12
50	9.95	0.28	14.5	0.05	2.30
250	0.16	0.07	1.72	0.051	0.53

Figures 10a and 11a demonstrate the significant impact of traps on the response magnitude. The I$_L$ value increases until it reaches a steady-state value with each subsequent measurement. The current–voltage characteristic's behavior is due to the gradual filling of traps. These traps capture electrons and do not participate in recombination processes during subsequent measurements.

The structures with interdigitated electrodes exhibit a monotonic dependence of current on voltage in the first measurement. However, the shape of the dependence of I$_L$ on U changes with subsequent measurements (Figures 10a and 11a). Deviation from the monotonic dependence of light current on voltage occurs at electric fields above a certain critical value E$_{cr}$ [28]. The critical field is defined as the electric field strength at which the drift length L$_E$ is equal to twice the hole diffusion length (L$_E$ = 2L$_p$) [29].

As the voltage increases, I$_L$ becomes dependent on voltage in the form of a curve with a maximum. The decrease in I$_L$ at voltages exceeding V$_{max}$ is due to the appearance of an internal electric field E$_{in}$, in the opposite direction of the external field E$_{ex}$. The voltage V$_{max}$, corresponding to the maximum value of the light current, increases with subsequent measurements (Figure 11b). The effect is explained by an increase in internal field strength

E_{in} with increasing photocurrent, and higher values of E_{ex} are required to compensate for it. It should be noted that the internal electric field plays a significant role at values of $I_L = 10^{-4}$–10^{-3} A.

The time dependences of $I_L(t)$ are characterized by two processes: fast and slow changes in current. The fast stage of change in I_L is determined by processes of generation of electrons from the valence band and excitation of electrons from trap centers located below the Fermi level into the conduction band. The capture of electrons and holes at trap centers occurs simultaneously with the generation process.

Slow processes are caused by recombination (mainly Shockley–Reed–Hall recombination) and capture by traps. The recombination rate $\Delta n/\tau_n$ depends on the concentration of free holes in the valence band. An increase in the concentration of free holes in the valence band increases the rate of recombination. Thus, the stationary state is established faster, determined by the radiation intensity and the voltage on the structure. In turn, trapped p_t holes do not participate in recombination. An increase in p_t leads to a decrease in the recombination rate and an increase in the response times τ_r and recovery times τ_d. Thus, trapped holes cause high photocurrent values.

It is assumed that free holes are captured not only by defects in the bulk of the gallium oxide film, but also by surface states at the metal/Ga_2O_3 interface. Table 3 shows the response and recovery times for structures with different interelectrode distances. The P parameter determines the perimeter of the contacts used. Times τ_r and τ_d were found as the average value measured for each of the pulses shown in Figure 14.

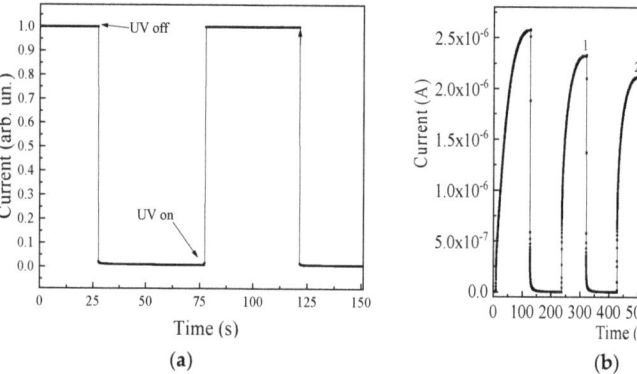

Figure 14. Time dependence of I_L for structures with two electrodes (**a**) and with interdigitated electrodes d = 50 μm (**b**) under UV exposure with λ = 254 nm and applied bias U = 1 V.

From Table 3 and Figure 14, we can draw the following conclusions:

1. The response τ_r and recovery times τ_d of the photocurrent in structures with interdigitated electrodes are described by the following expression [30]:

$$I = I_s + A \times \exp(-t/\tau_{r1}) + B \times \exp(-t/\tau_{r2}), \quad (5)$$

$$I = I_s + A \times \exp(-t/\tau_{d1}) + B \times \exp(-t/\tau_{d2}), \quad (6)$$

where I_s represents the steady-state current and A and B are the fitting constants. τ_{r1} and τ_{r2} are fast and slow components of response times, whereas τ_{d1} and τ_{d2} denote the fast and slow components of recovery times;

2. Structures with interdigitated electrodes have recovery times that are an order of magnitude shorter than their response times;
3. The maximum value of I_L decreases with each successive UV exposure (Figure 14b);

4. The minimum values of τ_r and τ_d correspond to structures with two electrodes, which have the lowest responsivity (Table 3).

The differences in responsivity values and time parameters between structures of the first and second types are explained by the different concentrations of traps that take part in the processes of generation and recombination.

Considering the metal/Ga_2O_3 interface, it is important to take into account the surface states, whose number depends on the perimeter of the contacts W. The Table 3 shows that the lowest W values were obtained for two-electrode samples. At a low concentration of surface states, a large fraction of the generated holes remain free and are capable of participating in recombination. Thus, there is a low concentration of electrons remaining in the conduction band. Consequently, the response values R (Table 2) are relatively small, and the values of τ_r and τ_d are fast (Table 3, Figure 14a).

Structures with large W values (d = 5 μm) exhibit maximum responsivity R and the longest τ_r and τ_d times (Tables 2 and 3). Increasing the interelectrode distance to 50 μm leads to a fourfold decrease in W. As a result, responsivity is reduced by one order of magnitude, and recovery times are cut in half.

Studies of isotypic and anisotypic Ga_2O_3 semiconductor heterostructures confirm the crucial role of free holes in responding to far-range UV radiation [31,32]. The presence of a source of free holes in the valence band of the semiconductor provides a high recombination rate and low values of τ_r compared to similar data for samples with interdigitated electrodes.

5. Conclusions

The electrical and photoelectric characteristics of thin-film M/Ga_2O_3/M structures with two parallel electrodes (d = 250 μm) and interdigitated contacts with interelectrode distances d = 50, 30, 10, and 5 μm were studied.

The dark currents of the studied structures, regardless of interelectrode distance, were caused by currents limited by the space charge in the semiconductor. Traps were exponentially distributed in the bandgap of gallium oxide. The concentration of trap centers, estimated from experimental data, was $(4-6) \cdot 10^{17}$ cm^{-3}.

The relationship between the magnitude of responsivity and the time characteristics for structures with different electrode topologies was considered. It was shown that the response and time parameters τ_r and τ_d are determined by the concentration of free holes p participating in recombination processes. However, the presence of free holes depends on the concentration of traps that can capture the holes. For the first time, it was proposed to take into account the capture of holes by the surface states of N_{ts} at the metal/Ga_2O_3 interface in addition to their capture by trap centers in the bulk of the oxide film. The concentration of N_{ts} depends on the topology of the electrodes, in other words, on the perimeter.

Author Contributions: Conceptualization, V.M.K. and A.V.T.; methodology, V.M.K., A.V.T. and A.V.K.; formal analysis, V.M.K., A.V.T. and P.M.K.; investigation, V.M.K., A.V.T., A.V.K. and E.V.Z.; resources, V.M.K., A.V.K. and E.V.Z.; data curation, V.M.K. and A.V.T.; writing—original draft preparation, V.M.K. and A.V.T.; writing—review and editing, V.M.K. and A.V.T.; visualization, V.M.K., P.M.K. and A.V.T.; supervision, V.M.K.; project administration, V.M.K.; funding acquisition, V.M.K. All authors have read and agreed to the published version of the manuscript.

Funding: This work was supported by a grant under the Decree of the Government of the Russian Federation No. 220 of 9 April 2010 (Agreement No. 075-15-2022-1132 of 1 July 2022).

Data Availability Statement: All data that support the findings of this study are included within the article.

Conflicts of Interest: The authors declare no conflicts of interest. The funders had no role in the design of the study; in the collection, analyses, or interpretation of data; in the writing of the manuscript, or in the decision to publish the results.

References

1. Osipov, A.V.; Sharafudinov, S.S.; Kremleva, A.V.; Osipova, E.V.; Smirnov, A.M.; Kukushkin, S.A. Phase transformations in gallium oxide layers. *Tech. Phys. Lett.* **2023**, *49*, 4–7.
2. Al-Hardan, N.H.; Abdul Hamid, M.A.; Jalar, A.; Firdaus-Raih, M. Unleashing the potential of gallium oxide: A paradigm shift in optoelectronic applications for image sensing and neuromorphic computing applications. *Mater. Today Phys.* **2023**, *38*, 101279. [CrossRef]
3. Xiao, Y.; Yang, S.; Cheng, L.; Zhou, Y.; Qian, Y. Research progress of solar-blind UV photodetectors based on amorphous gallium oxide. *Opto-Electron. Eng.* **2023**, *50*, 230005.
4. Fei, Z.; Chen, Z.; Chen, W.; Chen, S.; Wu, Z.; Lu, X.; Wang, G.; Liang, J.; Pei, Y. ε-Ga_2O_3 thin films grown by metal-organic chemical vapor deposition and its application as solar-blind photodetectors. *J. Alloys Compd.* **2022**, *925*, 166632. [CrossRef]
5. Yakovlev, N.N.; Nikolaev, V.I.; Stepanov, S.I.; Almaev, A.V.; Pechnikov, A.I.; Chernikov, E.V.; Kushnarev, B.O. Effect of Oxygen on the Electrical Conductivity of Pt-Contacted α-Ga_2O_3/ε(κ)-Ga_2O_3 MSM Structures on Patterned Sapphire Substrates. *IEEE Sens. J.* **2021**, *21*, 14636–14644. [CrossRef]
6. Fornari, R.; Pavesi, M.; Montedoro, V.; Klimm, D.; Mezzadri, F.; Cora, I.; Pécz, B.; Boschi, F.; Parisini, A.; Baraldi, A.; et al. Thermal stability of ε-Ga_2O_3 polymorph. *Acta Mater.* **2017**, *140*, 411–416. [CrossRef]
7. Galazka, Z. β-Ga_2O_3 for wide-bandgap electronics and optoelectronics. *Semicond. Sci. Technol.* **2018**, *33*, 113001. [CrossRef]
8. Varshney, U.; Sharma, A.; Vashishtha, P.; Goswami, L.; Gupta, G. Ga_2O_3/GaN Heterointerface-Based Self-Driven Broad-Band Ultraviolet Photodetectors with High Responsivity. *ACS Appl. Electron. Mater.* **2022**, *4*, 5641–5651. [CrossRef]
9. Wang, Y.; Li, S.; Cao, J.; Jiang, Y.; Zhang, Y.; Tang, W.; Wu, Z. Improved response speed of β-Ga_2O_3 solar-blind photodetectors by optimizing illumination and bias. *Mater. Des.* **2022**, *221*, 110917. [CrossRef]
10. Almaev, A.; Nikolaev, V.; Kopyev, V.; Shapenkov, S.; Yakovlev, N.; Kushnarev, B. Solar-Blind Ultraviolet Detectors Based on High-Quality HVPE α-Ga_2O_3 Films With Giant Responsivity. *IEEE Sens. J.* **2023**, *23*, 19245–19255. [CrossRef]
11. Polyakov, A.; Nikolaev, V.; Yakimov, E.; Ren, F.; Pearton, S.; Kim, J. Deep level defect states in β-, α-, and ε-Ga_2O_3 crystals and films: Impact on device performance. *J. Vac. Sci. Technol. A* **2022**, *40*, 020804. [CrossRef]
12. Labed, M.; Sengouga, N.; Prasad, C.; Henini, M.; Rim, Y. On the nature of majority and minority traps in β-Ga_2O_3: A review. *Mater. Today Phys.* **2023**, *36*, 101155. [CrossRef]
13. Traoré, A.; Okumura, H.; Sakurai, T. Photoconductivity buildup and decay kinetics in unintentionally doped β-Ga_2O_3. *Jpn. J. Appl. Phys.* **2022**, *61*, 091002. [CrossRef]
14. Kananen, B.E.; Giles, N.C.; Halliburton, L.E.; Foundos, G.K.; Chang, K.B.; Stevens, K.T. Self-trapped holes in β-Ga_2O_3 crystals. *J. Appl. Phys.* **2017**, *122*, 215703. [CrossRef]
15. Qin, Y.; Long, S.; Dong, H.; He, Q.; Jian, G.; Zhang, Y.; Hou, X.; Tan, P.; Zhang, Z.; Lv, H.; et al. Review of deep ultraviolet photodetector based on gallium oxide. *Chinese Phys. B* **2019**, *28*, 018501. [CrossRef]
16. Kalygina, V.M.; Tsymbalov, A.V.; Almaev, A.V.; Kushnarev, B.O.; Oleinik, V.L.; Petrova, J.S. Influence of White Light on the Photoelectric Characteristics of UV Detectors Based on β-Ga_2O_3. *IEEE Sens. J.* **2023**, *23*, 15530–15536. [CrossRef]
17. Yang, X.; Du, K.; He, L.; Wang, D. Fabrication and optoelectronic properties of Ga_2O_3/Eu epitaxial films on nanoporous GaN distributed Bragg reflectors. *J. Mater. Sci.* **2020**, *55*, 8231–8240. [CrossRef]
18. Winkler, N.; Wibowo, R.; Kautek, W.; Ligorio, G.; List-Kratochvil, E.; Dimopoulos, T. Nanocrystalline Ga_2O_3 films deposited by spray pyrolysis from water-based solutions on glass and TCO substrates. *J. Mater. Chem. C* **2019**, *7*, 69–77. [CrossRef]
19. Lampert, M.A.; Mark, P. *Current Injection in Solids*; Academic Press: New York, NY, USA, 1970; pp. 332–338.
20. Lovejoy, T.C.; Chen, R.; Zheng, X.; Villora, E.G.; Shimamura, K.; Yoshikawa, H.; Yamashita, Y.; Ueda, S.; Kobayashi, K.; Dunham, S.T.; et al. Band bending and surface defects in β-Ga_2O_3. *Appl. Phys. Lett.* **2012**, *100*, 181602. [CrossRef]
21. Ghadi, H.; McGlone, J.F.; Jackson, C.M.; Farzana, E.; Feng, Z.A.; Bhuiyan, U.; Zhao, H.; Arehart, A.R.; Ringel, S.A. Full bandgap defect state characterization of β-Ga_2O_3 grown by metal organic chemical vapor deposition. *APL Mater.* **2020**, *8*, 021111. [CrossRef]
22. Wang, Z.; Chen, X.; Ren, F.-F.; Gu, S.; Ye, J. Deep-level defects in gallium oxide. *J. Phys. D Appl. Phys.* **2021**, *54*, 043002. [CrossRef]
23. Chen, X.; Ren, F.-F.; Ye, J.; Gu, S. Gallium oxide-based solar-blind ultraviolet photodetectors. *Semicond. Sci. Technol.* **2020**, *35*, 023001. [CrossRef]
24. Hou, X.; Zou, Y.; Ding, M.; Qin, Y.; Zhang, Z.; Ma, X.; Tan, P.; Yu, S.; Zhou, X.; Zhao, X.; et al. Review of polymorphous Ga_2O_3 materials and their solar-blind photodetector applications. *J. Phys. D Appl. Phys.* **2021**, *54*, 043001. [CrossRef]
25. Greczynski, G.; Hultman, L. X-ray photoelectron spectroscopy: Towards reliable binding energy referencing. *Prog. Mater. Sci.* **2020**, *107*, 100591. [CrossRef]
26. Fares, C.; Ren, F.; Pearton, S.J. Temperature-Dependent Electrical Characteristics of β-Ga_2O_3 Diodes with W Schottky Contacts up to 500 °C. *ECS J. Solid State Sci. Technol.* **2019**, *8*, Q3007. [CrossRef]
27. Huan, Y.W.; Sun, S.M.; Gu, C.J.; Liu, W.J.; Ding, S.J.; Yu, H.Y.; Xia, C.T.; Zhang, D.W. Recent Advances in β-Ga_2O_3-Metal Contacts. *Nanoscale Res. Lett.* **2018**, *13*, 246. [CrossRef] [PubMed]
28. Borelli, C.; Bosio, A.; Parisini, A.; Pavesi, M.; Vantaggio, S.; Fornary, R. Electronic properties and photo-gain of UV-C photodetectors based on high-resistivity orthorhombic κ-Ga_2O_3 epilayers. *Mater. Sci. Eng. B* **2022**, *286*, 116056. [CrossRef]
29. Shalimova, K.V. *Semiconductor Physics*; Energy: Moscow, Russia, 1976; pp. 257–262.
30. Guo, D.; Liu, H.; Li, P.; Wu, Z.; Wang, S.; Cui, C.; Li, C.; Tang, W. Zero-Power-Consumption Solar-Blind Photodetector Based on β-Ga_2O_3/NSTO Heterojunction. *ACS Appl. Mater. Interfaces* **2017**, *9*, 1619–1628. [CrossRef] [PubMed]

31. Kalygina, V.; Podzyvalov, S.; Yudin, N.; Slyunko, E.; Zinoviev, M.; Kuznetsov, V.; Lysenko, A.; Kalsin, A.; Kopiev, V.; Kushnarev, B.; et al. Effect of UV and IR Radiation on the Electrical Characteristics of $Ga_2O_3/ZnGeP_2$ Hetero-Structures. *Crystals* **2023**, *13*, 1203. [CrossRef]
32. Chen, X.; Xu, Y.; Zhou, D.; Yang, S.; Ren, F.-F.; Lu, H.; Tang, K.; Gu, S.; Zhang, R.; Zheng, Y.; et al. Solar-Blind Photodetector with High Avalanche Gains and Bias-Tunable Detecting Functionality Based on Metastable Phase α-Ga_2O_3/ZnO Isotype Heterostructures. *ACS Appl. Mater. Interfaces* **2017**, *9*, 36997–37005. [CrossRef] [PubMed]

Disclaimer/Publisher's Note: The statements, opinions and data contained in all publications are solely those of the individual author(s) and contributor(s) and not of MDPI and/or the editor(s). MDPI and/or the editor(s) disclaim responsibility for any injury to people or property resulting from any ideas, methods, instructions or products referred to in the content.

Article

Strain-Modulated Electronic Transport Properties in Two-Dimensional Green Phosphorene with Different Edge Morphologies

Shuo Li and Hai Yang *

School of Physics Science and Technology, Kunming University, Kunming 650214, China
* Correspondence: kmyangh@kmu.edu.cn

Abstract: Based on two-dimensional green phosphorene, we designed two molecular electronic devices with zigzag (Type 1) and whisker-like (Type 2) configurations. By combining density functional theory (DFT) and non-equilibrium Green's function (NEGF), we investigated the electronic properties of Types 1 and 2. Type 1 exhibits an interesting negative differential resistance (NDR), while the current characteristics of Type 2 show linear growth in the current–voltage curve. We studied the electronic transport properties of Type 1 under uniaxial strain modulation and find that strained devices also exhibit a NDR effect, and the peak-to-valley ratio of device could be controlled by varying the strain intensity. These results show that the transport properties of green phosphorene with different edge configuration are different, and the zigzag edge have adjustable negative differential resistance properties.

Keywords: first-principles calculation; negative differential resistance; green phosphorene

Citation: Li, S.; Yang, H. Strain-Modulated Electronic Transport Properties in Two-Dimensional Green Phosphorene with Different Edge Morphologies. *Crystals* **2024**, *14*, 239. https://doi.org/10.3390/cryst14030239

Academic Editor: Ye Zhu

Received: 15 December 2023
Revised: 16 January 2024
Accepted: 29 January 2024
Published: 29 February 2024

Copyright: © 2024 by the authors. Licensee MDPI, Basel, Switzerland. This article is an open access article distributed under the terms and conditions of the Creative Commons Attribution (CC BY) license (https://creativecommons.org/licenses/by/4.0/).

1. Introduction

In recent years, the development of molecular devices has led to significant advancements in both simulational and experimental aspects. Molecular electronic devices exhibit various physical phenomena, such as molecular rectification, spin filtering, and NDR [1–5]. These physical effects hold practical value in molecular devices.

After 2004, graphene was successfully isolated, offering excellent electrical properties and a two-dimensional structure as a new material for molecular electronic device electrodes [6–11]. Similar to the preparation of graphene [12–16], two-dimensional black phosphene was successfully stripped in experiments in 2014 [17–20]. Phosphorene possesses anisotropic in-plane characteristics, with a direct bandgap of 0.3–2 eV as its thickness decreases from bulk to monolayer. Its carrier mobility is 10^3 cm^2/Vs, which indicates that phosphene has good electron transport properties. Due to its two-dimensional structure and exceptional electrical properties, phosphorene has emerged as a promising material in nanoelectronics and nanophotonics. In addition, phosphorus has a variety of allotropes, such as white phosphorus, blue phosphorus, and red phosphorus. On the basis of first-principles calculations, Han et al. designed a new allotrope of phosphorus, which they named green Phosphorus [21], which has a direct band gap of 0.7 to 2.4 eV and strong anisotropy in optical and transport properties [22].

Strain modulation is an important method that can change the electrical, optical, and magnetic properties of materials by applying force to them. The study of Liu et al. mentioned that a moderate strain can induce an antiferromagnetic to ferromagnetic phase transition, driving monolayer MnB to a ferromagnetic metal with Weyl Dirac nodal loops [23]. Graphene, as a typical representative of two-dimensional materials, has excellent mechanical strength and outstanding electrical properties. By introducing strain, the electrical properties of graphene, such as band gap and carrier mobility, can be adjusted [24]. Transition metal dichalcogenides (TMDs) [25] are a class of materials with significant spin-orbit coupling effects, and their electronic structure and spin polarization can be controlled through strain engineering. Strain modulation [26,27] controls electronic transport properties and energy bands of electrons. In the works of Ren [28,29], the change of edge morphology also has a large effect on the electronic properties of the material. He proposes that the electronic properties of the isolated regions are different when cutting blue phosphene. In addition, they discuss the differences in mechanical, electronic, and magnetic properties of various nanoribbons cut along typical crystal orientations from single-layer C3N sheets. Considering the many effects of strain and edge morphology on two-dimensional material properties, we choose to apply strain engineering and shear on two-dimensional material green phosphene, and discuss the effects of strain and edge morphology on its electronic properties.

In this work, we studied the structural transport and differences of single-layer green phosphene devices with Zigzag edge configuration (Type 1) and whisker edge configuration (Type 2) by first principles calculation and the non-equilibrium Green function method [30–37]. We select this Type 1 configuration to analyze its response to strain to study their electron transport properties. The results show that Type 1 structures have adjustable negative differential resistance properties under strain. This important property comes from the structural properties under strain conditions, which help in the design of electronic devices. Then, the corresponding mechanism of structural characteristic change under negative differential resistance and the pressure strain is analyzed.

2. Materials and Methods

The considered model is shown in Figure 1 and consists of several single layers of green phosphorene repeated in the z-direction. It has been demonstrated that single-layer green phosphorene without hydrogen termination at the edges is a stable structure [38]. We consider two different edge morphologies of monolayer green phosphene: one is zigzag, and the other is whisker, which are labeled Type 1 and Type 2, respectively. Type 1 is the existing structure, and Type 2 is obtained by cutting on the basis of Type 1. The molecular device consists of three parts: the first part includes the left and right electrodes, the second part is the electrode extension region, and the third part is the central scattering region. In the simulation calculations, the central scattering region contains 60 phosphorus atoms. In the molecular device structure, We perform n-type doping on the left and right electrodes as well as the electrode extension region to generate a non-equilibrium state in the device system, with a doping concentration of 1×10^{19} e/cm^3. The doping concentration in our simulated device structure corresponds to doping only a few electrons at the scale of our device. During the device experiment, the device's electrodes need to make contact with metal electrodes, which also leads to electron injection. This effectively simulates the realistic experimental environment. After geometric optimization, in Type 1, the bond lengths between adjacent phosphorus atoms are measured to be 2.23 Å and 2.26 Å, as shown in Figure 1a. In Type 2, the bond lengths between adjacent phosphorus atoms are 2.25 Å and 2.26 Å, as shown in Figure 1b. This bond length is basically the same as described in the literature [39]. The entire structure adopts an x-z coplanar configuration, with the transport direction along the z-direction.

We perform the geometric optimization and calculation of transport properties for Types 1 and 2, using the developed first-principles calculation software package (ATK) [40], which combines DFT with NEGF theory. In the calculations of the energy bands, we employ the Perdew–Burke–Ernzerhof (PBE) exchange-correlation functional within the generalized gradient approximation (GGA). The grid cutoff value for the electrostatic potential of 285 Ry is chosen, and the Fermi function is at a temperature of 300 K. A Monkhorst–Pack k-point grid of $1 \times 1 \times 285$ was chosen for the calculations. The lattice parameters and atomic positions were fully relaxed, with energy and force convergence criteria set to 10^{-5} eV and 0.05 eV/Å, respectively. A vacuum spacing of at least 15 Å was added in the direction perpendicular to the electronic transport plane to avoid interactions caused by periodic boundary conditions. For Type 1 and Type 2, we employ the NEGF method to calculate their transmission eigenvalues spectra and current–voltage (I–V) curves. When a voltage is applied, the electrochemical potentials of the left and right electrodes move accordingly. The current at a specific bias voltage, V_b, is obtained using the Landauer–Büttiker formula:

$$I(V) = \frac{2e}{h} \int_{\mu_L}^{\mu_R} T(E, V_b)[f_L(E - \mu_L) - f_R(E - \mu_R)]dE, \quad (1)$$

in this equation, $\mu_{L(R)} = E_F \pm eV_b/2$ represents the electrochemical potentials of the left and right electrodes, where the potential difference between the electrodes is equal to the bias voltage V_b. $f_L(E - \mu_L)$ and $f_R(E - \mu_R)$ are the Fermi distribution functions of the left and right electrodes at equilibrium. $T(E, V_b)$ is the electron transmission probability function, and it is calculated using the retarded Green's function $G(E, V_b)$:

$$T(E, V_b) = T_r[\Gamma_L(E, V_b)G(E, V_b)\Gamma_R(E, V_b)G^\dagger(E, V_b)], \quad (2)$$

in the above equation, $\Gamma_{L(R)} = i(\Sigma_{L(R)} - \Sigma_{L(R)}^\dagger)$ represents the broadening matrices of the left (right) electrode, where $\Sigma_{L(R)}$ is the coupling self-energy between the semi-infinite left (right) electrode and the central region.

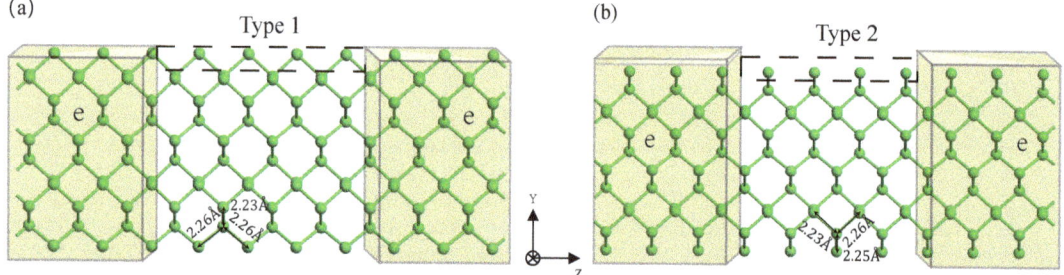

Figure 1. (**a**) Zigzag edge morphology of green phosphorus. (**b**) Whisker edge morphology of green phosphorus. N-type doping is carried out within the green box, which includes electrodes and electrode extension regions, with a doping concentration of 1×10^{19} e/cm^3.

3. Results

Based on the above considerations, we calculate the band structures of the left and right electrodes of Types 1 and 2, as shown in Figure 2a,c, which exhibit metal abundance. Besides, we also calculate their partial state density and total state density, as shown in Figure 2b,d, where the solid red line represents the state density of the upper and lower edge atoms of Types 1 and 2, the blue line represents the state density of all phosphorus atoms except the edge atoms, and the black line represents the state density of all atoms. Near the Fermi level, whether in Type 1 or Type 2, the red line representing the density of states (DOS) is higher than the blue line. This indicates that the DOS contributed by the edge atoms is larger, and the electrons are more localized at the edge atoms. For Type 1, because the phosphorus atoms at the edge have only one unformed bond, Type 2 has two unformed bonds, resulting in a high local density of electrons at the edge of Type 2. As one can see from the red line in the figure, DOS of Type 1 is 35 eV^{-1}, while DOS of Type 2 reaches 67 eV^{-1}. From the observation of the band structure near the Fermi level in Figure 2a,c, we find that two energy bands in Type 1 passing through the Fermi level are distributed in the conduction band region near the Γ point, and in the valence band region near the Z point, while Type 2 has four bands close to the Fermi level, which is an important aspect. It must affect their response to strain and transport characteristics.

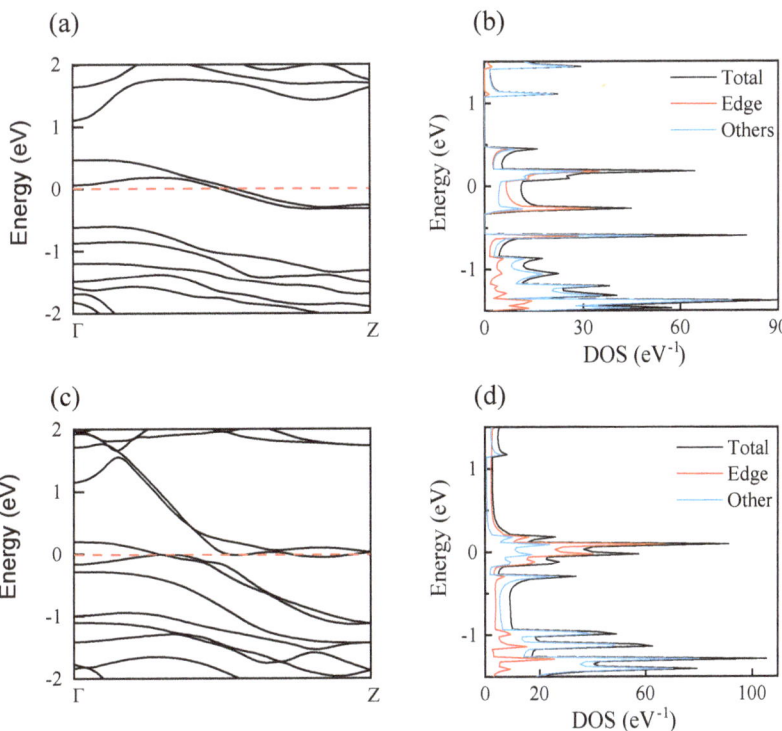

Figure 2. (**a**,**c**) Band structures of the left electrode. (**b**,**d**) Total density of states and partial density of states for Type 1 and Type 2, respectively.

After discussing the electronic structure of Type 1 and Type 2, in order to verify their differences in electron transport caused by energy band structures, we construct electronic devices and analyze their transport characteristics. The current–voltage (I–V) curves calculated using the Landauer–Büttiker Formula (1) are shown in Figure 3a,b. The electron transport properties of the two devices with different edge morphology are obviously distinct. Among them, the current passing through Type 1 gradually increases with the bias, reaching a maximum value of 34 µA at the bias voltage of 0.3 V, and then, when the bias voltage increases, the current passing through Type 1 drops to the lowest value of 0.23 µA.

As can be seen from the illustration in Figure 3a, the minimum current of the device is not zero. In contrast, the current through Type 2 increases linearly as the bias voltage increases. The electron transport characteristics of the electronic devices also confirm our previous theoretical analysis. In order to more clearly compare the electron transport properties of the two systems, we calculated their transmission spectra under the bias window with Formula (2), and the transmission spectra of the two structures are shown in Figure 3c,d. From the transmission spectrum of Type 1, we find that the transmittance of Type 1 has always been at a high level and observe that the transmission coefficient below the Fermi level starts to drop from 0.3 V and then rises to about 1.0 V, which also explains the reason for the NDR. For Type 2, in the bias window, there is a large number of transmission values exceeding 1, and with the opening of the bias window, the transmission value also gradually increases, which results in the I–V curves maintaining a linear increase in current.

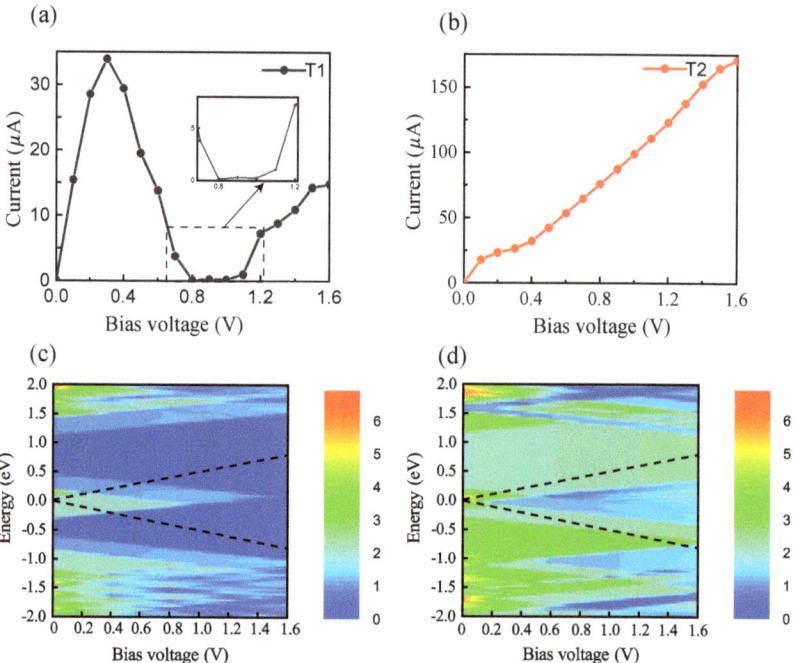

Figure 3. (**a**,**b**) I–V curves. (**c**,**d**) Transmission spectra for Type 1 and Type 2, respectively. The black dashed lines indicate the bias window.

Next, we discuss the response of Type 1 to strain and the effect of this regulation on the NDR effect. We define the strain as $\epsilon = (a - a_0)/a_0$, where a and a_0 are the lattice constants of the strained and the unstrained structures, respectively. When a strain occurs in one direction, the other direction is relaxed. First of all, Figure 4a,c,e display the band structure diagram, DOS diagram, and zero-bias transmission spectrum diagram when the compressive strain is −2%, −5%, and −8% on the Y-axis, respectively.

It can be seen from the Figure 4 that the energy bands of the three groups and their responses to transmission are different under different strain intensities. The energy band of the strain structure varies greatly compared with that of the original structure. We note that an energy band of the −2% strain structure crosses the Fermi level at the Γ point, where the energy gap between the conduction band and the valence band is small and increases rapidly. The energy gap at the Γ becomes smaller when the compressive strain is −5% and −8%. Similarly, from the DOS diagram, we also show that the contribution of each atom to the total electron orbital is also different under different compressive strains. When the compressive strain is −2%, the density of states at the Fermi level is 0. The DOS at Fermi level increases with the strain increase, and the peak value of −8% compression strain at 0.5 eV is higher than that at −5% compression strain. Besides, we note that the zero bias transmission spectra of three different strain degrees that their response to different strain degrees is corresponding to the band and DOS diagram. As can be seen from the figure, with the compression along the y direction, the transmission coefficient of the zero-bias transmission spectrum of several strain-configuration devices at the Fermi level exceeds 2, but the difference is that the transmission spectrum of the −5% compression strain has a very high peak value at 0.4 eV, while the transmission spectrum of the −8% compression strain has two high transmission coefficient peaks at this energy point. Therefore, we believe that the current transmission spectra of the device display distinct responses to different compressive strain degrees. Next, we calculate the transmission spectrum of the whole system under different bias voltages to verify this result.

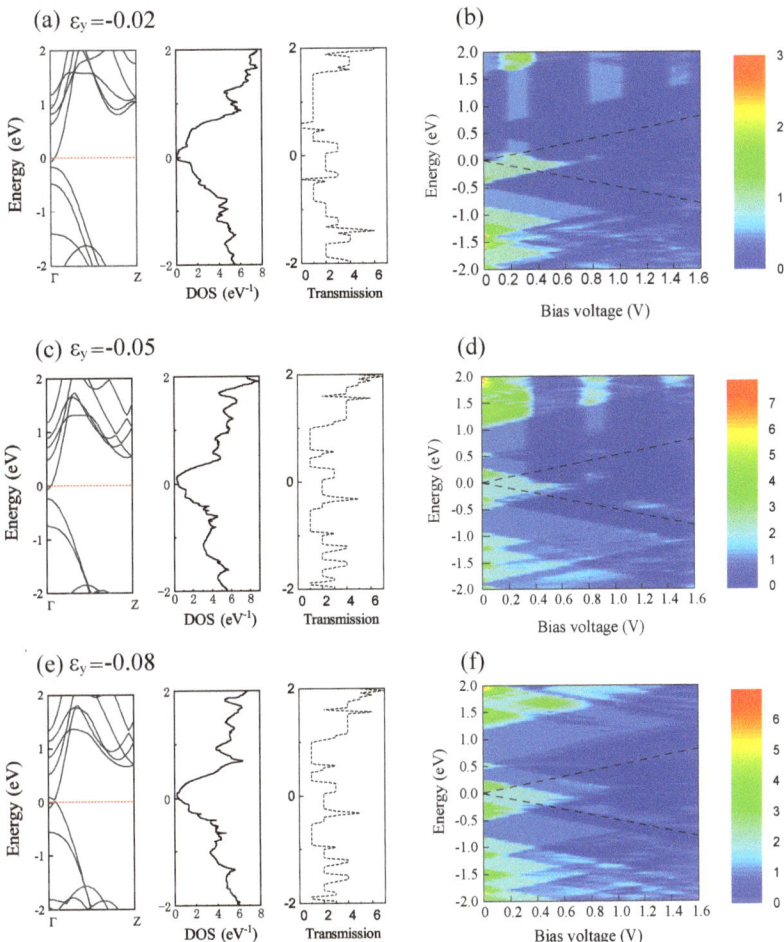

Figure 4. (**a**,**c**,**e**) Band structure on the left panel, DOS on the middle panel, and zero bias transmission spectra for the left electrode of Type 1 on the right panel, when the applied strain is $\epsilon_y = -0.02$, $\epsilon_y = -0.05$, and $\epsilon_y = -0.08$, respectively. The red dashed line represents the Fermi level. (**b**,**d**,**f**) The transmission spectra of Type 1 at strains of $\epsilon_y = -0.02$, $\epsilon_y = -0.05$, and $\epsilon_y = -0.08$, respectively. The black dashed line represents the bias window.

We calculated the transmission spectra under different strain conditions using Formula (2). Figure 4b,d,f shows the transmission spectra with compression strains of −2%, −5%, and −8%, respectively. The area between the two (black) dashed lines represents the bias window, with bias ranging from 0 to 1.6 V. As shown in Figure 4b, the transmittance of the device structure with a compression strain of −2% is always at a low level. In the bias window, the area with a transmission coefficient greater than 1 appears before the bias of 0.7 V, but after 0.4 V, the area covered by the part with a transmission coefficient greater than 1 in the bias window gradually decreases, indicating that the peak current is at 0.4 V. For Figure 4d with a compression strain of −5%, we note that the high transmission region appears before 0.5 V in the bias window. Unlike Figure 4b, most regions with a transmission coefficient greater than 1 appear between 0.5 V and 0.9 V, indicating that the current value of this part is higher than that of the structure with a compression strain of −2%. For Figure 4f with a compression strain of −8%, the peak transmission also appears at 0.4 V, and with the increase of bias, it also maintains a high transmission coefficient in most regions within the bias window.

From the transmission spectra of several structures, we obtained their I–V curves using Formula (1). As shown in Figure 5a, the variations in the I–V curves of different structures, represented by different colored lines, corresponded to the results we predicted from the transmittance spectra. First, we can clearly see that the peak current of the strain structure is significantly increased by about 15 µA compared with the original structure, and the current of these strain structures increases rapidly with the increase of bias voltage, reaching 0.4 V and then gradually decreases with the further increase of bias, and NDR effects emerge. In particular, the I–V curves of strain structures show different peak-to-valley current ratios. The peak-valley specific currents of $\epsilon_y = -0.02$, $\epsilon_y = -0.05$ and $\epsilon_y = -0.08$ are 9.27, 2.66, and 1.86, respectively. As shown in Figure 5b, we calculate DOS for several strain structures. Corresponding to the color of the I–V curve, the green line represents a device with a compressive strain of $\epsilon_y = -0.02$, the red line represents a device with a compressive strain of $\epsilon_y = -0.05$, and the blue line represents a device with a compressive strain of $\epsilon_y = -0.08$. We observe that with the increase of compressive strain, the peak value of DOS near the Fermi level of the device gradually increases, and the peak value increases. This result is also mirrored by the I–V diagram of the device under different strain conditions analyzed in the figure.

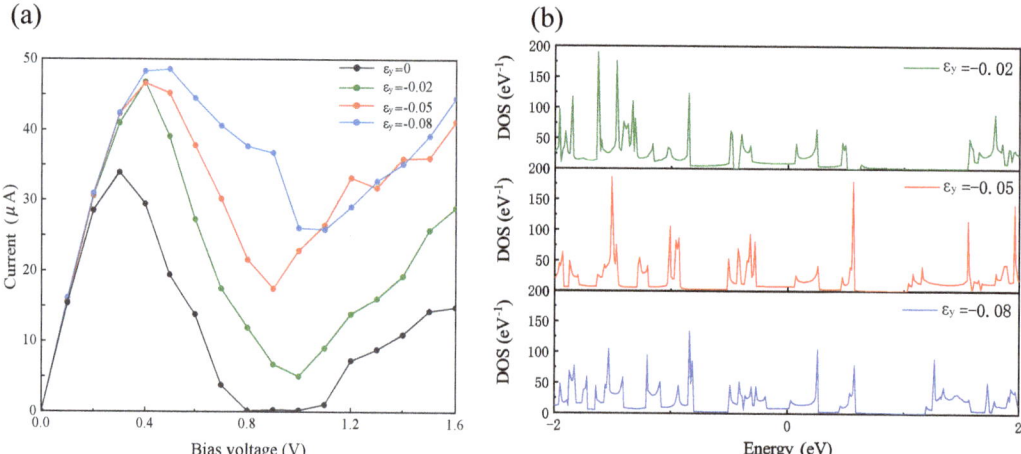

Figure 5. (a) I–V curves of Type 1 under different strain conditions. (b) DOS of the device under different strain conditions.

4. Conclusions

In summary, in this study we investigated the current transport characteristics of two different edge morphologies of green phosphorus devices. Firstly, we discussed the distinct electronic and transport properties of the zigzag edge (Type 1) and whisker-like (Type 2) configurations. Secondly, it was shown that the I–V characteristics of Type 1 exhibited a NDR effect, while the current of Type 2 linearly increased with bias. Thirdly, we also discussed the current transport characteristics of the Type 1 configuration under strain modulation. Fourthly, from the perspectives of the band structure, the DOS, and the transmission spectra, we analyzed the physical reasons for the transport differences under different strain modulations in detail. Finally, different strain intensities proved to result in different peak-to-valley ratios of the current. These results provide effective avenues for exploring the influence of edge morphology on device transport and studying the response between strain and NDR effect. These results share a guiding significance for the design of electronic nano-devices made of two-dimensional green phosphorene.

Author Contributions: Conceptualization, S.L. and H.Y.; methodology, S.L. and H.Y.; software, S.L. and H.Y.; resources, H.Y.; writing—original draft preparation, S.L.; writing—review and editing, H.Y.; funding acquisition, H.Y. All authors have read and agreed to the published version of the manuscript.

Funding: This work was supported by the National Nature Science Foundation of China (Grant No. 12164023), the Yunnan Local College Applied Basic Research Projects (Grant No. 2017FH001-001), and the Yunnan Local College Applied Basic Research Projects (Grant No. 2023Y0882).

Institutional Review Board Statement: Not applicable.

Informed Consent Statement: Not applicable.

Data Availability Statement: The data presented in this study are available on request from the corresponding author by the reason that the data has not been made publicly available.

Conflicts of Interest: The authors declare no conflict of interest.

References

1. Zhou, Y.; Zhang, D.; Zhang, J.; Ye, C.; Miao, X. Negative differential resistance behavior in phosphorus-doped armchair graphene nanoribbon junctions. *J. Appl. Phys.* **2014**, *115*, 073703. [CrossRef]
2. Hong, X.K.; Kuang, Y.W.; Qian, C.; Tao, Y.M.; Yu, H.L.; Zhang, D.B.; Liu, Y.S.; Feng, J.F.; Yang, X.F.; Wang, X.F. Axisymmetric All-Carbon Devices with High-Spin Filter Efficiency, Large-Spin Rectifying, and Strong-Spin Negative Differential Resistance Properties. *J. Phys. Chem. C* **2015**, *120*, 668–676. [CrossRef]
3. Wu, Q.; Liu, R.; Qiu, Z.; Li, D.; Li, J.; Wang, X.; Ding, G. Cr_3X_4 (X = Se, Te) monolayers as a new platform to realize robust spin filters, spin diodes and spin valves. *Phys. Chem. Chem. Phys.* **2022**, *24*, 24873–24880. [CrossRef]
4. Dong, X.; Peng, Z.; Chen, T.; Xu, L.; Ma, Z.; Liu, G.; Cen, K.; Xu, Z.; Zhou, G. Electronic structures and transport properties of low-dimensional GaN nanoderivatives: A first-principles study. *Appl. Surf. Sci.* **2021**, *561*, 150038. [CrossRef]
5. Huang, L.; Chen, S.Z.; Zeng, Y.J.; Wu, D.; Li, B.L.; Feng, Y.X.; Fan, Z.Q.; Tang, L.M.; Chen, K.Q. Switchable Spin Filters in Magnetic Molecular Junctions Based on Quantum Interference. *Adv. Electron. Mater.* **2020**, *6*, 2000689. [CrossRef]
6. Luisier, M. Atomistic simulation of transport phenomena in nanoelectronic devices. *Chem. Soc. Rev.* **2014**, *43*, 4357–4367. [CrossRef]
7. Zhou, Y.; Zheng, X.; Cheng, Z.-Q.; Chen, K.-Q. Current Superposition Law Realized in Molecular Devices Connected in Parallel. *J. Phys. Chem. C* **2019**, *123*, 10462–10468. [CrossRef]
8. Chen, Q.; Zheng, X.; Jiang, P.; Zhou, Y.-H.; Zhang, L.; Zeng, Z. Electric field induced tunable half-metallicity in an A-type antiferromagnetic bilayer $LaBr_2$. *Phys. Rev. B* **2022**, *106*, 245423. [CrossRef]
9. He, X.; Chen, J.; Li, S.; Lin, M.; Wang, Y.; Zheng, Y.; Lu, H. The tunability of electronic and transport properties of $InSe/MoSe_2$ van der Waals heterostructure: A first-principles study. *Surf. Interfaces* **2023**, *36*, 102634. [CrossRef]
10. Wang, X.; Feng, Z.; Rong, J.; Zhang, Y.; Zhong, Y.; Feng, J.; Yu, X.; Zhan, Z. Planar net-τ: A new high-performance metallic carbon anode material for lithium-ion batteries. *Carbon* **2019**, *142*, 438–444. [CrossRef]
11. Wang, Z.F.; Jin, K.-H.; Liu, F. Quantum spin Hall phase in 2D trigonal lattice. *Nat. Commun.* **2016**, *7*, 12746. [CrossRef] [PubMed]
12. Li, X.; Wang, X.; Zhang, L.; Lee, S.; Dai, H. Chemically Derived, Ultrasmooth Graphene Nanoribbon Semiconductors. *Science* **2008**, *319*, 1229–1232. [CrossRef]
13. Su, H.-P.; Shao, Z.-G. Lithium enhancing electronic transport properties of monolayer 6,6,12-graphyne from first principles. *Surf. Interfaces* **2021**, *22*, 100903. [CrossRef]
14. Wu, Q.-H.; Zhao, P.; Liu, H.-Y.; Liu, D.-S.; Chen, G. Odd–even dependence of rectifying behavior in carbon chains modified diphenyl–dimethyl molecule. *Chem. Phys. Lett.* **2014**, *62*, 605–606. [CrossRef]
15. Wu, X.; Xiao, S.; Quan, J.; Tian, C.; Gao, G. Perylene-based molecular device: Multifunctional spintronic and spin caloritronic applications. *Phys. Chem. Chem. Phys.* **2023**, *25*, 7354–7365. [CrossRef]
16. Kinikar, A.; Phanindra Sai, T.; Bhattacharyya, S.; Agarwala, A.; Biswas, T.; Sarker, S.K.; Krishnamurthy, H.R.; Jain, M.; Shenoy, V.B.; Ghosh, A. Quantized edge modes in atomic-scale point contacts in graphene. *Nat. Nanotechnol.* **2017**, *12*, 564–568. [CrossRef]
17. Wu, Q.; Shen, L.; Yang, M.; Cai, Y.; Huang, Z.; Feng, Y.P. Electronic and transport properties of phosphorene nanoribbons. *Phys. Rev. B* **2015**, *92*, 035436. [CrossRef]
18. Stegner, A.R.; Pereira, R.N.; Klein, K.; Lechner, R.; Dietmueller, R.; Brandt, M.S.; Stutzmann, M.; Wiggers, H. Electronic Transport in Phosphorus-Doped Silicon Nanocrystal Networks. *Phys. Rev. Lett.* **2008**, *100*, 026803. [CrossRef]
19. Jia, C.; Cao, L.; Zhou, X.; Zhou, B.; Zhou, G. Low-bias negative differential resistance in junction of a benzene between zigzag-edged phosphorene nanoribbons. *J. Phys. Condens. Matter* **2018**, *30*, 265301. [CrossRef] [PubMed]
20. Zheng, C.; Wu, K.; Jiang, K.; Yao, K.; Zhu, S.; Lu, Y. Perfect spin filtering effect, tunnel magnetoresistance and thermoelectric effect in metals-adsorbed blue phosphorene nanoribbons. *Physica B* **2022**, *626*, 413580. [CrossRef]

21. Han, W.H.; Kim, S.; Lee, I.-H.; Chang, K.J. Prediction of Green Phosphorus with Tunable Direct Band Gap and High Mobility. *J. Phys. Chem. Lett.* **2017**, *8*, 4627–4632. [CrossRef]
22. Qu, H.; Guo, S.; Zhou, W.; Wu, Z.; Cao, J.; Li, Z.; Zeng, H.; Zhang, S. Enhanced interband tunneling in two-dimensional tunneling transistors through anisotropic energy dispersion. *Phys. Rev. B* **2022**, *105*, 075413. [CrossRef]
23. Liu, C.; Fu, B.; Yin, H.; Zhang, G.; Dong, C. Strain-tunable magnetism and nodal loops in monolayer MnB. *Appl. Phys. Lett.* **2020**, *117*, 103101. [CrossRef]
24. Rezania, H.; Nourian, E.; Abdi, M.; Astinchap, B. Strain and magnetic field effects on the electronic and transport properties of γ-graphyne. *RSC Adv.* **2023**, *13*, 7988–7999. [CrossRef]
25. Roldán, R.; Castellanos-Gomez, A.; Cappelluti, E.; Guinea, F. Strain engineering in semiconducting two-dimensional crystals. *J. Phys. Condens. Matter* **2015**, *27*, 313201. [CrossRef]
26. Chen, Q.-Y. First-principles study on superconductive properties of compressive strain-engineered cryogenic superconducting heavy metal lead (Pb). *Commun. Theor. Phys.* **2021**, *73*, 035703. [CrossRef]
27. Yang, N.; Chen, Q.; Xu, Y.; Luo, J.; Yang, H.; Jin, G. Strain-modulated electronic transport in two-dimensional carbon allotropes. *AIP Adv.* **2022**, *12*, 045102. [CrossRef]
28. Ren, Y.; Cheng, F.; Zhou, X.-Y.; Chang, K.; Zhou, G.-H. Tunable mechanical, electronic and magnetic properties of monolayer C3N nanoribbons by external fields. *Carbon* **2019**, *143*, 14–20. [CrossRef]
29. Ren, Y.; Liu, P.; Zhou, B.; Zhou, X.; Zhou, G. Crystallographic Characterization of Black Phosphorene and its Application in Nanostructures. *Phys. Rev. Appl.* **2019**, *12*, 064025. [CrossRef]
30. Chen, J.; Zhang, Z.; Lu, H. Structure design and properties investigation of Bi_2O_2Se/graphene van der Waals heterojunction from first-principles study. *Surf. Interfaces* **2022**, *33*, 102289. [CrossRef]
31. Brandbyge, M.; Mozos, J.L.; Ordejón, P.; Taylor, J.; Stokbro, K. Density-functional method for nonequilibrium electron transport. *Phys. Rev. B* **2022**, *65*, 165401. [CrossRef]
32. Zhu, Y.; Li, H.; Chen, T.; Liu, D.; Zhou, Q. Investigation of the electronic and magnetic properties of low-dimensional $FeCl_2$ derivatives by first-principles calculations. *Vacuum* **2020**, *182*, 109694. [CrossRef]
33. Li, Y.-F.; Zhao, P.; Xu, Z.; Chen, G. Effect of linkage mode on the spin-polarized transport of a TPV radical-based molecular device. *Chem. Phys. Lett.* **2022**, *794*, 139515. [CrossRef]
34. Poljak, M.; Suligoj, T. Immunity of electronic and transport properties of phosphorene nanoribbons to edge defects. *Nano Res.* **2016**, *9*, 1723–1734. [CrossRef]
35. Zhang, W.; Basaran, C.; Ragab, T. Impact of geometry on transport properties of armchair graphene nanoribbon heterojunction. *Carbon* **2017**, *124*, 422–428. [CrossRef]
36. Wu, Q.; Ang, L.K. Giant tunneling magnetoresistance in atomically thin VSi_2N_4/$MoSi_2N_4$/VSi_2N_4 magnetic tunnel junction. *Appl. Phys. Lett.* **2022**, *120*, 022401. [CrossRef]
37. Chen, Q.-Y.; Liu, M.-Y.; Cao, C.; He, Y. Strain-tunable electronic and optical properties of novel anisotropic green phosphorene: A first-principles study. *Nanotechnology* **2019**, *30*, 335710. [CrossRef]
38. Han, X.; Stewart, H.M.; Shevlin, S.A.; Catlow, C.R.A.; Guo, Z.X. Strain and Orientation Modulated Bandgaps and Effective Masses of Phosphorene Nanoribbons. *Nano Lett.* **2014**, *14*, 4607–4614. [CrossRef]
39. Singsen, S.; Watwiangkham, A.; Ngamwongwan, L.; Fongkaew, I.; Jungthawan, S.; Suthirakun, S. Defect Engineering of Green Phosphorene Nanosheets for Detecting Volatile Organic Compounds: A Computational Approach. *ACS Appl. Nano Mater.* **2023**, *6*, 1496–1506. [CrossRef]
40. QuantumATK Version T-2022.03, Synopsys QuantumATK. Available online: http://www.quantumwise.com (accessed on 1 December 2023).

Disclaimer/Publisher's Note: The statements, opinions and data contained in all publications are solely those of the individual author(s) and contributor(s) and not of MDPI and/or the editor(s). MDPI and/or the editor(s) disclaim responsibility for any injury to people or property resulting from any ideas, methods, instructions or products referred to in the content.

Article

Thermal Evaporation Synthesis, Optical and Gas-Sensing Properties of ZnO Nanowires

Pham Hong Thach [1,2] and Tran Van Khai [1,2,*]

1. Faculty of Materials Technology, Ho Chi Minh City University of Technology (HCMUT), 268 Ly Thuong Kiet Street, District 10, Ho Chi Minh City 700000, Vietnam
2. Vietnam National University—Ho Chi Minh City (VNU-HCM), Linh Trung Ward, Thu Duc District, Ho Chi Minh City 700000, Vietnam
* Correspondence: tvkhai1509@hcmut.edu.vn; Tel.: +84-70-332-7675

Abstract: The purpose of this study is to synthesize and explore the relationship between the optical properties and gas-sensing performance of ZnO nanowires (NWs). Well-aligned ZnO nanowire (NW) arrays were synthesized on a silicon substrate using the thermal evaporation method without any catalyst or additive. The structures, surface morphologies, chemical compositions, and optical properties of the products were characterized using X-ray diffraction (XRD), field emission scanning electron microscopy (FESEM) together with energy-dispersive spectroscopy (EDS), high-resolution transmission electron microscopy (HRTEM), X-ray photoelectron spectroscopy (XPS), and photoluminescence (PL) spectroscopy, and their gas-sensing properties for NO_2 were examined. The results showed that single-crystalline ZnO NWs with high density grow uniformly and vertically on a Si substrate. The FESEM and TEM images indicate that ZnO NWs have an average diameter of roughly 135–160 nm with an average length of roughly 3.5 µm. The results from XRD confirm that the ZnO NWs have a hexagonal wurtzite structure with high crystalline quality and are highly oriented in the [0001] direction (i.e., along the c-axis). The deconvoluted O 1s peak at ~531.6 eV (29.4%) is assigned to the oxygen deficiency, indicating that the ZnO NWs contain very few oxygen vacancies. This observation is further confirmed by the PL analysis, which showed a sharp and high-intensity peak of ultraviolet (UV) emission with a suppressed deep-level (DL) emission (very high: $I_{UV}/I_{DL} > 70$), indicating the excellent crystalline quality and good optical properties of the grown NWs. In addition, the gas-sensing properties of the as-prepared ZnO NWs were investigated. The results indicated that under an operating temperature of 200 °C, the sensor based on ZnO NWs is able to detect the lowest concentration of 1.57 ppm of NO_2 gas.

Keywords: sensing; ZnO; thermal evaporation; nanowires; semiconductor

1. Introduction

Nitrogen oxide (NO_2) is a strong oxidizing gas that has become one of the extremely harmful gases that derive from combustion of fossil fuels and car exhaust [1,2]. Exposure to NO_2 gas results in serious damage to lung tissues and reduces the fixation of O_2 on red blood corpuscles, and the gas also contributes to acid rain as well as generating increased ozone in the lower atmosphere [3]. Nowadays, due to the increase in the numbers of cars, power plants, and factories, the concentration of NO_2 gas in the air atmosphere has increased rapidly [4]. According to health and safety guidelines, humans should not be exposed to more than 2.5 ppm of NO_2 [5,6]. Therefore, the development of gas sensors for the precise detection of NO_2 is of great importance for environmental monitoring and human health protection.

In recent years, metal oxide semiconductors (MOSs) such as SnO_2, ZnO, In_2O_3, WO_3, Fe_2O_3, NiO, Cu_2O, etc., have attracted tremendous attention in the field of gas sensing because of their low cost, simple fabrication methods, long life, high stability, and superior

sensing performance [7–13]. Among them, zinc oxide (ZnO) is an n-type semiconductor with a wide direct band gap of 3.37 eV (at 300 K), possessing a large exciton binding energy (60 meV) and high electron mobility (100–200 cm^2 V^{-1} s^{-1}) [14,15], making it suitable not only for gas sensing [16] but also for potential applications in solar cells [17], electronic and optoelectronic devices [18], and catalysts [19]. In fact, due to its advantages of high sensitivity, good oxidation resistance, great chemical–thermal stability, environmental friendliness, and low cost, nanosized ZnO with different morphologies has been widely employed to detect various gases including H_2, H_2S, NO_2, NH_3, CH_4, CO, ethanol, and acetone [16].

It is well known that gas-sensing performance strongly depends upon structural parameters including crystal size, specific surface area, microstructure, crystallographic planes and the crystallinity of sensing materials [20]. For instance, a decrease in crystal size, an increase in surface-to-volume ratio, and high crystallinity are required to enhance the sensitivity of gas sensors. Since sensing reactions take place mainly on the surface of the sensitive material [21–24], the control of the size of the semiconductor materials is one of the first requirements for enhancing the sensitivity of the sensor. Apart from these factors, morphology also plays a critical role in governing sensing performance [25–29]. Therefore, many efforts have been focused on the synthesis of ZnO nanostructures with different morphologies, such as nanoparticles [30], nanorods [31], nanowires [32], nanotubes [33], nanofibers [34], nanoflowers [35], and hierarchical nanostructures [36]. These nanostructured ZnO materials have higher surface-to-volume ratios compared with the bulk material, which leads to there being more active sites for gas absorption and facilitates charge transfer, thereby improving sensing performance. Recently, numerous efforts have been made to investigate the relationship between the morphology and sensing properties of ZnO nanostructures [37–41]. For example, Agarwal et al. [35] synthesized two types of ZnO nanostructures, ZnO nanorods and flower-like ZnO nanostructures, using a hydrothermal method. They reported that compared with nanorods, nanoflowers have a greater surface area and more surface defects. Due to these differences, flower-like ZnO is able to adsorb more target gas molecules, resulting in an enhanced gas response. Several researchers have studied the use of ZnO nanorods for monitoring H_2S gas, and their findings indicate an improved capacity to detect H_2S over other gases as well as showing an increased response and selectivity through the use of ZnO nanorods compared with bulk ZnO material [42,43]. In order to explore morphology-dependent gas-sensing properties, Zhang et al. [44] prepared ITO materials with various morphologies, including film, nanoparticles, nanorods, and nanowires, using a sputtering method. The results of this study indicated that compared with the other materials, the nanowires possess a larger specific surface area, a greater number of oxygen vacancies, and well-defined electron transport pathways, so their sensing properties are the best. Furthermore, by adjusting the density of nanowires, their sensing performance for ethanol gas was greatly enhanced. Compared with other morphologies of ZnO nanostructures, ZnO NWs have recently drawn a large amount of interest because of their distinctive geometric characteristics, high surface-to-volume ratio, well-defined electron transportation direction, and lower agglomeration tendency. These features are able to enhance the electron flow and affect the reaction between surface-adsorbed oxygen and gas molecules, thereby improving the sensing performance [45–48]. Up to now, there have been many reports on the gas-sensing behavior of ZnO NWs for various gases, which have exhibited excellent sensing performance [49–52]. The majority of these ZnO gas sensors, however, are based on a single NW or tangled NW membrane with no alignment. The vertically well-aligned NW arrays can provide a simple matrix for studying the average effect of assembled NWs. However, so far, to the best of our knowledge, little or no work has been reported on the application of vertical ZnO NWs on the Si substrate for NO_2 sensing. Recently, various methods have been employed for the synthesis of ZnO NWs, including the hydrothermal [53], thermal evaporation [54,55], chemical vapor deposition (CVD) [49], physical vapor deposition [56], and metal organic chemical vapor deposition (MOCVD) methods [57,58]. Among these methods, thermal evaporation is frequently employed due to its ease of operation, environmental friendliness, ease of scaling up, relatively low-cost process, etc. Moreover, with the use of this technique, the as-prepared ZnO NWs show good quality [59,60]. Therefore, the purpose of this study is to synthesize and explore the relationship between the optical properties and gas-sensing performance of ZnO NWs using the thermal evaporation method.

2. Materials and Methods

2.1. Fabrication of ZnO Nanowires

The well-aligned ZnO NWs were grown via thermal evaporation deposition in a horizontal tube furnace (OTF-1200x-STM, MTI Corp., Richmond, CA, USA) with an inner diameter of 720 mm and a heating zone of 400 mm. Silicon wafers (Si (100)) with thickness of 500 μm (Okemetic Co., Ltd., Tokyo, Japan) and size of 1 cm × 1 cm were used as the substrate for the growth of ZnO NWs. Before the substrates were put inside the chamber for growth, they were chemically etched with H_3PO_4 (85%, Merck Chemicals Co., Ltd., Darmstadt, Germany) solution for 60 s to remove the native oxide layer (SiO_2), then sonicated with a mixture of acetone, methanol, and deionized water in a sonication bath (Elma Select 60) for 10 min, and finally dried by air. The cleaned Si (1 cm × 1 cm) substrates were placed on top of an alumina boat with a diameter and length of 1.5 and 4 cm, respectively, containing about 0.25 g high-quality metallic zinc powder (75 μm, 99.99%, Sigma-Aldrich, Burlington, MA, USA), and then put at the center of the furnace, where the temperature is highest. The vertical distance between the zinc source and the substrate was about 2.5 mm, with a downstream separation of 15 mm. The temperature inside the furnace was raised from 20 °C to the reaction temperature of 620 °C at a rate of 20 °C per min with Ar (99.99%, Bao Khanh Co. Ltd., Ha Noi, Vietnam) at a flow rate of 350 mL/min The zinc source was thermally vaporized to grow ZnO NWs at atmospheric pressure under Ar at a flow rate of 350 mL/min for 90 min at 620 °C. After the reaction, the equipment was switched off and quartz tube was cooled naturally to room temperature under Ar at a flow rate of 100 mL/min.

2.2. Material Characterization

The crystal structure and phase composition of synthesized samples were determined via X-ray diffraction using a Bruker D8 Advanced diffractometer (Rigaku, Tokyo, Japan) with CuKα radiation at λ = 1.54178 Å, operating at 40 kV voltage and 200 mA current. Data were taken for the 2θ range of 25 to 75 degrees with a step of 0.02 degree and a scan step time of 25 s. The surface morphology and compositions of the ZnO NWs were examined using a scanning electron microscope (JEOL JSM-5900 LV SEM, Tokyo, Japan) operated at an accelerating voltage of 20 kV and equipped with an energy-dispersive spectroscopy (EDS) microanalysis. The microstructure of the NWs was observed with a high-resolution electron microscope (HRTEM, JEM-2010, JEOL Ltd., Tokyo, Japan) operated at an accelerating voltage of 200 kV. The elemental components and bonding configuration of the ZnO samples were investigated using an X-ray photoelectron spectroscope (XPS, VG Multilab ESCA 2000 system, East Grinstead UK) with a monochromatized Al K X-ray source (hν = 1486.6 eV). Data analysis was carried out with XPSPEAK41 software using mixed Gaussian–Lorentzian functions after Shirley background correction. The photoluminescence (PL) measurements were performed with a laser Raman spectrometer (HORIBA Jobin Yvon, iHR550) at room temperature, using a He-Cd laser-line with an excitation wavelength of 325 nm and at a laser power of 25 mW. For the sensing measurements, a thin (~100 nm) Au film was deposited on the ZnO NW samples via direct current (DC) magnetron sputtering to form electrodes using an interdigital electrode mask. The electrode pattern consisted of seven Au electrode fingers, each of them being 7 mm long and 0.5 mm wide, with 0.5 mm spacing. The gas-sensing characteristics were determined with a Keithley source-meter model no. 2400 connected to computer. Two mass flow controllers (MFC-3660) were applied to adjust the NO_2 gas concentration and carrier gas (dry air). A precise concentration of NO_2 was produced via dilution from the standard concentration of NO$_2$ (0.1% NO$_2$ + 99.9 N$_2$), as shown in Table 1. The concentration of the diluted NO$_2$ was calculated using Equation (1):

$$C_{dulited} = C_o \frac{F1}{F1 + F2} \quad (1)$$

where F_1 and F_2 are flow rate of NO$_2$ and carrier dry air, respectively, and C_0 is 0.1%. Total gas flow (target gas and dry air) was kept constant at 500 ppm. The resistance of the sensor in dry air or in the target gas was measured at 200 °C when a potential difference of 1 V was applied between the Au (~100 nm) electrodes. Sensitivity is one of the important factors in determining gas-sensing performance. Mathematically, sensitivity (S) of the ZnO NW sensor is defined as (R_g/R_a), where R_a and R_g are the electrical resistances of the sensor in air and target gas, respectively. The response time ($t_{response}$) is defined as the period in which the electrical resistance of the sensor reaches 90% of the response value upon exposure to the target gas, while the recovery time ($t_{recovery}$) is defined as the period in which the electrical resistance of the sensor returns to 10% of the response value after the target gas is removed [61]. The sensing characteristics were investigated at NO$_2$ concentrations of 2–10 pm, operating at temperature of 200 °C under atmospheric pressure.

Table 1. Volumetric flow rates of the target gas (NO$_2$), dry air, and standard concentration of NO$_2$ (Co = 0.1%).

MFC-1 NO$_2$ (sccm)	MFC-2 Dry Air (sccm)	NO$_2$ Concentration in Tube: $C_{dulited}$ (ppm)
1	499	2
3	498	6
5	495	10

3. Results and Discussion

The crystalline structure and phase composition of the synthesized samples were examined using powder X-ray diffraction. A typical XRD pattern of the obtained ZnO NWs is shown in Figure 1. From the pattern, it can be seen that all diffraction peaks at 2θ~31.8, 34.6, 36.4, 47.6, 56.7, 62.8, 67.8, and 72.5 can be indexed to the reflection from the $(10\bar{1}0)$, (0002), $(10\bar{1}1)$, $(10\bar{1}2)$, $(11\bar{2}0)$, $(10\bar{1}3)$, $(11\bar{2}2)$, and (0004) crystal planes of the hexagonal wurtzite ZnO structure with the cell constants of a = 3.249 Å and c = 5.206 Å (JCPDS card number: 36-1451, P6$_{3mc}$ space group) [62,63]. The lattice constants of the prepared ZnO NWs can be calculated as a = 3.249 Å and c = 5.185 Å with the c/a ratio = 1.5957. The unit cell volume for deposited ZnO NWs was estimated by using [64]: $V = \frac{\sqrt{3}}{2}a^2c$, where a and c are the lattice constants of the ZnO NWs. Accordingly, the calculated value of the unit cell volume for ZnO NWs is 47.4 Å^3. Therefore, the lattice parameters (a = b, c, V) estimated for synthesized NWs are in good agreement with the reported crystallographic data for ZnO [62,63,65]. The strongest peak intensity at the (0002) plane suggests that the ZnO NWs are preferentially oriented in the c-axis direction or in a direction perpendicular to the substrate surface. No traces of zinc, impurities, or substrates are detected from this pattern, indicating that the ZnO NW samples were composed of a pure ZnO hexagonal phase. It is widely known that the hexagonal crystal structure of ZnO is composed of an alternating arrangement of tetrahedrally coordinated Zn^{2+} and O^{2-} ions along the c-axis [66]. The most common crystal faces found on the one-dimensional ZnO nanostructures are polar Zn-terminated (0001) and O-terminated $(000\bar{1})$ as well as nonpolar $(01\bar{1}0)$ and $(2\bar{1}\bar{1}0)$ facets [67]. Therefore, adjusting the growth rate along the different directions can induce anisotropic growth of ZnO. Kinetic studies have shown that incoming precursor species tend to preferentially adsorb on the polar surfaces to decrease the surface energy because the polar surfaces are less thermodynamically stable. Accordingly, the higher surface energy facets grow faster than the others, and they usually have a smaller surface area. The sequence of the surface energy (E) of the facets in different crystallographic directions of ZnO is commonly E[0001] > E$[10\bar{1}1]$ > E$[10\bar{1}0]$ > E$[10\bar{1}\bar{1}]$ > E$[000\bar{1}]$, which is the same order as the crystal growth velocity, indicating the fastest growth rate in the [0001] direction [68]. Thus, in the case of ZnO NWs, the fastest growth rate is along the c-axis, resulting in the anisotropic growth of nanowires. Meanwhile, ZnO NWs with large areas of nonpolar $(01\bar{1}0)$ and $(2\bar{1}\bar{1}0)$ facets are more energetically favorable [67]. This explanation is supported by the XRD results, in which the (0002) peak intensity is much higher than the others, indicating that ZnO NWs were formed along the c-axis in the [0001] direction (i.e., with predominantly exposed polar (0001) facets).

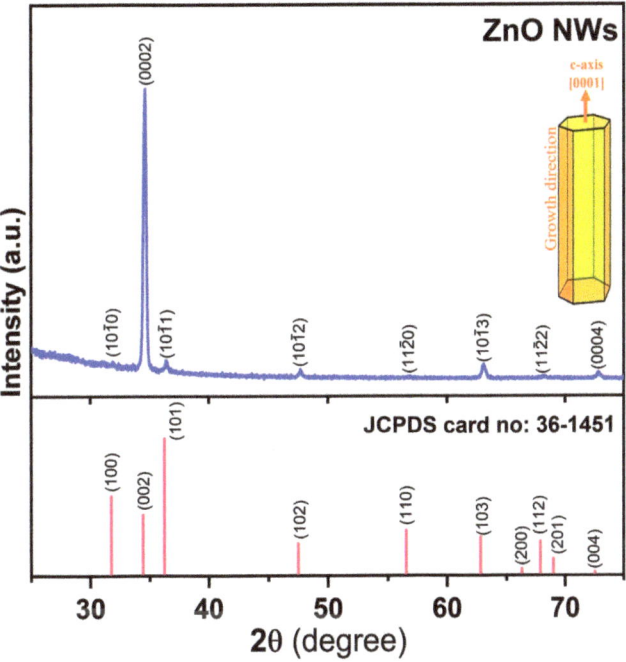

Figure 1. XRD pattern of ZnO NWs.

The morphology of the as-prepared samples was investigated using FESEM. Figure 2a–d show typical FESEM images of the obtained ZnO NWs at different magnifications. These sample ZnO NWs were synthesized at a temperature of 620 °C with a synthesis time of 90 min. Figure 2a,b show the low-magnification FESEM images of the deposited ZnO NWs and confirm that large-area, uniform, and dense NWs are vertically grown across the surface of the substrate. The NWs possess a smooth and clean surface along their entire lengths. Figure 2c,d reveal high-magnification images of ZnO NWs, indicating that the top of the NWs is completely flat with a perfect hexagonal structure and the diameter of the NWs is in the range of 160–200 nm with an area density of ~6.85/µm². Figure 3 illustrates the typical EDS spectrum of the as-grown ZnO NWs, indicating that there is no evidence of metallic elements in the NWs, and the as-prepared ZnO NWs exclusively include zinc and oxygen components. The existence of a Si peak in the spectrum is attributable to the Si substrate.

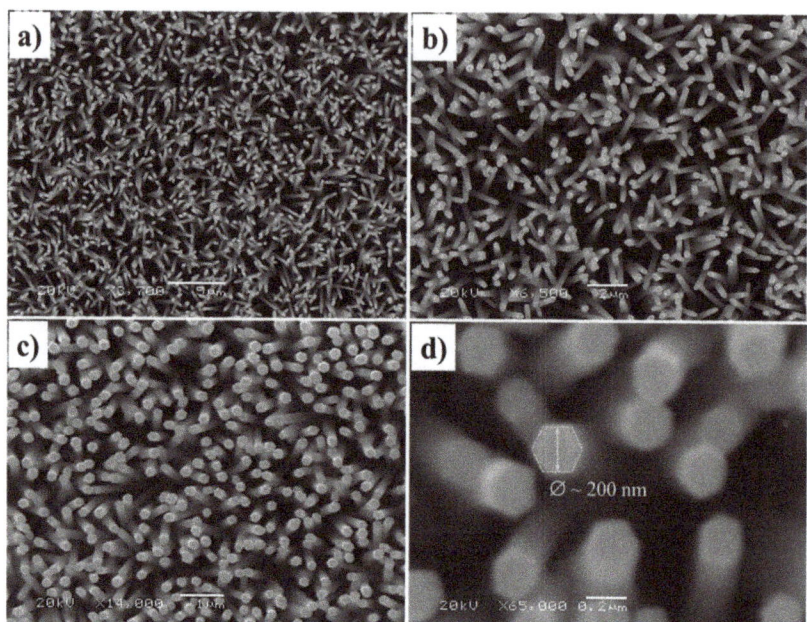

Figure 2. FESEM images with different magnifications—(**a**) 3700×; (**b**) 6500×; (**c**) 14,000×; and (**d**) 65,000×—of ZnO NWs.

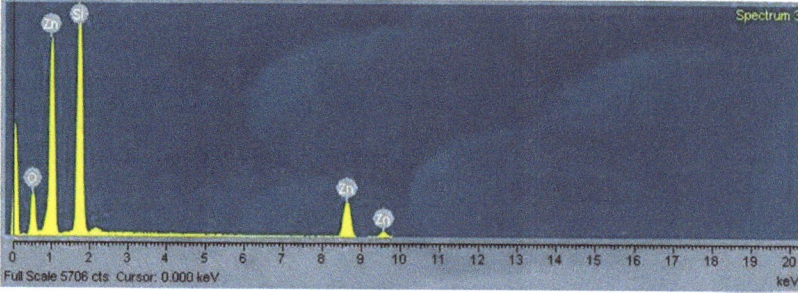

Figure 3. The EDS spectrum of ZnO NWs.

Detailed structural features of the as-prepared ZnO NWs are shown in Figure 4a,b. Figure 4a shows that the obtained ZnO NWs possess a very smooth surface and a relatively straight morphology with an average diameter of ~135–160 nm and a length of ~1.5–3.5 μm. A typical HRTEM image is shown in Figure 4b. It can be clearly observed that the ZnO NW contains no defects such as dislocations or stacking faults, indicating that it has a clean surface structure with excellent quality. The inset in Figure 4b is a lattice-resolved HRTEM image of a segment of a single ZnO NW. It can be clearly seen that the ZnO crystal lattice is well oriented and there are no noticeable structural defects throughout the region. This indicates that the obtained ZnO NWs are structurally uniform and defect-free. However, it should not be assumed that the as-synthesized ZnO NW arrays are free of other defects, such as vacancies and gaps, which may be invisible in HRTEM observations. As seen from the inset in Figure 4b, the measured lattice spacing of the ZnO NW was around 0.26 nm, corresponding to the distance between two (0002) crystal planes, indicating that the ZnO NWs are grown in the [0001] direction. This is consistent with the XRD results.

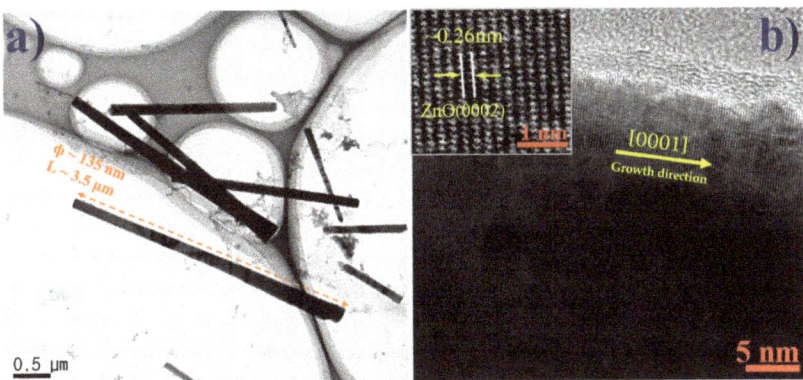

Figure 4. TEM images of the synthesized ZnO NWs: (**a**) conventional TEM; (**b**) HRTEM; the inset is a high-magnification TEM image revealing that the measured lattice spacing of 0.26 nm confirms the growth of ZnO NW in the [0001] direction.

The surface chemical compositions and their corresponding valence states of the as-prepared ZnO NWs were further investigated via XPS, and the results are shown in Figure 5a–d. All of the binding energies in the XPS analysis were corrected for specimen charging by reflecting them to the C1s peak position at ~284.6 eV [69]. Figure 5a exhibits the wide-scan XPS spectrum, from which the peaks located at ~284.4, ~533.2, and ~1021.5–1045.5 eV, corresponding to the C1s, O1s, and Zn2p core levels, respectively, can be observed clearly. Here, the peaks for Zn and O are expected to be from the ZnO material, whereas the presence of the C peak is due to the adventitious carbon contamination as well as the chemisorbed CO_2 on the surface of samples when exposed to air after the growth, before they are transferred to the XPS chamber (often observed in the XPS spectra of ZnO available in the literature) [70,71]. A more detailed investigation of the chemical states of Zn2p, O1s, and C1s was carried out, taking into consideration the high-resolution scans from core-level XPS, as shown in Figure 5b–d. The XPS spectra were fitted by Gaussian–Lorentzian functions after Shirley background correction using XPSPEAK41 software. Shown in Figure 5b is the high-resolution C1s core-level XPS spectrum, which can be fitted into four components. The main peak is centered at ~284.7 eV and represents C=C bonds due to the adventitious carbon, while the peaks located at ~285.7, ~286.6, and ~288.7 eV can belong to the C-OH group, C-O group, and carbonate phase (CO_2), respectively. As recorded, the high-resolution O1s core-level XPS spectrum (Figure 5c) can be deconvoluted into three peaks with binding energies of ~530.3, ~531.6, and ~532.4 eV. The peak located at ~530.3 eV (~24%) is attributed to O^{2-} species in the lattice (O_{Lat}) of the wurtzite structure of hexagonal ZnO [72–76], whereas the peak centered at ~531.6 eV (~29.4%) is assigned to the O^{2-} state of oxygen defects such as oxygen vacancies (V_O), oxygen interstitials (O_i), or surface oxygen atoms, suggesting the formation of the nonstoichiometric ZnO. Finally, the peaks at ~532.4 eV (~46.6%) are typically related to the chemisorbed or dissociated (O_C) oxygen species on the surface, such as O_2, H_2O, and CO_2 [77–79]. As shown in Figure 5d, the high-resolution Zn2p core-level XPS spectrum can be deconvoluted into two peaks at binding energies of ~1022.4 eV and ~1045.5 eV, which are attributed to the spin–orbits of $Zn2p_{3/2}$ and $Zn2p_{1/2}$ of tetrahedral Zn^{2+}, respectively, confirming that the oxidation state of Zn is +2 in the ZnO NWs [80,81]. The binding energy splitting between these two components (ΔBE) is 23.1 eV, which is consistent with previous studies [72].

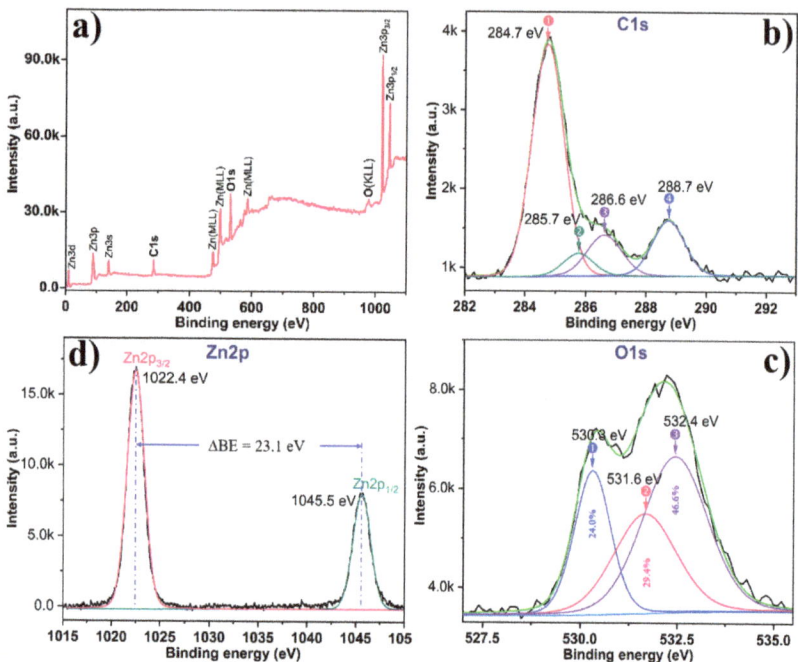

Figure 5. (**a**) Wide-scan XPS spectrum and high-resolution scan XPS of (**b**) C1s, (**c**) O1s, and (**d**) Zn2p of ZnO NWs. C1s peak position at ~284.6 eV is used as a binding energy reference to correct the XPS spectra in case of any charging effect.

The room-temperature PL spectrum of the ZnO NWs is presented in Figure 6. The PL spectrum was deconvoluted via Gaussian–Lorentzian fitting. The PL spectrum is very useful for analyzing any possible point defects in the NWs. Figure 6 shows that ZnO NWs have a strong and narrow ultraviolet (UV) emission peak at ~379.5 nm (~3.27 eV) as well as a broad and weak visible emission band that peaks at ~491 nm (~2.53 eV). The UV emission is attributed to the recombination of free excitons between the conduction and valence bands and is called near-band-edge emission (NBE) [82]. On the other hand, the visible emission band is the so-called deep-level (DL) emission of ZnO, which consists of the emissions associated with different types of DL defects in ZnO as well as surface states of ZnO NWs, such as oxygen vacancies (V_O) [83,84], zinc interstitials (Zn_i) [85], oxygen interstitials (O_i), zinc vacancies (V_{Zn}), and antisite oxygen (O_{Zn}) [86]. Additionally, there is a tiny PL peak at ~390 nm (~3.18 eV), which is caused by the electronic transition from Zn_i states to the valence band maximum [87–89]. The intensity ratio between the UV and DL peaks, I_{UV}/I_{DL}, is commonly used to estimate the defect density in ZnO [90]. Among various synthesis techniques, the ZnO NWs grown through thermal evaporation frequently reveal a very intense exciton emission [91–93]. The intensity ratio I_{UV}/I_{DL} recorded from our samples is more than 77, which is higher than those reported in most of the literature (see Table 2) [91–99], indicating that the prepared ZnO NWs have a high crystalline quality. The poor DL emission indicates a low concentration of DL defects in ZnO NWs produced by thermal evaporation.

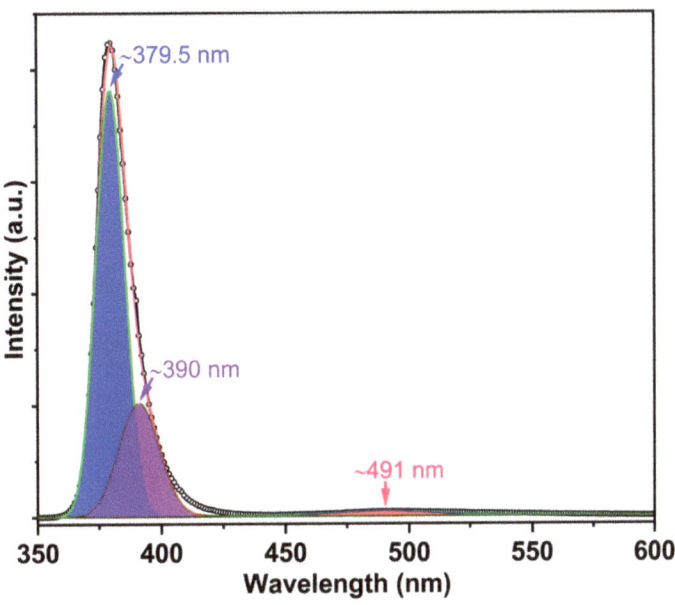

Figure 6. Room-temperate PL spectrum of as-prepared ZnO NWs.

Table 2. Comparison of intensity ratios I_{UV}/I_{DL} of ZnO nanostructures synthesized using different methods.

Morphology	Synthesis Method	I_{UV}/I_{DL}	References
ZnO NWs	Pulsed laser deposition	~11.9–45.4	[94]
ZnO nanorods	Hydrothermal	~0.3	[95]
ZnO nanorods	Hydrothermal	~0.6	[96]
ZnO nanorods	Hydrothermal	>19	[97]
ZnO nanorods	Hydrothermal	~0.6–10	[98]
ZnO nanorods	Thermal decomposition	~1.8	[99]
ZnO NWs	Physical vapor deposition	2.5–4.7	[90]
ZnO NWs	Thermal evaporation	~1.4	[91]
ZnO nanocolumns	Thermal evaporation	~2.4	[92]
ZnO NWs	Thermal evaporation	~16.2	[93]
ZnO NWs	Thermal evaporation	>77	Present work

The intrinsic point defects in nanostructured ZnO have been demonstrated to play a crucial role in determining the performance of electronic devices [100–103], in terms of doping control, free charge density, minority carrier lifetime, and luminescence efficiency [104]. For instance, it has been shown that by increasing the number of surface defects in ZnO NWs, highly sensitive UV sensors may be produced [102]. These surface defects can serve as trapping or recombination centers for the charge carriers and lead to the formation of a surface depletion region. A greater width of the depletion layer at the NW surface leads to higher UV sensitivity. However, a density of defects on the NW surface that is too high may cause negative effects on nanogenerator devices' performance [100,101]. It has been reported that ZnO nanostructures with high crystalline quality are known to improve photocatalytic activity due to the enhanced separation efficiency of photogenerated electron–hole pairs [105,106]. In the study of nanostructured ZnO-based UV sensors, high electron mobility is considered to contribute to enhancing the photocurrent [107–109]. High electron mobility may be attained by improving the crystallinity of ZnO. Therefore, precisely controlling the quality of the produced ZnO nanomaterials is critical in the production of a high-performance electronic device. As mentioned above, the optical behavior of the ZnO NWs is studied by analyzing the defect-related deep-level (DL) emission and near-band-edge (NBE) UV emission in the PL spectrum of NWs. Two main features noted in the PL spectrum are located at ~379.5 eV (~3.27 eV) and ~491 nm (~2.53 eV), and respectively labeled as ultraviolet (UV) emission (I_{UV}) and deep-level defect emission (I_{DL}) peaks. The DL emissions are usually caused by oxygen and zinc vacancies (V_O and V_{Zn}) and interstitials (O_i and Zn_i), antisites (O_{Zn} and Zn_O), and hydrogen impurities [110]. The ratio I_{UV}/I_{DL} can provide a qualitative indicator of the number of radiation defects in the produced nanomaterial. The higher the I_{UV}/I_{DL} ratio, the lower the number of point defects. Therefore, controlling the defect states and entirely suppressing defect-related emissions have become the most important issues for improving UV emission efficiency. DL emission was believed to cause a decrease in UV emission intensity. Many studies have reported that ZnO nanostructures obtained through wet chemical methods [95–98] usually exhibit a low UV emission with relatively strong visible emission (I_{UV}/I_{DL}: ~0.3÷19), indicating the presence of a large number of defects in the ZnO nanostructures. This is attributed to the insufficient reaction caused by low synthesis temperature and residual precursors and/or the formation of by-products during the synthesis process. This restricts the application of these ZnO nanostructures in optoelectronic devices as well as biochemical sensors. Unlike those prepared via other methods, the ZnO NWs prepared via thermal evaporation show a very high ratio of I_{UV}/I_{DL}, over 77. This confirms the high crystallinity of the as-grown NWs and the existence of a very low number of defects. These properties, along with the lack of grain boundaries and the long-distance order, make ZnO NWs suitable for applications in electronics, optoelectronics, photonics, solar cells, photocatalysis, etc. [111].

Figure 7a indicates typical resistance curves of ZnO NW sensors for 2, 6, and 10 ppm NO_2 gas. The NO_2 sensing was measured at 200 °C, the moderate working temperature of ZnO-based sensor devices [46,112]. After exposure to NO_2 gas, the resistance of the sensor increased, with the n-type behavior of the sensor being derived from the intrinsic n-type behavior of ZnO. The resistance increased upon exposure to NO_2 gas and approximately recovered to the initial state after removal of NO_2; variations in resistance were found to be reversible behaviors. Also, as predicted, the sensor sensitivities (S) increase when NO_2 gas concentrations increase. As seen in Table 3, the sensitivity of ZnO NW sensors changes as a function of NO_2 concentration. As the NO_2 concentration rises from 2 to 10 ppm, the sensitivity rises from 1.02 to 1.15. Figure 7b displays a typical response and recovery curve of the ZnO NW sensor towards 2 ppm NO_2 when operating at 200 °C. When the sensor is exposed to 2 ppm of NO_2 gas, an apparent increase in resistance is detected. In this case, the sensor response and recovery times are recorded to be 340 and 760 s, respectively. It was found that $t_{recovery}$ is larger than $t_{response}$; this could be due to the fact that NO_2 is strongly bonded to oxygen vacancies on the ZnO NW surface, with almost three times the bond strength of lattice oxygen ($E_{ads} = -0.30$ eV) [113]; as a result, its desorption occurs at a slower rate.

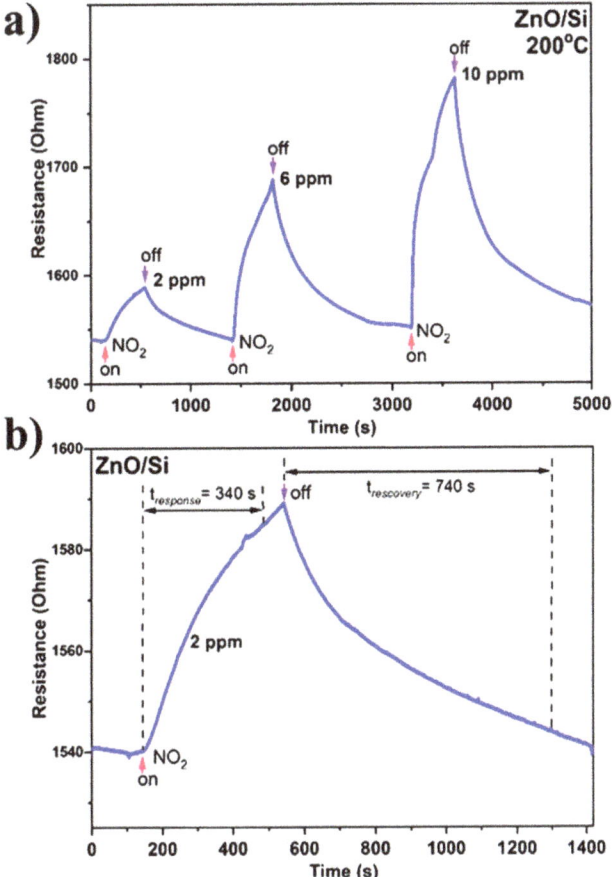

Figure 7. (a) The changes in the sensor resistance of the as-prepared ZnO NWs in various concentrations of NO_2 at operating temperature of 200 °C. (b) Typical response and recovery times of the ZnO NW sensor when exposed to 2 ppm NO_2 at 200 °C.

Table 3. Sensitivities measured at different NO_2 concentrations for the ZnO NW sensor at temperature of 200 °C.

NO_2 Conc. (ppm)	Sensitivity (S)
2	1.02
6	1.09
10	1.15

The NO_2 gas-sensing mechanism of ZnO NWs (n-type semiconductor) is mainly based on the electric resistance change when NO_2 gases make contact with the surface of the ZnO NWs. In particular, in the air atmosphere, oxygen molecules act as electron acceptors and are chemically adsorbed on the surface of ZnO NWs (see Equation (2)) [114–116]. By capturing electrons from the ZnO conduction band, oxygen molecules ionize on the surface material to form chemisorbed oxygen ions. As a result, an electron-depleted region develops on the ZnO surface, and a potential barrier is formed due to the decrease in electron density, leading to a decrease in the number of electrons in the conduction band of ZnO and an increase in resistance [117]. This is called the baseline resistance in air. The types of oxygen species adsorbed (O^{2-}, O^-, and O_2^-) on the ZnO surface depend on the ambient temperature, as can be expressed in Equations (3)–(5) below [118]:

$$O_{2(gas)} \leftrightarrow O_{2(ads)} \tag{2}$$

$$O_{2(ads)} + e^-_{(surf)} \leftrightarrow O_2^-{}_{(ads)}, T < 100\ °C \tag{3}$$

$$O_2^-{}_{(ads)} + e^-_{(surf)} \leftrightarrow 2O^-{}_{(ads)}, 100\ °C < T < 300\ °C \tag{4}$$

$$O^-{}_{(ads)} + e^-_{(surf)} \leftrightarrow O^{2-}{}_{(ads)}, T > 300\ °C \tag{5}$$

$$O^{2-}{}_{(ads)} \leftrightarrow O^{2-}{}_{(Lat)} \tag{6}$$

At low temperatures, oxygen is commonly adsorbed in its molecular state O_2^-, whereas at high temperatures it dissociates as atomic O^- and O^{2-}. Thus, there are two types of oxygen species on the ZnO NW surface; one type is oxygen adsorption and the other is lattice oxygen ($O^{2-}{}_{(Lat)}$) (see Equation (6)). In this work, the sensors were operated at a low temperature of 200 °C. When the sensor is exposed to the atmosphere of NO_2 (strong oxidizing gas), the NO_2 molecules not only withdraw the electrons from the ZnO conduction band, but also interact with the oxygen species adsorbed on the ZnO surface because of their stronger electrophilic properties (2.28 eV) compared with oxygen (0.43 eV) [119–121]. This results in the increase in the width of the electron depletion layer and junction potential barriers, which eventually increase the sensor resistance and generate a sensing response [122]. The absorption of NO_2 on the ZnO surface is complicated; however, it can be summarized by Equations (7)–(9) [123–127]. When NO_2 gas supply is stopped and the sensor is exposed to air again, NO can readily react with dissociated oxygen species and releases the electrons back to the conduction band, causing NO_2 to be released into the air and reducing the resistance (see Equations (10)–(13)).

$$NO_{2(gas)} \leftrightarrow NO_{2(ads)} \tag{7}$$

$$NO_{2(ads)} + O_2^-{}_{(ads)} + e^-_{(surf)} \leftrightarrow 2O^-{}_{(ads)} + NO_2^-{}_{(ads)}, T < 100\ °C \tag{8}$$

$$NO_2^-{}_{(ads)} + O^-{}_{(ads)} + 2e^-_{(surf)} \leftrightarrow 2O^{2-}{}_{(ads)} + NO_{(gas)}, 100\ °C < T < 300\ °C \tag{9}$$

$$O^{2-}{}_{(ads)} + 1/2O_2 \leftrightarrow 2O^-{}_{(ads)} \tag{10}$$

$$2O^-{}_{(ads)} \leftrightarrow O_2^-{}_{(ads)} + e \tag{11}$$

$$O_2^-{}_{(ads)} \leftrightarrow O_{2(gas)} + e \tag{12}$$

$$NO_{(gas)} + O^-{}_{(ads)} \leftrightarrow NO_{2(gas)} + e \tag{13}$$

It is known that point defects on ZnO surfaces play an important role in gas sensor applications since they strongly influence electrical properties. For example, it has been demonstrated that ZnO with a high density of oxygen vacancies has high electrical conductivity [128,129]. Furthermore, it has been demonstrated that increasing the number of defects or oxygen vacancies on the surface of sensing materials considerably improves reactivity, thereby enhancing the sensing performance [129,130]. It was believed that oxygen vacancies could act as favorable adsorption sites for NO_2 molecules [131], leading to an increase in the electrostatic interaction of reactive NO_2 molecules with the ZnO NW surface and consequently resulting in an increase in gas sensitivity. Indeed, several theoretical results suggest that surface defects such as oxygen vacancies can control the chemical/electronic properties and adsorption characteristics of metal oxide surfaces [132,133]. Very recently, using density functional theory (DFT) calculations, An et al. [113] determined that the adsorption energy of NO_2 on the oxygen vacancy site is $E_{ads} = -0.98$ eV, which is three times larger than that on the perfect site ($E_{ads} = -0.30$ eV), and hence, it was expected that the charge transfer from the oxygen vacancy site to the NO_2 adsorbate would be larger than that from the perfect site to the NO_2 adsorbate. This means that oxygen vacancies form stronger bonds with NO_2 molecules, attracting more charge from the ZnO surface than would occur with an oxygen-vacancy-free ZnO surface. This finding is consistent with the close correlation between the concentration of oxygen-vacancy-related defects and the sensitivity of ZnO-based NO_2 sensors [129,130]. Since NO_2 is a strong oxidizing gas with high electrophilicity (~2.28 eV), it directly chemisorbs on oxygen vacancy sites present on the sensor surface, such that the adsorbed NO_2 molecules can readily dissociate into O^- adatoms and release NO gas, following the reaction shown in Equation (14) [134]. In such a case, one of the O atoms of NO_2 fills the oxygen vacancy site, and the weakly bonded NO can desorb from the surface [113]. In addition to this, the NO_2 adsorption can also occur on lattice oxygen ($O^{2-}_{(Lat)}$), which further contributes to the sensing performance.

$$NO_{2(ads)} + V_{\ddot{O}} \leftrightarrow (V_{\ddot{O}} - O^-_{(ads)}) + NO_{(gas)} \tag{14}$$

It is evident from the XPS and PL results (see Figures 5 and 6) that these ZnO NWs contained very few oxygen vacancies and very little deep interstitial oxygen on the surface, which may be considered as a factor contributing to the sensing performance. The more oxygen vacancies there are in the material, the more sensitive the sensor is. It should be noted here that the as-grown ZnO NWs possess excellent optical properties (i.e., a very high I_{UV}/I_{DL} ratio); however, they can suffer from low sensitivity when the gas concentration is at the ppb level. Recently, much effort has been carried out to correlate the gas sensitivity with the point defects commonly present in ZnO [129,130,135–137].

In the gas sensors field, it has been indicated that the specific surface area is the predominant factor in determining the sensing performance of a material. Generally, the increase in the specific surface area of the material is favorable for increasing the number of the adsorbed target gas molecules (e.g., NO_2), thereby improving sensing performance. The specific surface area of the material is usually dependent on its morphological structure [138], which can be adjusted according to the method and conditions of preparation [40,41,139]. ZnO nanomaterials with a variety of sizes and shapes (including nanoparticles, nanorods, nanowires, nanotubes, nanofibers, nanoflowers, and hierarchical nanostructures) can be fabricated, and accordingly, their electrical and chemical properties as well as their specific surface area may be adjusted [139,140]. This is compatible with the gas-sensing characteristics of the material; the same materials with different morphologies can produce different gas-sensing properties [141–145]. It is anticipated that in addition to size/diameter [137] the length of nanowires can also have an impact on gas-sensing performance. For instance, Shooshtari et al. [146] investigated the effect of ZnO nanowire length on the sensing properties of CNT-ZnO samples. They indicated that as the lengths of the ZnO NWs increased from 0.5 to 1.5 um, the sensor response increased up to 30%, due to the adsorption site's effects. Furthermore, by increasing the aspect ratio of ZnO nanorods (length/diameter), the response to liquefied petroleum gas is greatly enhanced [31].

At present, the characteristic parameters of gas sensors, such as sensitivity, selectivity, stability, response time, recovery time, analyte concentration, working temperature, power consumption, and detection limit, have been reported many times in the literature [147–152]. Among them, selectivity is commonly used to evaluate the sensing performance of materials. A sensor is called selective if it only reacts with the target analyte and not with other components in a mixture. Hence, high selectivity confirms that the sensor gives precise information about the presence and concentration of gases. The selectivity of micro/nanostructured ZnO-based sensors can be calculated using Equation (15), as follows [150]:

$$K = \frac{S_{target}}{S_{interference}} \tag{15}$$

where S_{target} and $S_{interference}$ are the responses of a sensor towards target gas and interfering gas, respectively. Chougule et al. [150] fabricated a gas sensor based on ZnO film for detecting NO_2, H_2S, and NH_3 and revealed high selectivity for NO_2 over H_2S in comparison with NH_3. In another study, researchers selected seven different gases, including ethanol, toluene, methanol, NH_3, acetone, CO, and isopropanol, as target gases to investigate the gas response of a single-crystalline ZnO NW sensor [153]. They demonstrated that the ZnO NW sensor exhibits a higher response to ethanol vapor as compared with the other interference gases. The findings of their study indicated that the ZnO NW sensor had superior selectivity towards ethanol vapor. The selectivity of the sensor based on our ZnO NWs/Si-substrate samples is expected to be presented in another work (in preparation). The limit of detection (LOD) is considered an important measure that reflects the sensing performance. For high-performance sensor applications, a sensor should be able to detect extremely low concentrations of the gases. The lowest concentration of an analyte or gas that can be detected by a sensor under operating conditions is called its limit of detection. The LOD is calculated using the following Formula (16) [154]:

$$LOD = 3\frac{S_D}{Slope} \quad (16)$$

where S_D is the intercept standard deviation, as given below (17):

$$\text{Standard deviation }(S_D) = \text{Standard error} \times \sqrt{N} \quad (17)$$

(here, N is the number of data points).

To calculate the LOD of the sensor, we plotted a graph of the sensitivity curve of the NO_2 concentrations. Figure 8 displays the linear fitting curve to NO_2 concentrations ranging from 2 to 10 ppm, indicating that the gas sensor has high linearity for NO_2 concentrations ranging from 2 to 10 ppm. The calculated slope and standard error values are 0.01625 and 0.00493, respectively. According to the definition of LOD, the value of LOD for NO_2 at 200 °C can be calculated using Equation (16):

$$LOD = 3\frac{S_D}{Slope} = 3\frac{0.00493 \times \sqrt{3}}{0.01625} = 1.57 \text{ ppm.}$$

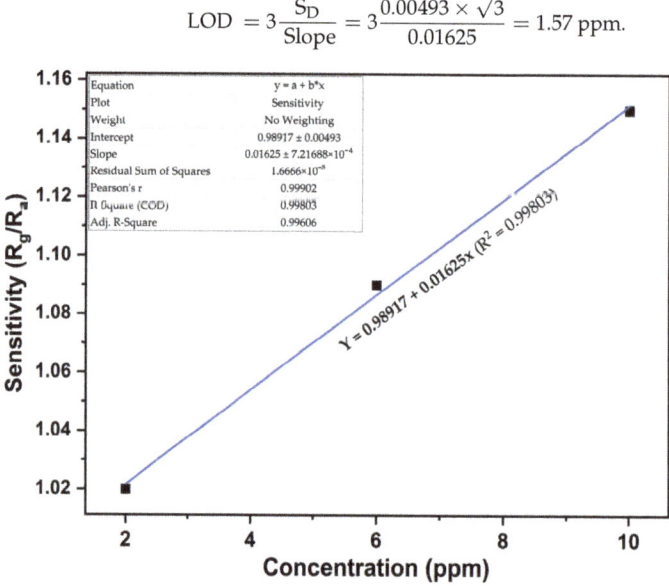

Figure 8. Sensitivity curve of ZnO NWs with different concentrations of NO_2 at 200 °C.

This result is comparable with previous findings for ZnO nanorod-based gas sensors, in which the lowest detection limits were estimated as 10 ppb at 250 °C [155], 50 ppb at 300 °C [156], and 100 ppb at 250 °C [151] for NO_2 gas.

4. Conclusions

In summary, well-aligned ZnO NWs were synthesized via the simple thermal evaporation route without using any catalyst or additive. The obtained samples were analyzed through various techniques such as XRD, FESEM, TEM, EDS, XPS, and photoluminescence spectroscopy. The FESEM and TEM images indicated that the as-prepared ZnO NWs are single-crystalline with a diameter in the range of 135–160 nm and an average length of roughly 3.5 μm, and grow vertically on the Si substrates with homogeneous distribution. The XRD and HRTEM results demonstrated that the ZnO NWs have a hexagonal wurtzite structure and are grown preferentially along the c-axis (0002). The XPS and PL results confirmed that the as-grown ZnO NWs are of high crystalline quality with very few oxygen defects on the surface. The room-temperature PL spectrum of the as-prepared ZnO NWs exhibited a sharp and strong UV emission (~380 nm) with a weak deep-level emission (~491 nm), indicating good optical properties. These properties make ZnO NWs attractive for application in fabricating high-performance optoelectronic devices. Additionally, the ZnO NWs were used as a sensor, which exhibited good sensing performance for NO_2 gas at an operating temperature of 200 °C; this was likely due to their high specific surface area and special atomic structure according to all the characteristics presented above.

Author Contributions: Conceptualization, T.V.K. and P.H.T.; methodology, T.V.K. and P.H.T.; software, T.V.K.; validation, T.V.K. and P.H.T.; formal analysis, P.H.T.; investigation, T.V.K.; resources, T.V.K. and P.H.T.; data curation, T.V.K.; writing—original draft preparation, T.V.K. and P.H.T.; writing—review and editing, T.V.K.; visualization, T.V.K. and P.H.T.; supervision, T.V.K.; project administration, T.V.K.; funding acquisition, T.V.K. All authors have read and agreed to the published version of the manuscript.

Funding: This research received no external funding.

Data Availability Statement: Not applicable.

Acknowledgments: We acknowledge the support of time and facilities from Ho Chi Minh City University of Technology (HCMUT), VNU-HCM for this study.

Conflicts of Interest: The authors declare no conflict of interest.

References

1. Khaniabadi, Y.O.; Goudarzi, G.; Daryanoosh, S.M.; Borgini, A.; Tittarelli, A.; De Marco, A. Exposure to PM10, NO_2, and O_3 and Impacts on Human Health. *Environ. Sci. Pollut. Res. Int.* **2017**, *24*, 2781–2789. [CrossRef] [PubMed]
2. Kwon, Y.J.; Mirzaei, A.; Na, H.G.; Kang, S.Y.; Choi, M.S.; Bang, J.H.; Oum, W.; Kim, S.S.; Kim, H.W. Porous Si nanowires for highly selective room temperature NO_2 gas sensing. *Nanotechnology* **2018**, *29*, 294001. [CrossRef] [PubMed]
3. Patnaik, P. *A Comprehensive Guide to the Hazardous Properties of Chemical Substances*, 3rd ed.; John Wiley & Sons: New York, NY, USA, 2007; p. 405.
4. Brunet, J.; Talazac, L.; Battut, V.; Pauly, A.; Blanc, J.P.; Germain, J.P.; Pellier, S.; Soulier, C. Evaluation of Atmospheric Pollution by Two Semiconductor Gas Sensors. *Thin Solid Films* **2001**, *391*, 308–313. [CrossRef]
5. National Research Council. *Emergency and Continuous Exposure Guidance Levels for Selected Submarine Contaminants*; National Academies Press: Washington, DC, USA, 2009; Volume 3.
6. Drewniak, S.; Drewniak, Ł.; Pustelny, T. Mechanisms of NO_2 Detection in Hybrid Structures Containing Reduced Graphene Oxide. *Sensors* **2022**, *22*, 5316. [CrossRef] [PubMed]
7. Kamble, D.L.; Harale, N.S.; Patil, V.L.; Patil, P.S.; Kadam, L.D. Characterization and NO_2 gas sensing properties of spray pyrolyzed SnO_2 thin films. *J. Anal. Appl. Pyrolysis* **2017**, *127*, 38–46. [CrossRef]
8. Sun, X.; Hao, H.; Ji, H.; Li, X.; Cai, S.; Zheng, C. Nanocasting Synthesis of In_2O_3 with Appropriate Mesostructured Ordering and Enhanced Gas-Sensing Property. *ACS Appl. Mater. Interfaces* **2014**, *6*, 401–409. [CrossRef]
9. Kim, J.W.; Porte, Y.; Ko, K.Y.; Kim, H.; Myoung, J.M. Micropatternable Double-Faced ZnO Nanoflowers for Flexible Gas Sensor. *ACS Appl. Mater. Interfaces* **2017**, *9*, 32876–32886. [CrossRef]
10. Boyadjiev, S.I.; Georgieva, V.; Stefan, N.; Stan, G.E.; Mihailescu, N.; Visan, A.; Mihailescu, I.N.; Besleaga, C.; Szilagyi, I.M. Characterization of PLD grown WO_3 thin films for gas sensing. *Appl. Surf. Sci.* **2017**, *417*, 218–223. [CrossRef]

11. Capone, S.; Benkovicova, M.; Forleo, A.; Jergel, M.; Manera, M.G.; Siffalovic, P.; Taurino, A.; Majkova, E.; Siciliano, P.; Vavra, I.; et al. Palladium/gamma-Fe$_2$O$_3$ nanoparticle mixtures for acetone and NO$_2$ gas sensors. *Sens. Actuators B* **2017**, *243*, 895–903. [CrossRef]
12. Luan, V.; Tien, H.N.; Hur, S.H.; Han, J.H.; Lee, W. Three dimensional porous nitrogen-doped NiO nanostructures as highly sensitive NO$_2$ sensors. *Nanomaterials* **2017**, *7*, 313. [CrossRef]
13. Deng, S.; Tjoa, V.; Fan, H.; Tan, H.; Sayle, D.C.; Olivo, M.; Mhaisalkar, S.; Wei, J.; Sow, C.H. Reduced Graphene Oxide Conjugated Cu$_2$O Nanowire Mesocrystals for High-Performance NO$_2$ Gas Sensor. *J. Am. Chem. Soc.* **2012**, *134*, 4905–4917. [CrossRef]
14. Kumar, R.; Kumar, G.; Umar, A. Zinc Oxide Nanomaterials for Photocatalytic Degradation of Methyl Orange: A Review. *Nanosci. Nanotechnol. Lett.* **2014**, *6*, 631–650. [CrossRef]
15. Hahn, Y.B. Zinc oxide nanostructures and their applications. *Korean J. Chem. Eng.* **2011**, *28*, 1797–1813. [CrossRef]
16. Kumar, R.; Al-Dossary, O.; Kumar, G.; Umar, A. Zinc Oxide Nanostructures for NO$_2$ Gas–Sensor Applications: A Review. *Nano-Miro Lett.* **2015**, *7*, 97–120. [CrossRef] [PubMed]
17. Wibowo, A.; Marsudi, M.A.; Amal, M.I.; Ananda, M.B.; Stephanie, R.; Ardy, H.; Diguna, L.J. ZnO Nanostructured Materials for Emerging Solar Cell Applications. *RSC Adv.* **2020**, *10*, 42838–42859. [CrossRef]
18. Wang, Z.L. ZnO Nanowire and Nanobelt Platform for Nanotechnology. *Mater. Sci. Eng. R* **2009**, *64*, 33–71. [CrossRef]
19. Pascariu, P.; Tudose, I.V.; Suchea, M.; Koudoumas, E.; Fifere, N.; Airinei, A. Preparation and Characterization of Ni, Co Doped ZnO Nanoparticles for Photocatalytic Applications. *Appl. Surf. Sci.* **2018**, *448*, 481–488. [CrossRef]
20. Wang, C.; Yin, L.; Zhang, L.; Xiang, D.; Gao, R. Metal Oxide Gas Sensors: Sensitivity and Influencing Factors. *Sensors* **2010**, *10*, 2088–2106. [CrossRef]
21. Liu, L.; Li, S.; Zhuang, J.; Wang, L.; Zhang, J.; Li, H.; Liu, Z.; Han, Y.; Jiang, X.; Zhang, P. Improved selective acetone sensing properties of Co-Doped ZnO nanofibers by electrospinning. *Sens. Actuators B* **2011**, *155*, 782–788. [CrossRef]
22. Tain, S.; Yang, F.; Zeng, D.; Xie, C. Solution-Processed Gas Sensors Based on ZnO Nanorods Array with an Exposed (0001) Facet for Enhanced Gas-Sensing Properties. *J. Phys. Chem. C* **2012**, *116*, 10586–10591.
23. Batzill, M.; Diebold, U. The Surface and Materials Science of Tin Oxide. *Prog. Surf. Sci.* **2005**, *79*, 47–154. [CrossRef]
24. Yamazoe, N.; Shimanoe, K. Theory of power laws for semiconductor gas sensors. *Sens. Actuators B Chem.* **2002**, *128*, 566–573. [CrossRef]
25. Fioravanti, A.; Marani, P.; Morandi, S.; Lettieri, S.; Mazzocchi, M.; Sacerdoti, M.; Carotta, M.C. Growth Mechanisms of ZnO Micro-Nanomorphologies and Their Role in Enhancing Gas Sensing Properties. *Sensors* **2021**, *21*, 1331. [CrossRef] [PubMed]
26. Rai, P.; Kwak, W.K.; Yu, Y.T. Solvothermal synthesis of ZnO nanostructures and their morphology-dependent gas-sensing properties. *ACS Appl. Mater. Interfaces* **2013**, *5*, 3026–3032. [CrossRef] [PubMed]
27. Obereggera, S.P.; Jonesb, O.A.H.; Spencer, M.J.S. Effect of nanostructuring of ZnO for gas sensing of nitrogen dioxide. *Comput. Mater. Sci.* **2017**, *132*, 104–115. [CrossRef]
28. Carotta, M.C.; Cervi, A.; di Natale, V.; Gherardi, S.; Giberti, A.; Guidi, V.; Puzzovio, D.; Vendemiati, B.; Martinelli, G.; Sacerdoti, M.; et al. ZnO gas sensors: A comparison between nanoparticles and nanotetrapods-based thick films. *Sens. Actuators B Chem.* **2009**, *137*, 164–169. [CrossRef]
29. Öztürk, S.; Kılınç, N.; Taşaltın, N.; Öztürk, Z.Z. A comparative study on the NO$_2$ gas sensing properties of ZnO thin films, nanowires and nanorods. *Thin Solid Films* **2011**, *520*, 932–938. [CrossRef]
30. Bai, S.; Hu, J.; Li, D.; Luo, R.; Chen, A.; Liu, C.C. Quantum-sized ZnO nanoparticles: Synthesis, characterization and sensing properties for NO$_2$. *J. Mater. Chem.* **2011**, *21*, 12288–12294. [CrossRef]
31. Gurav, K.V.; Gang, M.G.; Shin, S.W.; Patil, U.M.; Deshmukh, P.R.; Agawane, G.L.; Suryawanshi, M.P.; Pawar, S.M.; Patil, P.S.; Lokhande, C.D.; et al. Gas sensing properties of hydrothermally grown ZnO nanorods with different aspect ratios. *Sens. Actuators B Chem.* **2014**, *190*, 439–445. [CrossRef]
32. Khai, T.V.; Thu, L.V.; Ha, L.T.T.; Thanh, V.M.; Lam, T.D. Structural, optical and gas sensing properties of vertically well-aligned ZnO nanowires grown on graphene/Si substrate by thermal evaporation method. *Mater. Charact.* **2018**, *141*, 296–317. [CrossRef]
33. Huang, B.; Zhao, C.; Zhang, M.; Zhang, Z.; Xie, E.; Zhou, J.; Han, W. Doping effect of In$_2$O$_3$ on structural and ethanol-sensing characteristics of ZnO nanotubes fabricated by electrospinning. *Appl. Surf. Sci.* **2015**, *349*, 615–621. [CrossRef]
34. Cao, Y.; Zou, X.; Wang, X.; Qian, J.; Bai, N.; Li, G.D. Effective detection of trace amount of explosive nitro-compounds by ZnO nanofibers with hollow structure. *Sens. Actuators B Chem.* **2016**, *232*, 564–570. [CrossRef]
35. Agarwal, S.; Rai, P.; Gatell, E.N.; Llobet, E.; Güell, F.; Kumar, M.; Awasthi, K. Gas sensing properties of ZnO nanostructures (flowers/rods) synthesized by hydrothermal method. *Sens. Actuators B Chem.* **2019**, *292*, 24–31. [CrossRef]
36. Liu, X.; Zhang, J.; Wang, L.; Yang, T.; Guo, X.; Wu, S.; Wang, S. 3D hierarchically porous ZnO structures and their functionalization by au nanoparticles for gas sensors. *J. Mater. Chem.* **2011**, *21*, 349–356. [CrossRef]
37. Shouli, B.; Liangyuan, C.; Dianqing, L.; Wensheng, Y.; Pengcheng, Y.; Zhiyong, L.; Aifan, C.; Liu, C.C. Different morphologies of ZnO nanorods and their sensing property. *Sens. Actuators B Chem.* **2010**, *146*, 129–137. [CrossRef]
38. Iversen, K.J.; Spencer, M.J.S. Effect of ZnO Nanostructure Morphology on the Sensing of H$_2$S Gas. *J. Phys. Chem. C* **2013**, *117*, 26106–26118. [CrossRef]
39. Zhang, S.; Nguyen, S.T.; Nguyen, T.H.; Yang, W.; Noh, J.S. Effect of the Morphology of Solution-Grown ZnO Nanostructures on Gas-Sensing Properties. *J. Am. Ceram. Soc.* **2017**, *100*, 5629–5637. [CrossRef]
40. Gupta, S.K.; Joshi, A.; Kaur, M. Development of gas sensors using ZnO nanostructures. *J. Chem. Sci.* **2010**, *122*, 57–62. [CrossRef]

41. Wei, S.; Wang, S.; Zhang, Y.; Zhou, M. Different morphologies of ZnO and their ethanol sensing property. *Sens. Actuators B Chem.* **2014**, *192*, 480–487. [CrossRef]
42. Cao, Y.; Jia, D.; Wang, R.; Luo, J. Rapid One-Step Room-Temperature Solid-State Synthesis and Formation Mechanism of Zno Nanorods as H_2S-Sensing Materials. *Solid-State Electron.* **2013**, *82*, 67–71. [CrossRef]
43. Wang, C.; Chu, X.; Wu, M. Detection of H_2S Down to ppb Levels at Room Temperature Using Sensors Based on ZnO Nanorods. *Sens. Actuators B Chem.* **2006**, *113*, 320–323. [CrossRef]
44. Zhang, Y.; Li, Q.; Tian, Z.; Hu, P.; Qin, X.; Yun, F. Gas-sensing properties of ITO materials with different morphologies prepared by sputtering. *SN Appl. Sci.* **2020**, *2*, 264. [CrossRef]
45. An, S.; Park, S.; Ko, H.; Jin, C.; Lee, W.I.; Lee, C. Enhanced Gas Sensing Properties of Branched ZnO Nanowires. *Thin Solid Films* **2013**, *547*, 241–245. [CrossRef]
46. Ahn, M.W.; Park, K.S.; Heo, J.H.; Kim, D.W.; Choi, K.J.; Park, J.G. On-chip fabrication of ZnO-nanowire gas sensor with high gas sensitivity. *Sens. Actuators B Chem.* **2009**, *138*, 168–173. [CrossRef]
47. Waclawik, E.R.; Chang, J.; Ponzoni, A.; Concina, I.; Zappa, D.; Comini, E.; Motta, N.; Faglia, G.; Sberveglieri Beilstein, G. Enhanced NO_2 gas sensor response by chemical modification of nanowire surfaces. *J. Nanotechnol.* **2012**, *3*, 368–377.
48. Cardoza-Contreras, M.N.; Romo-Herrera, J.M.; Ríos, L.A.; García-Gutiérrez, R.; Zepeda, T.A.; Contreras, O.E. Single ZnO Nanowire-Based Gas Sensors to Detect Low Concentrations of Hydrogen. *Sensors* **2015**, *16*, 30539–30544. [CrossRef]
49. Lupan, O.; Emelchenko, G.A.; Ursaki, V.V.; Chai, G.; Redkin, A.N.; Gruzintsev, A.N.; Tiginyanu, I.M.; Chow, L.; Ono, L.K.; Roldan Cuenya, B.; et al. Synthesis and Characterization of ZnO Nanowires for Nanosensor Applications. *Mater. Res. Bull.* **2010**, *45*, 1026–1032. [CrossRef]
50. Wan, Q.; Li, Q.H.; Chen, Y.J.; Wang, T.H.; He, X.L.; Li, J.P.; Lin, C.L. Fabrication and Ethanol Sensing Characteristics of ZnO Nanowire Gas Sensors. *Appl. Phys. Lett.* **2004**, *84*, 3654–3656. [CrossRef]
51. Wu, W.Y.; Ting, J.M.; Huang, P.J. Electrospun ZnO nanowires as gas sensors for ethanol detection. *Nanoscale Res. Lett.* **2009**, *4*, 513–517. [CrossRef]
52. Hieu, N.V.; Chien, N.D. Low-temperature growth and ethanol-sensing characteristics of quasi-one-dimensional ZnO nanostructures. *Phys. B Condens. Matter* **2008**, *403*, 50–56. [CrossRef]
53. Pivert, M.L.; Martin, N.; Leprince-Wang, Y. Hydrothermally Grown ZnO Nanostructures for Water Purification via Photocatalysis. *Crystals* **2022**, *12*, 308. [CrossRef]
54. Li, S.; Zhang, X.; Yan, B.; Yu, T. Growth mechanism and diameter control of well-aligned small-diameter ZnO nanowire arrays synthesized by a catalyst-free thermal evaporation method. *Nanotechnology* **2009**, *20*, 495604. [CrossRef] [PubMed]
55. Xing, Y.J.; Xi, Z.H.; Zhang, X.D.; Song, J.H.; Wang, R.M.; Xu, J.; Xue, Z.Q.; Yu, D.P. Thermal evaporation synthesis of zinc oxide nanowires. *Appl. Phys. A* **2005**, *80*, 1527–1530. [CrossRef]
56. Wang, L.; Zhang, X.; Zhao, S.; Zhou, G.; Zhou, Y.; Qi, J. Synthesis of well-aligned ZnO nanowires by simple physical vapor deposition on c-oriented ZnO thin films without catalysts or additives. *Appl. Phys. Lett.* **2005**, *86*, 024108. [CrossRef]
57. Ashraf, S.; Jones, A.C.; Bacsa, J.; Steiner, A.; Chalker, P.R.; Beahan, P.; Hindley, S.; Odedra, R.; Williams, P.A.; Heys, P.N. MOCVD of vertically aligned ZnO nanowires using bidentate ether adducts of dimethylzinc. *Chem. Vap. Depos.* **2011**, *17*, 45–53. [CrossRef]
58. Woong, L.; Min-Chang, J.; Jae-Min, M. Fabrication and application potential of ZnO nanowires grown on GaAs(002) substrates by metal–organic chemical vapour deposition. *Nanotechnology* **2004**, *15*, 254.
59. Khai, T.V.; Maneeratanasarn, P.; Choi, B.G.; Ham, H.; Shim, K.B. Diameter-and density-controlled synthesis of well-aligned ZnO nanowire arrays and their properties using a thermal evaporation technique. *Phys. Status Solidi A* **2012**, *209*, 1498–1510. [CrossRef]
60. Khai, T.V.; Thu, L.V.; Lam, T.D. Vertically well-aligned ZnO nanowire arrays directly synthesized from Zn vapor deposition without catalyst. *J. Electron. Mater.* **2016**, *45*, 2601–2607. [CrossRef]
61. Yang, Z.; Dou, X.; Zhang, S.; Guo, L.; Zu, B.; Wu, Z.; Zeng, H. A High-Performance Nitro-Explosives Schottky Sensor Boosted by Interface Modulation. *Adv. Funct. Mater.* **2015**, *25*, 4039–4048. [CrossRef]
62. Ham, H.; Shen, G.; Cho, J.; Lee, T.J.; Seo, S.; Lee, G. Vertically Aligned ZnO Nanowires Produced by a Catalyst-Free Thermal Evaporation Method and Their Field Emission Properties. *Chem. Phys. Lett.* **2005**, *404*, 69–73. [CrossRef]
63. Kavakebi, M.; Jamali-sheini, F. Ultrasonic Synthesis of Zn-Doped CdO Nanostructures and Their Optoelectronic Properties. *Trans. Nonferrous Met. Soc. China* **2018**, *28*, 2255–2264. [CrossRef]
64. Cullity, B.D.; Stock, S.R. *Elements of X-ray Diffraction*; Pearson Education Limited: London, UK, 2014.
65. Thool, G.S.; Singh, A.K.; Singh, R.S.; Gupta, A.; Susan, A.B.H. Facile synthesis of flat crystal ZnO thin films by solution growth method: A micro-structural investigation. *J. Saudi Chem. Soc.* **2014**, *18*, 712–721. [CrossRef]
66. Samadi, M.; Zirak, M.; Naseri, A.; Khorashadizade, E.; Moshfegh, A.Z. Recent Progress on Doped ZnO Nanostructures for Visible-Light Photocatalysis. *Thin Solid Films* **2016**, *605*, 2–19. [CrossRef]
67. Wang, Z.L. Zinc oxide nanostructures: Growth, properties and applications. *J. Phys. Condens. Matter* **2004**, *16*, R829–R858.
68. Xu, L.; Hu, Y.L.; Pelligra, C.; Chen, C.H.; Jin, L.; Huang, H.; Sithambaram, S.; Aindow, M.; Joesten, R.; Suib, S.L. ZnO with Different Morphologies Synthesized by Solvothermal Methods for Enhanced Photocatalytic Activity. *Chem. Mater.* **2009**, *21*, 2875–2885. [CrossRef]
69. Wagner, C.; Gale, L.; Raymond, R. Two-dimensional chemical state plots: A standardized data set for use in identifying chemical states by X-ray photoelectron spectroscopy. *Anal. Chem.* **1979**, *51*, 466–482. [CrossRef]

70. Saha, T.; Achath Mohanan, A.; Swamy, V.; Guo, N.; Ramakrishnan, N. An Optimal Thermal Evaporation Synthesis of C-Axis Oriented ZnO Nanowires with Excellent UV Sensing and Emission Characteristics. *Mater. Res. Bull.* **2016**, *77*, 147–154. [CrossRef]
71. Sahai, A.; Goswami, N. Probing the Dominance of Interstitial Oxygen Defects in ZnO Nanoparticles through Structural and Optical Characterizations. *Ceram. Int.* **2014**, *40*, 14569–14578. [CrossRef]
72. Al-Gaashani, R.; Radiman, S.; Daud, A.R.; Tabet, N.; Al-Douri, Y. XPS and Optical Studies of Different Morphologies of ZnO Nanostructures Prepared by Microwave Methods. *Ceram. Int.* **2013**, *39*, 2283–2292. [CrossRef]
73. Das, J.; Pradhan, S.K.; Sahu, D.R.; Mishra, D.K.; Sarangi, S.N.; Nayak, B.B.; Verma, S.; Roul, B.K. Micro-Raman and XPS Studies of Pure ZnO Ceramics. *Phys. B Condens. Matter* **2010**, *405*, 2492–2497. [CrossRef]
74. Zheng, J.H.; Jiang, Q.; Lian, J.S. Synthesis and Optical Properties of Flower-like ZnO Nanorods by Thermal Evaporation Method. *Appl. Surf. Sci.* **2011**, *257*, 5083–5087. [CrossRef]
75. Li, C.C.; Du, Z.F.; Li, L.M.; Yu, H.C.; Wan, Q.; Wang, T.H. Surface-Depletion Controlled Gas Sensing of ZnO Nanorods Grown at Room Temperature. *Appl. Phys. Lett.* **2007**, *91*, 032101. [CrossRef]
76. Ghosh, B.; Ray, S.C.; Pattanaik, S.; Sarma, S.; Mishra, D.K.; Pontsho, M.; Pong, W.F. Tuning of the Electronic Structure and Magnetic Properties of Xenon Ion Implanted Zinc Oxide. *J. Phys. D Appl. Phys.* **2018**, *51*, 095304. [CrossRef]
77. Tam, K.H.; Cheung, C.K.; Leung, Y.H.; Djurišic, A.B.; Ling, C.C.; Beling, C.D.; Fung, S.; Kwok, W.M.; Chan, W.K.; Phillips, D.L.; et al. Defects in ZnO Nanorods Prepared by a Hydrothermal Method. *J. Phys. Chem. B* **2006**, *110*, 20865–20871. [CrossRef]
78. Karamat, S.; Rawat, R.S.; Lee, P.; Tan, T.; Ramanujan, R. Structural, Elemental, Optical and Magnetic Study of Fe Doped ZnO and Impurity Phase Formation. *Prog. Nat. Sci. Mater.* **2014**, *24*, 142–149. [CrossRef]
79. Chen, Y.; Xu, X.L.; Zhang, G.H.; Xue, H.; Ma, S.Y. Blue shift of optical band gap in Er-doped ZnO thin films deposited by direct current reactive magnetron sputtering technique. *Physica E Low Dimens. Syst. Nanostruct.* **2010**, *42*, 1713–1716. [CrossRef]
80. Chen, H.; Liu, W.; Qin, Z. ZnO/ZnFe$_2$O$_4$ Nanocomposite as a Broad-Spectrum Photo-Fenton-like Photocatalyst with near-Infrared Activity. *Catal. Sci. Technol.* **2017**, *7*, 2236–2244. [CrossRef]
81. Papari, G.P.; Silvestri, B.; Vitiello, G.; de Stefano, L.; Rea, I.; Luciani, G.; Aronne, A.; Andreone, A. Morphological, Structural, and Charge Transfer Properties of F-Doped ZnO: A Spectroscopic Investigation. *J. Phys. Chem. C* **2017**, *121*, 16012–16020. [CrossRef]
82. Lisachenko, A.A. Study of self-sensitization of wide-gap oxides to visible light by intrinsic defects: From Terenin to the present days. *J. Photochem. Photobiol. A Chem.* **2018**, *354*, 47–60. [CrossRef]
83. Studenikin, S.A.; Golego, N.; Cocivera, M. Fabrication of green and orange photoluminescent, undoped ZnO films using spray pyrolysis. *J. Appl. Phys.* **1998**, *84*, 2287–2294. [CrossRef]
84. Hsu, N.E.; Hung, W.K.; Chen, Y.F. Origin of Defect Emission Identified by Polarized Luminescence from Aligned ZnO Nanorods. *J. Appl. Phys.* **2004**, *96*, 4671–4673. [CrossRef]
85. Liu, M.; Kitai, A.H.; Mascher, P. Point Defects and Luminescence Centres in Zinc Oxide and Zinc Oxide Doped with Manganese. *J. Lumin.* **1992**, *54*, 35–42. [CrossRef]
86. Lin, B.; Fu, Z.; Jia, Y. Green luminescent center in undoped zinc oxide films deposited on silicon substrates. *Appl. Phys. Lett.* **2001**, *79*, 943–945. [CrossRef]
87. Ahn, C.H.; Kim, Y.Y.; Kim, D.C.; Mohanta, S.K.; Cho, H.K. A Comparative Analysis of Deep Level Emission in ZnO Layers Deposited by Various Methods. *J. Appl. Phys.* **2009**, *105*, 013502. [CrossRef]
88. Zhao, S.; Zhou, Y.; Zhao, K.; Liu, Z.; Han, P.; Wang, S.; Xiang, W.; Chen, Z.; Lü, H.; Cheng, B.; et al. Violet luminescence emitted from Ag-nanocluster doped ZnO thin films grown on fused quartz substrates by pulsed laser deposition. *Phys. B Condens. Matter* **2006**, *373*, 154–156. [CrossRef]
89. Khai, T.V.; Kwak, D.S.; Kwon, Y.J.; Shim, K.B.; Kim, H.W. Catalyst-free thermally-evaporated growth and optical properties of ZnO nanowires on Si, GaN and sapphire substrates. *Cryst. Res. Technol.* **2013**, *48*, 75–86. [CrossRef]
90. Yousefi, R.; Zak, A.K. Growth and characterization of ZnO nanowires grown on the Si(111) and Si(100) substrates: Optical properties and biaxial stress of nanowires. *Mater. Sci. Semicond.* **2011**, *14*, 170–174. [CrossRef]
91. Hassan, N.K.; Hashim, M.R.; Mahdi, M.A.; Allam, N.K. A catalyst-free growth of ZnO nanowires on Si (100) substrates: Effect of substrate position on morphological, structural and optical properties. *ECS J. Solid State Sci. Technol.* **2012**, *1*, P86–P89. [CrossRef]
92. Umar, A.; Hahn, Y.B. Aligned hexagonal coaxial-shaped ZnO nanocolumns on steel alloy by thermal evaporation. *Appl. Phys. Lett.* **2006**, *88*, 173120. [CrossRef]
93. Ahmad, N.F.; Yasui, K.; Hashim, A.M. Seed/catalyst-free growth of zinc oxide on graphene by thermal evaporation: Effects of substrate inclination angles and graphene thicknesses. *Nanoscale Res. Lett.* **2015**, *10*, 10. [CrossRef]
94. ElZein, B.; Yao, Y.; Barham, A.S.; Dogheche, E.; Jabbour, G.E. Toward the Growth of Self-Catalyzed ZnO Nanowires Perpendicular to the Surface of Silicon and Glass Substrates, by Pulsed Laser Deposition. *Materials* **2020**, *13*, 4427. [CrossRef] [PubMed]
95. Han, F.; Yang, S.; Jing, W.; Jiang, K.C.; Jiang, Z.; Liu, H.; Li, L. Surface Plasmon Enhanced Photoluminescence of ZnO Nanorods by Capping Reduced Graphene Oxide Sheets. *Opt. Express* **2014**, *22*, 11436–11445. [CrossRef] [PubMed]
96. Khai, T.V.; Long, L.N.; Khoi, N.H.T.; Hoc Thang, N. Effects of Hydrothermal Reaction Time on the Structure and Optical Properties of ZnO/Graphene Oxide Nanocomposites. *Crystals* **2022**, *12*, 1825. [CrossRef]
97. Li, L.; Yao, C.; Wu, L.; Jiang, K.; Hu, Z.; Xu, N.; Sun, J.; Wu, J. ZnS Covering of ZnO Nanorods for Enhancing UV Emission from ZnO. *J. Phys. Chem. C* **2021**, *125*, 13732–13740. [CrossRef]

98. Yin, Y.; Sun, Y.; Yu, M.; Liu, X.; Yang, B.; Liu, D.; Liu, S.; Cao, W.; Ashfold, M.N.R. Controlling the hydrothermal growth and the properties of ZnO nanorod arrays by pre-treating the seed layer. *RSC Adv.* **2014**, *4*, 44452. [CrossRef]
99. Shakti, N.; Prakash, A.; Mandal, T.; Katiyar, M. Processing Temperature Dependent Morphological and Optical Properties of ZnO Nanorods. *Mater. Sci. Semicond.* **2014**, *20*, 55–60. [CrossRef]
100. Opoku, C.; Dahiya, A.S.; Cayrel, F.; Poulin-Vittrant, G.; Alquier, D.; Camara, N. Fabrication of Field-Effect Transistors and Functional Nanogenerators Using Hydrothermally Grown ZnO Nanowires. *RSC Adv.* **2015**, *5*, 69925–69931. [CrossRef]
101. Sohn, J.I.; Cha, S.N.; Song, B.G.; Lee, S.; Kim, S.M.; Ku, J.; Kim, H.J.; Park, Y.J.; Choi, B.L.; Wang, Z.L.; et al. Engineering of Efficiency Limiting Free Carriers and an Interfacial Energy Barrier for an Enhancing Piezoelectric Generation. *Energy Environ. Sci.* **2012**, *6*, 97–104. [CrossRef]
102. Kushwaha, A.; Aslam, M. Defect Induced High Photocurrent in Solution Grown Vertically Aligned ZnO Nanowire Array Films. *J. Appl. Phys.* **2012**, *112*, 054316. [CrossRef]
103. Das, S.N.; Kar, J.P.; Choi, J.H.; Lee, T.I.; Moon, K.J.; Myoung, J.M. Fabrication and Characterization of ZnO Single Nanowire-Based Hydrogen Sensor. *J. Phys. Chem. C* **2010**, *114*, 1689–1693. [CrossRef]
104. Selim, F.A.; Weber, M.H.; Solodovnikov, D.; Lynn, K.G. Nature of native defects in ZnO. *Phys. Rev. Lett.* **2007**, *99*, 085502. [CrossRef] [PubMed]
105. Wang, Y.; Li, X.; Wang, N.; Quan, X.; Chen, Y. Controllable synthesis of ZnO nanoflowers and their morphology-dependent photocatalytic activities. *Sep. Purif. Technol.* **2008**, *62*, 727–732. [CrossRef]
106. Baradaran, M.; Ghodsi, F.; Bittencourt, C.; Llobet, E. The role of Al concentration on improving the photocatalytic performance of nanostructured ZnO/ZnO:Al/ZnO multilayer thin films. *J. Alloys Compd.* **2019**, *788*, 289–301. [CrossRef]
107. Goldberger, J.; Sirbuly, D.J.; Law, M.; Yang, P. ZnO Nanowire Transistors. *J. Phys. Chem. B* **2005**, *109*, 9–14. [CrossRef] [PubMed]
108. Liu, J.; Wu, W.; Bai, S.; Qin, Y. Synthesis of High Crystallinity ZnO Nanowire Array on Polymer Substrate and Flexible Fiber-Based Sensor. *ACS Appl. Mater. Interfaces* **2011**, *3*, 4197–4200. [CrossRef] [PubMed]
109. Das, S.N.; Moon, K.J.; Kar, J.P.; Choi, J.H.; Xiong, J.; Lee, T.I.; Myoung, J.M. ZnO Single Nanowire-Based UV Detectors. *Appl. Phys. Lett.* **2010**, *97*, 022103. [CrossRef]
110. Janotti, A.; Van de Walle, C.G. Native Point Defects in ZnO. *Phys. Rev. B* **2007**, *76*, 165202. [CrossRef]
111. Ding, M.; Guo, Z.; Zhou, L.; Fang, X.; Zhang, L.; Zeng, L.; Xie, L.; Zhao, H. One-dimensional Zinc oxide nanomaterials for application in high-performance advanced opolectronic devices. *Crystals* **2018**, *8*, 223. [CrossRef]
112. Nisha, R.; Madhusoodanan, K.N.; Vimalkumar, T.V.; Vijayakumar, K.P. Gas sensing application of nanocrystalline zinc oxide thin films prepared by spray pyrolysis. *Bull. Mater. Sci.* **2015**, *38*, 583–591.
113. An, W.; Wu, X.; Zeng, X.C. Adsorption of O_2, H_2, CO, NH_3, and NO_2 on ZnO Nanotube: A Density Functional Theory Study. *J. Phys. Chem. C* **2008**, *112*, 5747–5755. [CrossRef]
114. Mirzaei, A.; Kim, J.-H.; Kim, H.W.; Kim, S.S. Resistive-Based Gas Sensors for Detection of Benzene, Toluene and Xylene (BTX) Gases: A Review. *J. Mater. Chem. C* **2018**, *6*, 4342–4370. [CrossRef]
115. Mirzaei, A.; Kim, J.H.; Kim, H.W.; Kim, S.S. How Shell Thickness Can Affect the Gas Sensing Properties of Nanostructured Materials: Survey of Literature. *Sens. Actuators B Chem.* **2018**, *258*, 270–294. [CrossRef]
116. Mirzaei, A.; Leonardi, S.G.; Neri, G. Detection of Hazardous Volatile Organic Compounds (VOCs) by Metal Oxide Nanostructures-Based Gas Sensors: A Review. *Ceram. Int.* **2016**, *42*, 15119–15141. [CrossRef]
117. Al-Hardan, N.H.; Aziz, A.A.; Abdullah, M.J.; Ahmed, N.M. Conductometric Gas Sensing Based on ZnO Thin Films: An Impedance Spectroscopy Study. *ECS J. Solid State Sci. Technol.* **2018**, *7*, P487–P490. [CrossRef]
118. Ding, J.; Chen, S.; Han, N.; Shi, Y.; Hu, P.; Li, H.; Wang, J. Aerosol assisted chemical vapour deposition of nanostructured ZnO thin films for NO_2 and ethanol monitoring. *Ceram. Int.* **2020**, *46*, 15152–15158. [CrossRef]
119. Chen, E.S.; Wentworth, W.E.; Chen, E.C.M. The electron affinities of NO and O_2. *J. Mol. Struct.* **2002**, *606*, 1–7. [CrossRef]
120. Zhou, Z.; Gao, H.; Liu, R.; Du, B. Study on the structure and property for the $NO_2 + NO_2^-$ electron transfer system. *J. Mol. Struct. THEOCHEM* **2001**, *545*, 179–186. [CrossRef]
121. Šulka, M.; Pitoňák, M.; Neogrády, P.; Urban, M. Electron Affinity of the O_2 Molecule: CCSD(T) Calculations Using the Optimized Virtual Orbitals Space Approach. *Int. J. Quantum Chem.* **2008**, *108*, 2159–2171. [CrossRef]
122. Ganbavle, V.V.; Inamdar, S.I.; Agawane, G.L.; Kim, J.H.; Rajpure, K.Y. Synthesis of Fast Response, Highly Sensitive and Selective Ni:ZnO Based NO_2 Sensor. *Chem. Eng. J.* **2016**, *286*, 36–47. [CrossRef]
123. Barreca, D.; Bekermann, D.; Comini, E.; Devi, A.; Fischer, R.A.; Gasparotto, A.; Maccato, C.; Sada, C.; Sberveglieri, G.; Tondello, E. Urchin-like ZnOnanorod arrays for gas sensing applications. *Cryst. Eng. Comm.* **2010**, *12*, 3419–3421. [CrossRef]
124. Zhu, L.; Zeng, W. Room-Temperature Gas Sensing of ZnO-Based Gas Sensor: A Review. *Sens. Actuators A Phys.* **2017**, *267*, 242–261. [CrossRef]
125. Xiao, Y.; Yang, T.; Chuai, M.; Xiao, B.; Zhang, M. Synthesis of ZnO nanosheet arrays with exposed (100) facets for gas sensing applications. *Phys. Chem. Chem. Phys.* **2016**, *18*, 325–330. [CrossRef] [PubMed]
126. Choi, M.S.; Na, H.G.; Mirzaei, A.; Bang, J.H.; Oum, W.; Han, S.; Choi, S.W.; Kim, M.; Jin, C.; Kim, S.S.; et al. Room-temperature NO_2 sensor based on electrochemically etched porous silicon. *J. Alloys Compd.* **2019**, *811*, 151975. [CrossRef]

127. Parellada-Monreal, L.; Castro-Hurtado, I.; Martínez Calderón, M.; Presmanes, L.; Mandayo, G.G. Laser-induced periodic surface structures on ZnO thin film for high response NO$_2$ detection. *Appl. Surf. Sci.* **2019**, *476*, 569–575. [CrossRef]
128. Chang, P.C.; Fan, Z.; Wang, D.; Tseng, W.Y.; Chiou, W.A.; Hong, J.; Lu, J.G. ZnO Nanowires Synthesized by Vapor Trapping CVD Method. *Chem. Mater.* **2004**, *16*, 5133–5137. [CrossRef]
129. Ahn, M.W.; Park, K.S.; Heo, J.H.; Park, J.G.; Kim, D.W.; Choi, K.J.; Lee, J.H.; Hong, S.H. Gas sensing properties of defect-controlled ZnO-nanowire gas sensor. *Appl. Phys. Lett.* **2008**, *93*, 263103–263105. [CrossRef]
130. Li, G.; Zhang, H.; Meng, L.; Sun, Z.; Chen, Z.; Huang, X.; Qin, Y. Adjustment of Oxygen Vacancy States in ZnO and Its Application in Ppb-Level NO$_2$ Gas Sensor. *Sci. Bull.* **2020**, *65*, 1650–1658. [CrossRef]
131. Li, Y.; Zu, B.; Guo, Y.; Li, K.; Zeng, H.; Dou, X. Surface Superoxide Complex Defects-Boosted Ultrasensitive Ppb-Level NO$_2$ Gas Sensors. *Small* **2016**, *12*, 1420–1424. [CrossRef]
132. Henrich, V.E.; Cox, P.A. *The Surface Science of Metal Oxides*; Cambridge University Press: Cambridge, UK, 1994.
133. Schaub, R.; Wahlström, E.; Rønnau, A.; Lagsgaard, E.; Stensgaard, I.; Besenbacher, F. Oxygen-Mediated Diffusion of Oxygen Vacancies on the TiO$_2$(110) Surface. *Science* **2003**, *299*, 377–379. [CrossRef]
134. Spencer, M.J.S.; Yarovsky, I. ZnO Nanostructures for Gas Sensing: Interaction of NO$_2$, NO, O, and N with the ZnO$(10\bar{1}0)$ Surface. *J. Phys. Chem. C* **2010**, *114*, 10881–10893. [CrossRef]
135. Zou, C.; Liang, F.; Xue, S. Synthesis and oxygen vacancy related NO$_2$ gas sensing properties of ZnO:Co nanorods arrays gown by a hydrothermal method. *Appl. Surf. Sci.* **2015**, *353*, 1061–1069. [CrossRef]
136. Lee, B.W.; Kim, T.S.; Goswami, S.K.; Oh, E. Gas Sensitivity and Point Defects in Sonochemically Grown ZnO Nanowires. *J. Korean Phys. Soc.* **2012**, *60*, 415–419. [CrossRef]
137. Lupan, O.; Ursaki, V.V.; Chai, G.; Chow, L.; Emelchenko, G.A.; Tiginyanu, I.M.; Gruzintsev, A.N.; Redkin, A.N. Selective Hydrogen Gas Nanosensor Using Individual ZnO Nanowire with Fast Response at Room Temperature. *Sens. Actuators B Chem.* **2010**, *144*, 56–66. [CrossRef]
138. Choi, S.W.; Park, J.Y.; Kim, S.S. Dependence of Gas Sensing Properties in ZnO Nanofibers on Size and Crystallinity of Nanograins. *J. Mater. Res.* **2011**, *26*, 1662–1665. [CrossRef]
139. Morandi, S.; Fioravanti, A.; Cerrato, G.; Lettieri, S.; Sacerdoti, M.; Carotta, M. Facile synthesis of ZnO nano-structures: Morphology influence on electronic properties. *Sens. Actuators B Chem.* **2017**, *249*, 581–589. [CrossRef]
140. Righettoni, M.; Amann, A.; Pratsinis, S.E. Breath Analysis by Nanostructured Metal Oxides as Chemo-Resistive Gas Sensors. *Mater. Today* **2015**, *18*, 163–171. [CrossRef]
141. Khatibani, A.B. Characterization and Ethanol Sensing Performance of Sol–Gel Derived Pure and Doped Zinc Oxide Thin Films. *J. Electron. Mater.* **2019**, *48*, 3784–3793. [CrossRef]
142. Perekrestov, V.; Latyshev, V.; Kornyushchenko, A.; Kosminska, Y. Formation, Charge Transfer, Structural and Morphological Characteristics of ZnO Fractal-Percolation Nanosystems. *J. Electron. Mater.* **2019**, *48*, 2788–2793. [CrossRef]
143. Gul, M.; Amin, M.; Abbas, M.; Ilyas, S.Z.; Shah, N.A. Synthesis and Characterization of Magnesium Doped ZnO Nanostructures: Methane (CH$_4$) Detection. *J. Mater. Sci. Mater. Electron.* **2019**, *30*, 5257–5265. [CrossRef]
144. Chetna; Kumar, R.; Garg, A.; Chowdhuri, A.; Jain, A.; Kapoor, A. A Novel Method of Electrochemically Growing ZnO Nanorods on Graphene Oxide as Substrate for Gas Sensing Applications. *Mater. Res. Express* **2019**, *6*, 075039. [CrossRef]
145. Aydin, H.; Yakuphanoglu, F.; Aydin, C. Al-Doped ZnO as a Multifunctional Nanomaterial: Structural, Morphological, Optical and Low-Temperature Gas Sensing Properties. *J. Alloys Compd.* **2019**, *773*, 802–811. [CrossRef]
146. Shooshtari, M.; Pahlavan, S.; Rahbarpour, S.; Ghafoorifard, H. Investigating organic vapor sensing properties of composite carbon nanotube-zinc oxide nanowire. *Chemosensors* **2022**, *10*, 205. [CrossRef]
147. Shi, L.; Naik, A.J.T.; Goodall, J.B.M.; Tighe, C.; Gruar, R.; Binions, R.; Parkin, I.; Darr, J. Highly sensitive ZnO nanorod- and nanoprism-based NO$_2$ gas sensors: Size and shape control using a continuous hydrothermal pilot plant. *Langmuir* **2013**, *29*, 10603–10609. [CrossRef] [PubMed]
148. Han, H.V.; Hoa, N.D.; Tong, P.V.; Nguyen, H.; Hieu, N.V. Single-Crystal Zinc Oxide Nanorods with Nanovoids as Highly Sensitive NO$_2$ Nanosensors. *Mater. Lett.* **2013**, *94*, 41–43. [CrossRef]
149. Park, J.; Oh, J.Y. Highly-sensitive NO$_2$ Detection of ZnO Nanorods Grown by a Sonochemical Process. *J. Korean Phys. Soc.* **2009**, *55*, 1119–1122. [CrossRef]
150. Chougule, M.; Sen, S.; Patil, V. Fabrication of Nanostructured ZnO Thin Film Sensor for NO$_2$ Monitoring. *Ceram. Int.* **2012**, *38*, 2685–2692. [CrossRef]
151. Oh, E.; Choi, H.Y.; Jung, S.H.; Cho, S.; Kim, J.C.; Lee, K.H.; Kang, S.W.; Kim, J.; Yun, J.Y.; Jeong, S.H. High-Performance NO$_2$ Gas Sensor Based on ZnO Nanorod Grown by Ultrasonic Irradiation. *Sens. Actuators B Chem.* **2009**, *141*, 239–243. [CrossRef]
152. Bochenkov, V.E.; Sergeev, G.B. Sensitivity, selectivity, and stability of gas-sensitive metal-oxide nanostructures. In *Metal Oxide Nanostructures and Their Applications*; Umar, A., Hang, Y.B., Eds.; American Scientific Publishers: Valencia, CA, USA, 2010; pp. 31–35.
153. Yuan, Z.; Yin, L.; Ding, H.; Huang, W.; Shuai, C.; Deng, J. One-step synthesis of single-crystalline ZnO nanowires for the application of gas sensor. *J. Mater. Sci. Mater. Electron.* **2018**, *29*, 11559–11565. [CrossRef]
154. Priya, S.; Halder, J.; Mandal, D.; Chowdhury, A.; Singh, T.; Chandra, A. Hierarchical SnO$_2$ nanostructures for potential VOC sensor. *J. Mater. Sci.* **2021**, *56*, 9883–9893. [CrossRef]

155. Sahin, Y.; Öztürk, S.; Kılınç, N.; Käsemen, A.; Erkovane, M.; Ozturk, Z.Z. Electrical conduction and NO_2 gas sensing properties of ZnO nanorods. *Appl. Surf. Sci.* **2014**, *303*, 90–96. [CrossRef]
156. Öztürk, S.; Kılınç, N.; Öztürk, Z.Z. Fabrication of ZnO nanorods for NO_2 sensor applications: Effect of dimensions and electrode position. *J. Alloys Compd.* **2013**, *581*, 196–201. [CrossRef]

Disclaimer/Publisher's Note: The statements, opinions and data contained in all publications are solely those of the individual author(s) and contributor(s) and not of MDPI and/or the editor(s). MDPI and/or the editor(s) disclaim responsibility for any injury to people or property resulting from any ideas, methods, instructions or products referred to in the content.

Article

Effect of UV and IR Radiation on the Electrical Characteristics of Ga$_2$O$_3$/ZnGeP$_2$ Hetero-Structures

Vera Kalygina [1], Sergey Podzyvalov [1], Nikolay Yudin [1,*], Elena Slyunko [1], Mikhail Zinoviev [1], Vladimir Kuznetsov [2], Alexey Lysenko [1], Andrey Kalsin [1], Victor Kopiev [1], Bogdan Kushnarev [1], Vladimir Oleinik [1], Houssain Baalbaki [1] and Pavel Yunin [3]

[1] Laboratory for Radiophysical and Optical Methods of Environmental Studies, National Research Tomsk State University, 634050 Tomsk, Russia; kalygina@ngs.ru (V.K.); cginen@yandex.ru (S.P.); elenohka266@mail.ru (E.S.); muxa9229@gmail.com (M.Z.); festality@yandex.ru (A.L.); andrejkalsin@gmail.com (A.K.); kopiev@mail.ru (V.K.); kushnarev@mail.ru (B.K.); oleinik@mail.ru (V.O.); houssainsyr1@gmail.com (H.B.)

[2] Institute of High Current Electronics, Akademichesky Ave. 2/3, 634055 Tomsk, Russia; robert_smith_93@mail.ru

[3] Institute of Physics of Microstructures RAS, Akademicheskaya St., 7, 603087 Nizhny Novgorod, Russia; yunin@ipmras.ru

* Correspondence: rach3@yandex.ru; Tel.: +7-996-938-71-32

Citation: Kalygina, V.; Podzyvalov, S.; Yudin, N.; Slyunko, E.; Zinoviev, M.; Kuznetsov, V.; Lysenko, A.; Kalsin, A.; Kopiev, V.; Kushnarev, B.; et al. Effect of UV and IR Radiation on the Electrical Characteristics of Ga$_2$O$_3$/ZnGeP$_2$ Hetero-Structures. *Crystals* **2023**, *13*, 1203. https://doi.org/10.3390/cryst13081203

Academic Editors: Andrew F. Zhou and Peter X. Feng

Received: 30 May 2023
Revised: 18 July 2023
Accepted: 19 July 2023
Published: 2 August 2023

Copyright: © 2023 by the authors. Licensee MDPI, Basel, Switzerland. This article is an open access article distributed under the terms and conditions of the Creative Commons Attribution (CC BY) license (https://creativecommons.org/licenses/by/4.0/).

Abstract: The data on electrical and photoelectric characteristics of Ga$_2$O$_3$/ZnGeP$_2$ hetero-structures formed by RF magnetron sputtering Ga$_2$O$_3$ target with a purity of (99.99%) were obtained. The samples are sensitive to UV radiation with a wavelength of λ = 254 nm and are able to work offline as detectors of short-wave radiation. Structures with Ga$_2$O$_3$ film that was not annealed at 400 °C show weak sensitivity to long-wavelength radiation, including white light and near-IR (λ = 808 and 1064 nm). After annealing in an air environment (400 °C, 30 min), ZnGeP$_2$ crystals in contact with Ga$_2$O$_3$ show n-type conductivity semiconductor properties, the sensitivity of Ga$_2$O$_3$/ZnGeP$_2$ hetero-structures increases in the UV and IR ranges; the photovoltaic effect is preserved. Under λ = 254 nm illumination, the open-circuit voltage is fixed at positive potentials on the electrode to Ga$_2$O$_3$, the short-circuit current increases by three orders of magnitude, and the responsivity increases by an order of magnitude. The structures detect the photovoltaic effect in the near-IR range and are able to work offline (remotely) as detectors of long-wavelength radiation.

Keywords: hetero-structure; UV radiation; IR range; response

1. Introduction

Technological progress is largely determined by obtaining high-quality semiconductor materials of complex composition. ZnGeP$_2$ (ZGP) is a semiconductor compound with a chalcopyrite structure.

This is a p-type semiconductor with a bandgap $E_g \approx 2.0$ eV at 300 K. ZGP crystals are characterized by a wide transparency region (0.65–12 µm), high second-order electrical susceptibility ($d_{36} = 75 \times 10^{-12}$ m/V), birefringence sufficient for phase matching, low-temperature dependence of refractive indices, relatively high thermal conductivity 0.18 W/(cm * K), resistance to high humidity and aggressive environments, large values of temperature, and angular and spectral phase-matching widths [1].

The electrical properties of ZnGeP$_2$ have been studied less than other A$_2$B$_4$C$_5^2$ compounds, which is explained by the experimental difficulties that arise when measuring transport effects in substances with high resistance. In [2], the results of studying a number of characteristics of ZGP crystals are presented. The Hall effect measurements showed that all samples had p-type conductivity. The concentration values for various crystals at T = 300° K and the types of impurities used are given in Table 1.

Table 1. Charge carrier concentration (p), mobility (U_p), and acceptor impurity activation energy (E_a) for various $ZnGeP_2$ samples.

Doping Impurity	Impurity Action	P (cm^{-3})	U_p (cm^2/V s)	E_a (eV)
-	-	1.5×10^{10}	60	0.65
Au	inactive	2.5×10^{10}	30	0.65
Cu	inactive	1.0×10^{10}	13	-
Se	acceptor	2.5×10^{14}	0.5	0.4
Ga	acceptor	8.7×10^{16}	0.18	0.31
In	acceptor	1.4×10^{16}	0.13	-

It can be seen from the results presented in [2] that the doping ZGP melt with group I chemical elements, practically, does not lead to a change in the hole concentration compared to the undoped crystal; however, the mobility of charge carriers decreases significantly. Group VI impurities cause an increase in the hole concentration by four orders of magnitude; they have the role of acceptors. The Hall mobility of holes decreased by approximately two orders of magnitude compared to undoped material, which probably indicated an increase in the concentration of impurity ions and structural lattice defects. A significant increase in the concentration of holes is observed when elements of group III, gallium, and indium are introduced into the melt. In these samples, a significant decrease in the value of the Hall mobility is observed, as in the case of sample 3p. Based on the results of the studies, it was determined that impurity elements of groups III and VI in the ZGP material are effective acceptor centers.

$ZnGeP_2$ samples grown under various conditions and subjected to subsequent high-temperature annealing always have a p-type conductivity (from $p_0 = 1 \times 10^{10}$ to 1×10^{16} cm^{-3}) and a hole mobility of 20–40 cm^2/V * s [3–5], which is explained by the presence of zinc vacancies [V_{Zn}] or germanium atoms at the sites of the phosphorus [Ge_P] sublattice. The invariability of the type of conductivity indicates high stability on the part of structural growth defects, which thus determines the properties of $ZnGeP_2$.

Analyzing the temperature dependence of the electro-physical parameters in [2], the values of the activation energy of the acceptor impurity were estimated by 0.2–0.7 eV. The activation energies of deep centers increase with a decrease in the density of free holes in the studied crystals. A similar effect was also observed in [2] when doping $ZnGeP_2$ with various impurities.

$ZnGeP_2$ triple compounds are nonlinear crystals and are widely used in optical parametric oscillators, amplifiers, and other techniques based on the effects of nonlinear optics [6] like sources of coherent radiation of the mid-IR range, which are used in the following ways:

- In LIDAR systems for remote detection and quantification of gases based on the effect of resonant absorption. The range from 3 to 10 µm is most effective [7];
- In medicine, e.g., minimally invasive neurosurgery (tissue ablation) with a wavelength of 6.45 µm, requiring single high-energy pulses [8];
- In the evaporation and deposition of thin-film polymers;
- In spectroscopy and other scientific applications where high spatial and temporal resolution is required, e.g., excitation and vibrational transitions in molecules or inter-band transitions in semiconductor structures;
- Special military and civilian applications.

In addition to the applications aforementioned, this work presents the results of studies of electrical and photoelectric characteristics of structures based on $ZnGeP_2$ and gallium oxide films, and their sensitivity to ultraviolet and infrared radiation.

Ga_2O_3 is a transparent conducting oxide (TCO) that forms several polymorphs, such as α, β, γ, δ, and ε [9–11], which can either be insulators or conductors depending on the growth conditions from different techniques. Each polymorph can exhibit differences in band alignment [12]. Polymorphs have slightly varying material properties in the crystal space group. Differences between polymorphs include their properties in the crystal space group and the coordination number of Ga^{3+} ions. Ga_2O_3 is an n-type semiconductor with a bandgap of $E_g = 4.8–5.3$ eV. In the band gap of Ga_2O_3, there are local levels of defects, mainly due to oxygen and gallium vacancies [13].

Gallium oxide has a number of practical properties, where Ga_2O_3 is a semiconductor of n-type conductivity with a high specific resistance ($\rho \approx 10^{13}$ $\Omega \cdot$cm) that allows us to create fluorescent capacitors with a wide set of wavelengths and gas sensors with high stability and speed of operation even in conditions of high humidity. The use of oxide films in semiconductor/Ga_2O_3 structures allowed for the development of various devices for detecting UV radiation in outer space, flame detectors, and tracking the state of ozone holes.

The properties of the materials and the characteristics of the semiconductor/Ga_2O_3 interfaces depend on the method of manufacture, the thickness of the deposited layers [14], the pressure of the gas mixture during manufacture [15,16], and subsequent technological processes [17,18]. In turn, the processes at these interfaces determine the electrical and optical characteristics of the structures when thin layers of gallium oxide are used. Often devices of short-wave range, according to technical specifications, work in continuous modes of UV radiation, including systems of protected short-range communication, UV astronomy, and monitoring of ozone holes. In this regard, it is important to study the detector behavior under UV continuous illumination. One of the promising areas of development of short-wave radiation detectors is devices that can operate as self-powered UV photodetectors. Such photodetectors have a simple design and, most importantly, involve direct integration with metal–dielectric–semiconductor (MDS) fabrication technology.

2. Preparation of ZGP Substrates

A wire-cutting machine was used to cut ZGP ingot into plates with dimensions of $6 \times 6 \times 1$ mm. The wire used in Figure 1 is a wire with a diameter of 0.08–0.35 mm. A wire was used to improve cutting, on which a diamond-containing layer with a grain size of 10–60 μm was galvanized. The main advantage of wire cutting is that this method makes it possible to obtain formed parts with minimal disruption of the crystal structure due to low thermodynamic stresses that occur in the zone of contact between the wire and the workpiece. When processing brittle materials, the force is about 0.15–2.00 N.

The plates obtained after cutting had non-flatness and non-parallelism, a significant thickness of the damaged layer, and a spread in thickness. Technological operations of grinding and polishing were sequentially carried out to obtain the specified values of the thickness and surface roughness.

The block of elements was attached to the faceplate using wax glue and fixing equipment. Grinding was realized by the free-lapping principle. The abrasive suspension is collected from distilled water and diamond powder of the selected grain size, which is selected from the given specification of the roughness class.

During the grinding process, the abrasive grains, which are not connected, form an abrasive layer between the ZIP plates and the grinder, and rolling the particles of both the processed material and the material of the grinder break out. On the processed plates, the material is removed only from the side facing the grinder; the second side is the base.

When rotating in the same direction as the grinder and the faceplate with a block of elements, the grains of the abrasive grains are pressed with one of their faces into the grinder, and with the other, into the surface of the plates to be machined. In places of contact with abrasive grains, cracks appear on the surface of the grinder and elements, the maximum depth of which depends on the nature and size of the fractions of the selected abrasive grains. When, because of repeated impacts of grains, the entire surface layer is covered with similar cracks, subsequent grain movements in the same places will lead to the extraction of fragments of the plate and form "gouges". A number of adjacent gouges form a characteristic rough surface.

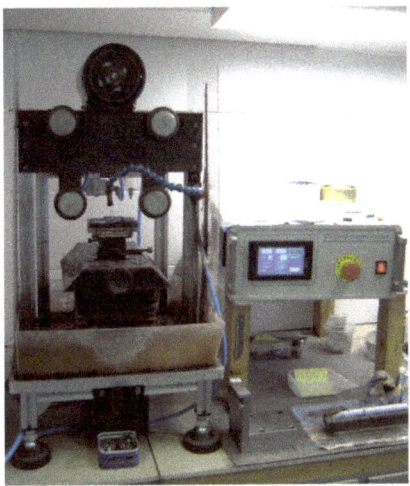

Figure 1. Photograph of a wire-cutting machine.

Polishing of optical elements is the final operation and aims to eliminate surface damage after grinding and to ensure high cleanliness (Figure 2) of the treated surface.

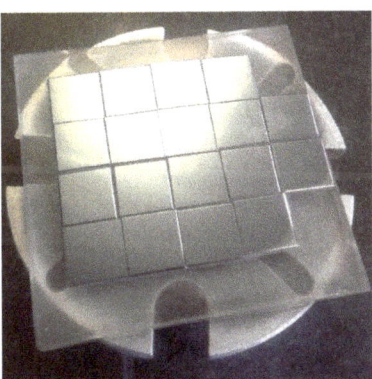

Figure 2. Photograph of polished ZGP plates glued to a faceplate.

The initial polishing of the working surfaces of all the studied samples was carried out on a 4-PD-200 polishing and finishing machine (SZOS, Republic of Belarus). The initial processing of the working surfaces of the samples consisted of polishing on a cambric-polishing pad using synthetic diamond powder ACM 0.5/0 (average grain size 270 nm). Next, the samples were additionally polished on a cambric polishing pad using synthetic diamond powder ACM 0.25/0. The samples were then polished on a resin polishing pad made from polishing resin using ACM 0.25/0 synthetic diamond powder.

3. Experimental Methodology

The $Ga_2O_3/ZnGeP_2$ structures were formed by depositing a Ga_2O_3 film by RF magnetron sputtering Ga_2O_3 (99.99%) target onto polished $ZnGeP_2$ substrates.

Thin layers of gallium oxide were obtained on an AUTO-500 unit (manufacturer: Edwards) in an Ar/O$_2$ gas mixture. The oxygen concentration in the mixture was maintained at (56.1 ± 0.5) vol.%. The distance between the target and the substrate was 70 mm. The pressure in the chamber during deposition was 0.7 Pa. During the deposition of the oxide film, the ZGP substrates were not heated. After the formation of the Ga$_2$O$_3$ film with a thickness of 150–200 nm, Pt electrodes were deposited on the Ga$_2$O$_3$ film through a metal mask, where the contact area to Ga$_2$O$_3$ is of s = 1.04 × 10^{-2} cm^2. A solid ohmic contact (In) was deposited onto the opposite surface of the ZnGeP$_2$ substrate. Schematic of the device structure is presented in Figure 3. Some of the samples were annealed in the air for 30 min at 400 °C.

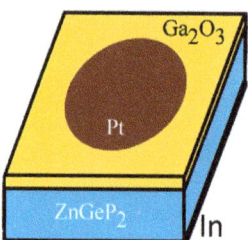

Figure 3. Schematic of device structure.

The current–voltage characteristics (I-V), when no radiation (dark) and under UV illumination, were measured at room temperature using a Keithley 2611B source meter. A VL-6.C krypton-fluorine lamp with a 254 nm filter was used as a source of UV radiation (UVr). The distance between the lamp and the sample was 1 cm, the radiation power was 8 × 10^{-6} W, and the incident radiation intensity was 0.78 mW/cm^2. When working in the IR range, lasers at 808 and 1064 nm were used. Regardless of the type of radiation, the samples were illuminated from the side of the oxide film.

4. Experimental Data and the Discussion

Figure 4 shows the data of X-ray diffraction analysis of a gallium oxide film deposited on a ZnGeP$_2$ substrate by RF magnetron sputtering Ga$_2$O$_3$ (99.99%) target. The absence of peaks in the curve for Ga$_2$O$_3$ indicates the amorphous phase of the oxide film.

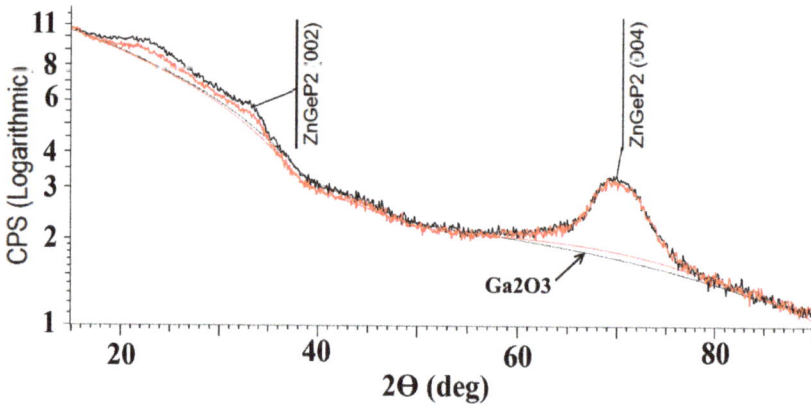

Figure 4. X-ray diffraction analysis of gallium oxide film on the ZnGeP$_2$ substrate.

4.1. Electrical and Photoelectric Characteristics of Structures without Annealing

Figure 5 shows a schematic representation of the energy diagram of the $Ga_2O_3/ZnGeP_2$ hetero-structure. Taking into account the electronic affinity of the materials in contact (χ_{Ga2O3} = 4.0 eV, χ_{ZGP} = 3.5 eV), the band offset of the conduction band is ΔE_c = 0.5 eV, and the valence band is ΔE_v = 3.7 eV (E_{g1} = 5.1 eV and E_{g2} =1.9 eV).

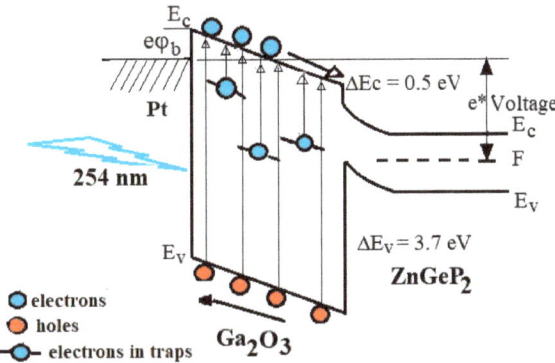

Figure 5. The schematic band diagram of the $Ga_2O_3/ZnGeP_2$ hetero-junction.

The current–voltage (I-V) characteristics of $Ga_2O_3/ZnGeP_2$ structures without annealing at 400 °C are not symmetrical with respect potential sign on the Pt electrode: at negative potentials on the contact to gallium oxide, the current increases faster than at positive values of V, which corresponds to direct currents (Figure 6a).

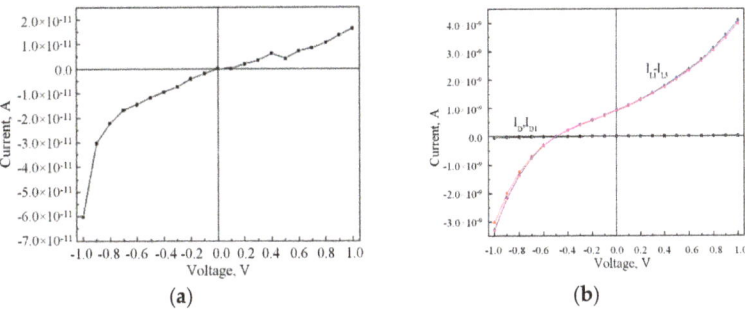

Figure 6. Current–voltage characteristics of the sample: (**a**) dark current (I_D); (**b**) dark currents (I_D, I_{D1}) and under UV continuous illumination (I_{L1}–I_{L5}); dark current measured immediately after turning off the UV is indicated as I_{D1}.

Under UV illumination, electrons from the valence band and from deep centers in the band gap of Ga_2O_3 are generated into the conduction band of gallium oxide, which leads to an increase in current in the structure. The current through the sample increases regardless of the potential sign on the Pt electrode (Figure 6b). The structures exhibit a photovoltaic effect, which allows to detect UV radiation offline. Open-circuit voltage V_{oc} = 0.5 V and short-circuit current Isc = 1 × 10^{-9} A. With repeated measurements of the I-V characteristics during UV continuous illumination, the curves of the dependence of I_L on V practically coincide (Figure 6b, curves I_{L1}–I_{L5}), which indicates the stable operation of the structures under UV continuous illumination.

The photocurrent I_{ph} (the difference between the current during UV radiation (I_L) and dark current (I_D)) is greater at V > 0 V. Figure 7 shows the dark I-V characteristics before (I_D) and after UV illumination (I_{D1}), the light I-V characteristics measured during

UV continuous illumination, and the dependence of the responsivity R as a function of voltage in the range $0 \leq V \leq 5$ V.

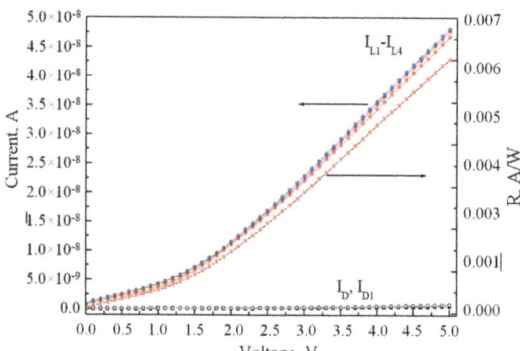

Figure 7. Dependencies of dark currents before (I_D) and after (I_{D1}) UV illumination; currents under UV continuous illumination (I_{L1}–I_{L4}) and response R on voltage.

After switching off UV radiation, the currents return to the original I_D values measured before ilumination. For comparison, Figure 8 shows the dark I-V characteristics measured before (I_D) and immediately after switching off the UV radiation (I_{D1}).

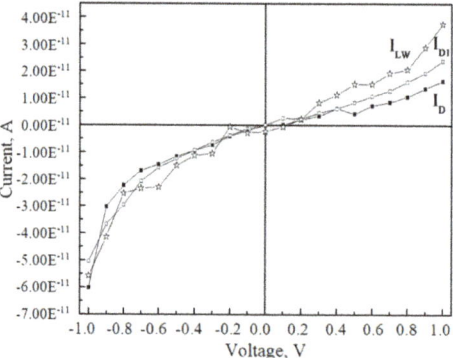

Figure 8. Dark I-V characteristics of the photodetector before (I_D) and after (I_{D1}) 254 nm illumination; I-V characteristics during white light illumination with power of 3×10^{-3} W (I_{LW}).

Close values of currents in most of the voltage range -1 V $\leq V \leq +1$ V indicate the absence of noticeable persistent conductivity in the samples. Analysis of the above data (Figures 3b and 6) shows that the currents during UV illumination (I_L) exceed the dark currents I_D by more than an order of magnitude. This allows to use $Ga_2O_3/ZnGeP_2$ structures as UV detectors. Detector efficiency is determined by several parameters, including dark current I_D, responsivity R, external quantum efficiency EQE, response time t_r, and recovery time td [19].

The responsivity is calculated using the formula [19]:

$$R = \frac{I_L - I_D}{P}, \tag{1}$$

P is the light power. In accordance with the growth of I_L, there is an increase in the response with an increase in the voltage on the structure. At positive potentials on the Pt electrode, the light current exceeds the dark current by orders of magnitude; therefore, it is more advantageous to use such structures as UV radiation detectors at positive voltages at contact with gallium oxide.

The external quantum efficiency EQE is calculated according to the following formula [19]:

$$\text{EQE} = \frac{hc}{e\lambda}R, \quad (2)$$

where h is Planck's constant, e is the electron charge, c is the speed of light, λ is the wavelength of incident light. In the range from 0 to 5 V, EQE increases from 4.7×10^{-4} to 2.8×10^{-2}.

Another important characteristic of the detector is the specific detectivity D* [cm × Hg$^{0.5}$ × W^{-1}], which characterizes the possibility of the photodetector to detect extremely small signals and is calculated by the formula [19]:

$$D^* = R\sqrt{\frac{s}{2eI_D}}. \quad (3)$$

where s is the area of the photosensitive region. The values of D* increase with increasing voltage on the sample, and at V ≥ 4.5 V the curve of the dependence of D* on V tends to saturation (Figure 9). The increase in the dark current I_D with increasing voltage causes the saturation region on the curve D* = f(V).

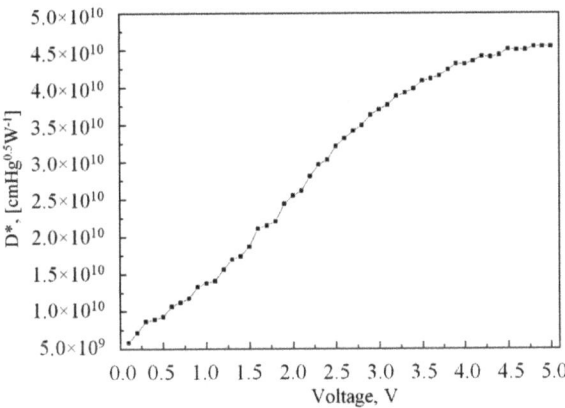

Figure 9. Dependence of specific detectivity D* on voltage.

A small difference between the dark currents before (I_D) and immediately after UV illumination (I_{D1}) (Figure 8) indicates the absence of noticeable persistent conductivity in the studied structures. Figure 10a shows the profile of the time dependence of the current of the Ga$_2$O$_3$/ZnGeP$_2$ structure for three pulses switching on and off the UV radiation. One of the pulses is shown in detail in Figure 10b. The response time t_r and recovery time t_d do not exceed 1 s (Figure 10).

The sensitivity of the samples to radiation with λ = 254 nm is explained by the absorption of light in the wide bandgap of Ga$_2$O$_3$ film.

Ga$_2$O$_3$/ZnGeP$_2$ structures with a gallium oxide film not annealed at 400 °C show weak sensitivity to long-wave radiation. At V = +5 V, the current I_L is 4.8×10^{-8} A at λ = 254 nm and a power of 8×10^{-6} W, and when illuminated with white light with a power of 3×10^{-3} W, the current I_{LW} = 7.9×10^{-10} A, which is about two orders of magnitude lower. However, if the oxide film is annealed at 400 °C in the air for 30 min, then the Ga$_2$O$_3$/ZnGeP$_2$ structures become sensitive to both UV and IR radiation.

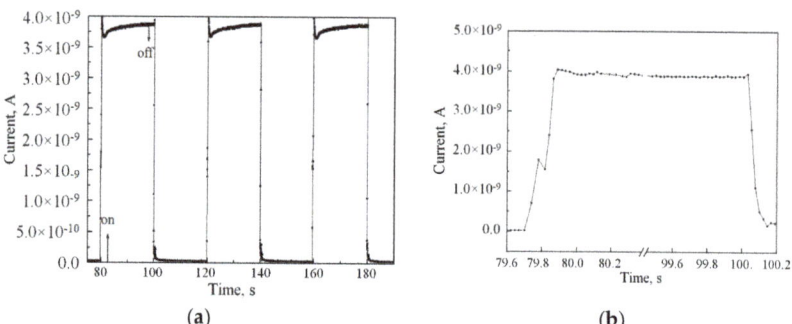

Figure 10. Time profile of the current when switching on/off the radiation with λ = 254 nm and V = 1 V (**a**); start and end of a single pulse on a more detailed scale (**b**).

4.2. Electrical and Photoelectric Characteristics of Ga_2O_3/$ZnGeP_2$ Structures after Annealing

Annealing leads to a significant change in the characteristics of Ga_2O_3/$ZnGeP_2$ structures: direct currents are observed at positive voltages on the Pt electrode; dark currents increase (Figure 11); and sensitivity to UV and IR radiation increases.

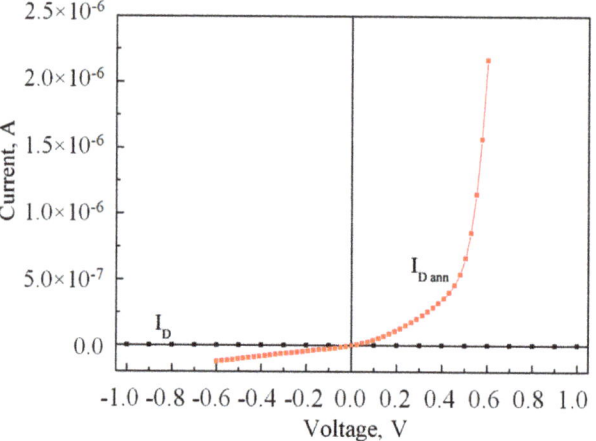

Figure 11. Dependences of dark currents of samples without (I_D) and after annealing ($I_{D\ ann}$) on voltage.

In structures with an annealed gallium oxide film, the sensitivity to UV and IR radiation increases; the photovoltaic effect is preserved; V_{oc} decreases to (0.3–0.5) V and is fixed at positive voltages; and short-circuit current increases by three orders of magnitude: $I_{sc} = (1.0–1.5) \times 10^{-6}$ A (Figure 12).

After annealing, the responsivity R to λ = 254 nm at all voltages increases by an order of magnitude; the maximum values of R are obtained at negative voltages on the Pt electrode (Figure 12); at V = 0 V, the specific detectivity of the structures is $D^* = 1 \times 10^{12}$ cm × $Hg^{0.5}$ × W^{-1}.

Lasers with wavelengths of 808 and 1064 nm and powers of 500 and 100 mW, respectively, were used as a source of IR radiation. Figure 10 shows the dark I-V characteristics and the light ones under λ = 808 nm (Figure 13a) and 1064 nm (Figure 13b) illumination.

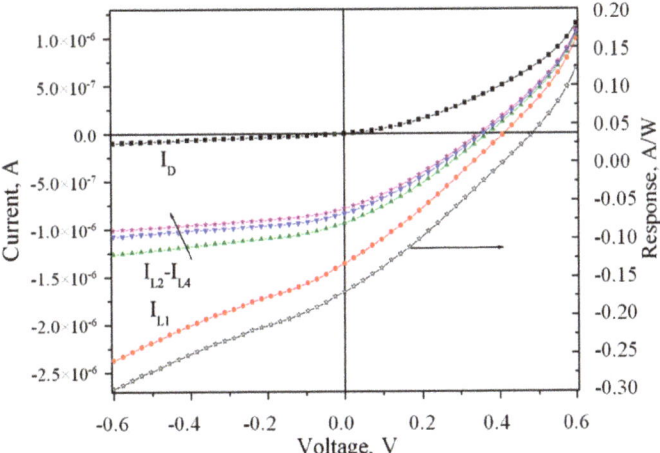

Figure 12. Dependences of dark (I_D), light I-V characteristics (I_{L1}–I_{L4}), responsivity R of $Ga_2O_3/ZnGeP_2$ structures at λ = 254 nm on voltage.

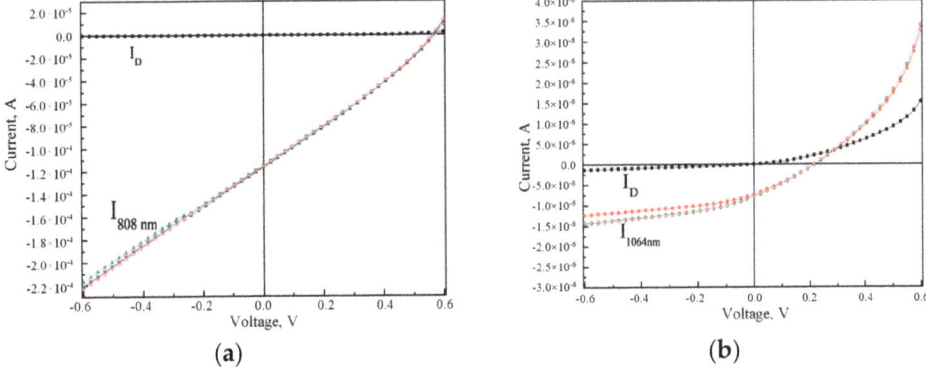

Figure 13. Dependences of dark (I_D) and light I-V characteristics of $Ga_2O_3/ZnGeP_2$ structure on voltage during illuminating: with λ = 808 nm (P = 500 mW) (**a**); with λ = 1064 nm (P = 100 mW) (**b**).

Samples with an annealed Ga_2O_3 film have a photovoltaic effect under IR illumination, and the open-circuit voltage is observed at positive values of V. Under λ = 808 nm illumination with P = 500 mW, the open-circuit voltage Voc = (0.57–0.60) V and current Isc = 0.12 mA (Figure 11a) are recorded. Thus, the results from above show these structures can function without an external power source, making it, therefore, a self-powered device, which has potentially important applications such as secure ultraviolet communication and space detection.

It can be assumed that the sensitivity of the $Ga_2O_3/ZnGeP_2$ structures to IR radiation with the annealed Ga_2O_3 film is due to the high transmittance T. However, the spectral dependences of T measured before and after annealing (Figure 14) refute this hypothesis. The question remains open, and research in this direction continues.

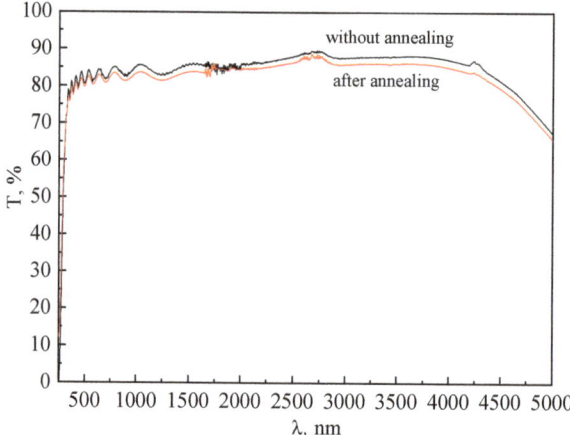

Figure 14. Transmission (T) spectra of Ga_2O_3 film without and after annealing.

5. Conclusions

The electrical and photoelectric characteristics of $Ga_2O_3/ZnGeP_2$ hetero-structures obtained by RF magnetron sputtering Ga_2O_3 (99.99%) target on polished $ZnGeP_2$ substrates are studied. The samples are sensitive to UV radiation with λ = 254 nm and are able to work as self-powered detectors of short-wave radiation. Structures with Ga_2O_3 film not annealed at 400 °C show weak sensitivity to long-wave radiation, including the near-IR range (λ = 808 and 1064 nm). Annealing of the oxide film in the air (400 °C, 30 min) leads to an increase in the sensitivity of hetero-structures in both the UV and IR ranges. These structures can be used as radiation detectors in the IR range with the ability to operate as self-powered devices.

Author Contributions: Conceptualization, V.K. (Vera Kalygina) and S.P.; methodology, S.P., M.Z. and N.Y.; software, E.S., V.K. (Vladimir Kuznetsov) and H.B.; validation, A.L. and A.K.; formal analysis, B.K., B.K. and V.O.; investigation, M.Z., S.P. and P.Y.; resources, V.K. (Victor Kopiev) data curation, E.S. and M.Z.; writing—original draft preparation, A.L.; writing—review and editing, A.L. and H.B.; visualization S.P.; supervision V.K. (Vera Kalygina) and S.P.; project administration, V.K. (Vera Kalygina); funding acquisition, N.Y. All authors have read and agreed to the published version of the manuscript.

Funding: This study was supported by the Tomsk State University Development Program (Priority 2030).

Institutional Review Board Statement: Not applicable.

Informed Consent Statement: Not applicable.

Data Availability Statement: Not applicable.

Conflicts of Interest: The authors declare no conflict of interest. The funders had no role in the design of the study; in the collection, analyses, or interpretation of data; in the writing of the manuscript, or in the decision to publish the results.

References

1. Kitaeva, G.K. Terahertz generation by means of optical lasers. *Laser Phys. Lett.* **2008**, *5*, 559–576. [CrossRef]
2. Grigoreva, V.S.; Prochukhan, V.D.; Rud, Y.V.; Yakovenko, A.A. Some electrical properties of high-resistance ZnGeP2 single crystals. *Phys. Status Solidi (a)* **1973**, *17*, K69–K73. [CrossRef]
3. Carnio, B.; Greig, S.; Firby, C.; Zawilski, K.; Schunemann, P.; Elezzabi, A. Terahertz electro-optic detection using a ⟨012⟩-cut chalcopyrite ZnGeP2 crystal. *Appl. Phys. Lett.* **2016**, *108*, 261109. [CrossRef]
4. Barnoski, M. *In the Integral Opticus/m. Barnoski, Ten, J. Goel et al. // In Order. M. Barnosky; Per. with English. in Order. T.A. Shmaonova*; Mir: Moscow, Russia, 1977.

5. Brudnyĭ, V.N.; Voevodin, V.G.; Grinyaev, S.N. Deep levels of intrinsic point defects and the nature of "anomalous" optical absorption in ZnGeP$_2$. *Phys. Solid State* **2006**, *48*, 1949–1961. [CrossRef]
6. Pearton, S.J.; Yang, J.; Cary, P.H.; Ren, F.; Kim, J.; Tadjer, M.J.; Mastro, M.A. A review of Ga$_2$O$_3$ materials, processing, and devices. *Appl. Phys. Rev.* **2018**, *5*, 011301. [CrossRef]
7. Lin, C.-H.; Lee, C.-T. Ga$_2$O$_3$-based solar-blind deep ultraviolet light-emitting diodes. *J. Lumin.* **2020**, *224*, 117326. [CrossRef]
8. Roy, R.; Hill, V.G.; Osborn, E.F. Polymorphism of Ga$_2$O$_3$ and the system Ga$_2$O$_3$–H$_2$O. *J. Am. Chem. Soc.* **1952**, *74*, 719–722. [CrossRef]
9. Zhang, D.; Dong, S. Challenges in band alignment between semiconducting materials: A case of rutile and anatase TiO$_2$. *Prog. Nat. Sci. Mater. Int.* **2019**, *29*, 277–284. [CrossRef]
10. Zhang, Z.; Farzana, E.; Arehart, A.R.; Ringel, S.A. Deep level defects throughout the bandgap of (010) β-Ga$_2$O$_3$ detected by optically and thermally stimulated defect spectroscopy. *Appl. Phys. Lett.* **2016**, *108*, 052105. [CrossRef]
11. Boyd, G.D.; Beuhler, E.; Stortz, F.G. Linear and nonlinear optical properties of ZnGeP$_2$ and CdSe. *Appl. Phys. Lett.* **1971**, *18*, 301–304. [CrossRef]
12. Romanovskii, O.; Sadovnikov, S.A.; Kharchenko, O.V.; Yakovlev, S.V. Development of near/mid IR differential absorption OPO lidar system for remote gas analysis of the atmosphere. In Proceedings of the Remote Sensing of Clouds and the Atmosphere XXIV, Strasbourg, France, 9–11 September 2019; Volume 11152, pp. 236–243.
13. Lv, Z.; Shen, Y.; Zong, N.; Bian, Q.; Wang, E.-P.; Chang, J.-Q.; Bo, Y.; Cui, D.-F.; Peng, Q.-J. 1.53 W all-solid-state nanosecond pulsed mid-infrared laser at 6.45 μm. *Opt. Lett.* **2022**, *47*, 1359. [CrossRef] [PubMed]
14. Wang, X.; Chen, Z.; Guo, D.; Zhang, X.; Wu, Z.; Li, P.; Tang, W. Optimizing the performance of a β-Ga$_2$O$_3$ solar-blind UV photodetector by compromising between photoabsorption and electric field distribution. *Opt. Mater. Express* **2018**, *8*, 2918–2927. [CrossRef]
15. Li, Z.; An, Z.; Xu, Y.; Cheng, Y.; Cheng, Y.; Chen, D.; Feng, Q.; Xu, S.; Zhang, J.; Zhang, C.; et al. Improving the production of high-performance solarblind b-Ga$_2$O$_3$ photodetectors by controlling the growth pressure. *J. Mater. Sci.* **2019**, *54*, 10335–10345. [CrossRef]
16. Lia, M.-Q.; Yanga, N.; Wanga, G.-G.; Zhanga, H.-Y.; Hana, J.-C. Highly preferred orientation of Ga$_2$O$_3$ films sputtered on SiC substrates for deep UV photodetector application. *Appl. Surf. Sci.* **2019**, *471*, 694–702. [CrossRef]
17. Kalygina, V.; Vishnikina, V.; Petrova, Y.; Prudaev, I.; Yaskevich, T. Photovoltaic characteristics of structures metal-Ga$_2$O$_3$-GaAs. *Phys. Technol. Semicond.* **2015**, *49*, 357–363.
18. Yu, J.; Lou, J.; Wang, Z.; Ji, S.; Chen, J.; Yu, M.; Peng, B.; Hu, Y.; Yuan, L.; Zhang, Y.; et al. Surface modification of β-Ga$_2$O$_3$ layer using Pt nanoparticles for improved deep UV photodetector performance. *J. Alloys Compd.* **2021**, *872*, 159508. [CrossRef]
19. Hou, X.; Zou, Y.; Ding, M.; Qin, Y.; Zhang, Z.; Ma, X.; Tan, P.; Yu, S.; Zhou, X.; Zhao, X.; et al. Review of polymorphous Ga$_2$O$_3$ materials and their solar-blind photodetector applications. *J. Phys. D Appl. Phys.* **2020**, *54*, 043001. [CrossRef]

Disclaimer/Publisher's Note: The statements, opinions and data contained in all publications are solely those of the individual author(s) and contributor(s) and not of MDPI and/or the editor(s). MDPI and/or the editor(s) disclaim responsibility for any injury to people or property resulting from any ideas, methods, instructions or products referred to in the content.

Article

Synthesis of Up-Conversion Fluorescence N-Doped Carbon Dots with High Selectivity and Sensitivity for Detection of Cu^{2+} Ions

Yuanyuan Xiong, Mengxiao Chen, Zhen Mao, Yiqing Deng, Jing He, Huaixuan Mu, Peini Li, Wangcai Zou and Qiang Zhao *

School of Chemical Engineering, Sichuan University, No. 24 South Section 1, Yihuan Road, Chengdu 610065, China; xiongyuanyuan@stu.scu.edu.cn (Y.X.); cmx961012@163.com (M.C.); maozhenjqk@163.com (Z.M.); dengyiqing@stu.scu.edu.cn (Y.D.); hejing7@stu.scu.edu.cn (J.H.); mhx@stu.scu.edu.cn (H.M.); lipeini@stu.scu.edu.cn (P.L.); zouwangcai@stu.scu.edu.cn (W.Z.)
* Correspondence: zhaoqiang@scu.edu.cn

Abstract: Carbon dots have drawn extensive attention in the detection of metal ions with good stability, excellent biocompatibility and low toxicity. Meanwhile, the quantum yield, response rate and the detection mechanism for Cu^{2+} ions are vital to their development and application. To obtain more selective and sensitive materials to detect Cu^{2+} ions, N-doped carbon dots (DN-CDs) were synthesized by a one-step hydrothermal method using citric acid as the carbon source and diethylenetriamine (DETA) as the nitrogen source. The obtained DN-CDs exhibited stable and intense blue light emission and special near-infrared up-conversion fluorescence at 820 nm, attributed to the effect of introducing N atoms into the structure of carbon dots. Due to the dynamic quenching of the DN-CDs by Cu^{2+} ions, the fluorescence intensity (λex = 820 nm) of DN-CDs was quantitatively decreased in the presence of Cu^{2+} ions. The DN-CDs had a rapid response within 3 min. The DN-CD system exhibited a linear relationship with a concentration range from 2.5 to 50 µM and low detection limit (LOD) of 42 nM. After careful investigation, an interesting conclusion was proposed: N-doped CDs with N/O = 1:1 or higher with relatively abundant N atoms prefer to detect Cu^{2+} ions while those with N/O = 1:2 or lower prefer to detect Fe^{3+} ions.

Keywords: carbon dots; up-conversion fluorescence; Cu^{2+} ions detection

Citation: Xiong, Y.; Chen, M.; Mao, Z.; Deng, Y.; He, J.; Mu, H.; Li, P.; Zou, W.; Zhao, Q. Synthesis of Up-Conversion Fluorescence N-Doped Carbon Dots with High Selectivity and Sensitivity for Detection of Cu^{2+} Ions. *Crystals* **2023**, *13*, 812. https://doi.org/10.3390/cryst13050812

Academic Editor: Paolo Olivero

Received: 14 April 2023
Revised: 8 May 2023
Accepted: 10 May 2023
Published: 13 May 2023

Copyright: © 2023 by the authors. Licensee MDPI, Basel, Switzerland. This article is an open access article distributed under the terms and conditions of the Creative Commons Attribution (CC BY) license (https://creativecommons.org/licenses/by/4.0/).

1. Introduction

Industrial pollutants pose serious threats to nature and human health, so their removal and detection are extremely important. At present, some effective materials have been developed for the removal of pollutants, such as chitosan-based polymer materials with excellent adsorption abilities for heavy metal ions [1]. Additionally, some effective analytical methods have been developed to detect chemical or biological drugs, metal ions and organic pollutants [2–5]. Analytical methods used to detect metal ions include potential/electrochemical sensors [6,7], fluorescence analysis [8], atomic absorption spectrometry (AAS) [9] and inductively coupled plasma mass spectrometry (ICP-MS) [10]. However, most of these approaches are not suitable for real-time monitoring due to their high cost, long assay times, special equipment, and laborious and complex experimental processes. Therefore, fluorescent sensors with low cost, simple operation, high sensitivity and good selectivity have gradually emerged [11]. Nowadays, various materials have been developed as fluorescent probes for environmental monitoring of metal ions, such as metal–organic frameworks (MOFs) [12,13], nanoclusters [14,15], organic fluorescent dyes [16] and semiconductor quantum dots. In particular, fluorescent carbon-based materials have attracted extensive attention due to their good chemical stability, excellent biocompatibility and low toxicity [17], distinguishing themselves from traditional fluorescent materials.

With the progress in scientific and technological research, more heavy metals remain in the human living environment. The excessive intake of metal ions will have extremely serious effects on human health [18,19]. Copper is one of the most important transition metals in the human body [20] and plays an extremely important role in biological processes, such as embryonic development, mitochondrial respiration, regulation of hemoglobin levels as well as hepatocyte and neuronal functions [21,22]. Copper ions are widely distributed in the natural environment, especially in industrial wastewater. Long-time exposure to copper ions will cause high toxicity to the human body, and is associated with neurodegenerative diseases, such as Wilson's syndrome, Alzheimer's disease and Parkinson's disease [23–25]. According to United States Environmental Protection Agency (USEPA) guidelines, drinkable water must not contain a Cu^{2+} content of ≥ 20 µM [26]. Therefore, the detection of copper ions in the environment requires a simple and practical detection method with high sensitivity and selectivity.

Carbon dots (CDs) are zero-dimensional carbon-based materials, mostly spherical nanoclusters with diameters of less than 10 nm, composed of amorphous or crystalline cores with sp^2 hybrid carbon atoms [27]. Owing to their small size, high photostability, resistance to photobleaching, low toxicity, good water solubility, excellent biocompatibility and ease of surface functionalization, carbon dots have become the focus of fluorescent probe materials as well as in cell labeling, optical bioimaging, drug delivery, storage applications and photocatalysis [28]. Metal ions, such as Hg^{2+}, Fe^{3+} and Cu^{2+}, can effectively quench the fluorescence of carbon dots [29–31]. Carbon dots can be obtained by thermal decomposition [32], hydrothermal carbonization [33], microwave-assisted pyrolysis [34], electrochemical methods [35], etc. Hydrothermal carbonization refers to heating water and carbon-rich substances in a closed container to trigger the carbonization reaction. Citric acid is the most used carbon-rich substance, which is easy to condense and induce to form carbon rings, and is also has atoms that can be easily doped. Meanwhile, the surface of the formed carbon dots has rich functional groups, such as hydroxyl and carboxyl groups. The hydrothermal method is the most widely used method, which has the characteristics of good feasibility; cheap, diverse carbon sources; no toxicity; and environmental friendliness [36]. However, carbon dots without modification usually have a low quantum yield (QY) and poor stability, which affect their practical application [37,38]. Additionally, the origin of fluorescence in carbon dots has not yet been clarified. The fluorescence emission is primarily caused by the competition between the optical centers, surface states and traps [39]. Up-conversion fluorescence means that the shorter wavelengths of light are emitted through longer excitation wavelengths [40]. This fluorescence is usually caused by the multiphoton absorption effect. The fluorescence of carbon dots can be altered by controlling size and shape, modifying the edges of the surfaces and doping with heteroatoms, due to the edge effect, surface effect and quantum confinement effect (QCE) [41]. Doping different elements leads to better performance, such fluorescence improvement and shifting of the fluorescence spectra. Doping with different heteroatoms can change the electronic properties and excite special optical properties through the passivation of the surface states or the introduction of defects [42,43]. Salehtabar et al. [44] prepared photo-luminescent, water-soluble, nitrogen-doped, highly fluorescent carbon dots based on the gum tragacanth polysaccharide and triethylenetetramine (TETA). Therefore, it is a better way to synthetize fluorescent probe materials based on carbon dots with a higher sensitivity, higher selectivity, high quantum yield and unique fluorescence performance.

Herein, we report a simple method for preparing nitrogen-doped carbon dots (DN-CDs) with intense down-conversion and unique up-conversion fluorescence using diethylenetriamine (DETA) and citric acid as the nitrogen and carbon sources, respectively (Scheme 1). The DN-CDs had good fluorescence properties, and excellent photobleaching and stability. We used it for rapid, efficient and selective detection of copper ions within 3 min with a detection limit (LOD) as low as 42 nM. Compared with other carbon dots, DN-CDs showed better and stable up-conversion fluorescence, which could convert low energy light into high energy radiation through a nonlinear optical process, avoiding interference with biological background luminescence and achieving good photon tissue penetration [45–48]. This would compensate for the disadvantages of down-conversion fluorescence carbon dots, and has great potential for detecting metals in organisms. The experimental results proved that the mechanism of fluorescence quenching through dynamic quenching. Additionally, combined with reports in the literature, the ratio of N and O atoms may be more critical on the premise that the N, O-based functional groups interact with Cu^{2+}. Two previously synthesized nitrogen-doped carbon dots were selected for comparison and preliminarily verified the proposed conclusion. This would be conducive to the selection and design of highly selective fluorescent probe materials for copper ions.

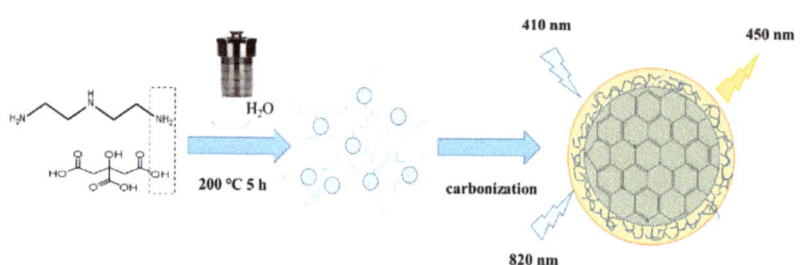

Scheme 1. Schematic illustration for DN-CD preparation.

2. Experimental

2.1. Chemicals and Materials

All chemicals were of analytical grade and were used directly without any further purification. Diethylenetriamine, quinine sulfate and all the metal nitrates, nicotinamide sulfate and chloride salts were purchased from Kelong Chemical Co., Ltd. (Chengdu, China). Citric acid and sulfuric acid were purchased from Xilong Chemical Co., Ltd. (Chengdu, China). Anhydrous ethanol was purchased from Changlian Chemical Co., Ltd. (Chengdu, China). N, N-Dimethylformamide (DMF) and Tetrabutylammonium Tetrafluoroborate were purchased from Dingsheng Chemical Co., Ltd. (Chengdu, China). l-Glutamic acid (l-Glu) and m-phenylenediamine (MPD) were purchased from Aladdin Industrial Co., Ltd. (Shanghai, China).

2.2. Preparation of N-CDs

N-doped carbon dots were synthesized by a simple one-step hydrothermal method. First, 0.44 g citric acid (CA) and a certain amount of diethylenetriamine (DETA) were added to 20 mL of ultra-pure water and sonicated for 20 min until the CA was dissolved. The prepared mixture was placed in a 50 mL PPL-lined stainless-steel autoclave and heated in an oven for a certain amount of time. The autoclave was cooled to room temperature after the reaction, and a slightly yellow solution was obtained. The solution was collected, dialyzed (MWCO = 500 Da) for 24 h to remove unreacted starting materials and then the dialysate was removed and internal solution was freeze-dried to obtain the final brown product. The optimal amount of DETA was determined by testing the varying molar ratios of CA:DETA (1:1, 1:2, 1:4, 1:8 and 1:10) with the reaction condition of 5 h and 180 °C. Additionally, the optimal reaction temperature was determined by testing different temperatures (150 °C, 180 °C, 200 °C, 220 °C and 240 °C) with a 1:4 molar ratio of CA:DETA and 5 h reaction time. Then, the reaction time was determined by testing different reaction times (4 h, 6 h, 8 h, 10 h and 12 h) with a 1:4 molar ratio of CA:DETA at 200 °C.

The two other carbon dots were prepared according to previous reports [49,50]. The specific preparation methods are detailed in the Supplementary Information.

2.3. Measurement of Quantum Yield

DN-CDs have similar absorption peaks (350 nm) and emission peaks (380–580 nm) as quinine acid; therefore, the QY could be measured using a relative method [51]. Quinine sulfate was selected as the reference compound (QY = 54% in 0.05 M H_2SO_4). The QY of our sample was calculated using Equation (1):

$$Y_1 = Y_2 \frac{S_1}{S_2} \frac{A_2}{A_1} \frac{n_1^2}{n_2^2} \quad (1)$$

where 'Y_1' is the QY of DN-CDs, 'Y_2' is the QY of quinine sulfate, 'S' is the peak area obtained from emission peak integration, 'A' is the absorbance intensity at λ(ex) of 350 nm, and 'n' is the refractive index (1.33 for both sample and reference). A series of solutions of DN-CDs and quinine sulfate were prepared until the absorbance intensities were below 0.05 at λex = 350 nm, which were recorded as A_1 and A_2, respectively. Then, the fluorescence spectra were measured and the emission peaks in the range of 400–700 nm were selected for integration to obtain the peak areas (S_1 and S_2). Finally, the QY for our DN-CDs was calculated using Equation (1).

2.4. Characterization

Transmission electron microscopy (TEM) was performed using an INSPECT F instrument (J&R Instrument Technology Co., Ltd., Shanghai, Chnia). X-ray photoelectron spectroscopy (XPS) results were obtained using XSAM 800 equipment (Kratos Analytical Ltd., Manchester, UK). The X-ray diffraction (XRD) patterns were identified using an X'Pert PRO diffractometer (PANalytical, Almelo, The Netherlands) with Cu Kα1 radiation, a step size of 0.01313° 2θ, and a counting time of 30 ms step-1 from 10° to 80°. FT-IR spectra were obtained using a Nicolet-6700 FT-IR spectrometer (Thermo Scientific, Madison, WI, USA). Fluorescence spectra were recorded using an F97-Pro fluorospectrophotometer (Lengguang Tech. Ltd., Shanghai, China). UV–Vis absorption spectra were recorded using a UV–vis spectrophotometer (Mapada Instruments Ltd., Shanghai, China). Fluorescence lifetime spectra were measured using F-7000 equipment. The redox properties of DN-CDs were examined by cyclic voltammetry which was carried out on a CH1660A electrochemical workstation.

2.5. Detection of Metal Ions

The PL spectra were measured to study the selectivity and sensitivity of the DN-CDs. The different chemical compounds with different metal ions (Na^+, K^+, Ca^{2+}, Fe^{2+}, Fe^{3+}, Cr^{3+}, Mg^{2+}, Mn^{2+}, Pb^{2+}, Zn^{2+}, Cd^{2+}) and anions (SO_4^{2-}, PO_4^{3-}, Cl^-, NO_3^-) were dissolved in DI-water to obtain a series of metal solutions with a concentration of 10 μM. 1 mL aqueous solution of metal ions and 1 mL aqueous solution of DN-CDs (20 μg/mL) were mixed. After a certain amount of time, PL intensity was detected three times to obtain the average value. To further study the sensitivity detection of Cu^{2+}, a series of $CuSO_4$ solutions with concentration of 0–500 μM were prepared and mixed with 1 mL DN-CDs (20 μg/mL). A Cu^{2+} solution was added into the solutions containing other metal ions to obtain a mixed solution with two metal ions to conduct interference tests.

2.6. Detection in Real Samples

We used the tap water from our laboratory, the water from Jinjiang River in Wuhou District, Chengdu, and the water from a lotus pond in the Wangjiang Campus of Sichuan University as the real water samples. These samples were filtered through a 0.22 μm membrane and centrifuged at 10,000 rpm for 5 min to remove any suspended materials. Then, 1 mL of the sample with a standard Cu^{2+} ion solution (10, 20 and 30 μM) was mixed with 1 mL of DN-CD solution for 4 min and the fluorescence intensity was measured.

3. Results and Discussion

3.1. Optical Properties of DN-CDs

In this study, by changing the ratio of CA and DETA, reaction temperature and reaction time, we developed a simple one-step synthesis method for carbon dots with a high quantum yield under down-conversion fluorescence. As shown in Figure S1, it could be seen that with the increase in DETA and reaction temperature, the quantum yield first increased and then decreased. Additionally, in Figure 1c, the quantum yield decreased and then increased slightly. Finally, carbon dots with a high quantum yield under down-conversion fluorescence were obtained. Additionally, the optimum conditions were found to be CA/DETA = 1:4, reaction time of 4 h, and reaction temperature of 200 °C. And we used these carbon dots, denoted as DN-CDs, to perform the further characterizations.

The optical properties of the DN-CDs were measured using UV-vis absorption and PL spectroscopy, as shown in Figure 1a. Typically, the absorption of carbon dots could be divided into two main regions: π—π* and n—π* transition [52]. The weak absorption peak at 241 nm was the π—π* transition of C=C, attributed to the absorption peak of the carbon atom in cyclic aromatic hydrocarbons [53]. The wide absorption peak at 350 nm indicated the n—π* transition of the nitrogen- or oxygen-containing functional groups on the surface of the DN-CDs, attributed to the doping with N atoms [54]. The bandgap energies were determined according to the Tauc plot (Equation (2)) as follows:

$$(ah\nu)^{\frac{1}{n}} = A(h\nu - E_g) \quad (2)$$

where 'a' is the absorption coefficient, 'h' is Planck's constant, 'ν' is the incident light frequency, 'E_g' is the bandgap energy, 'A' is a constant and 'v' is a variable dependent on the type of bandgap [55]. As shown in Figure 1b, the bandgap of DN-CDs was 3.21 eV.

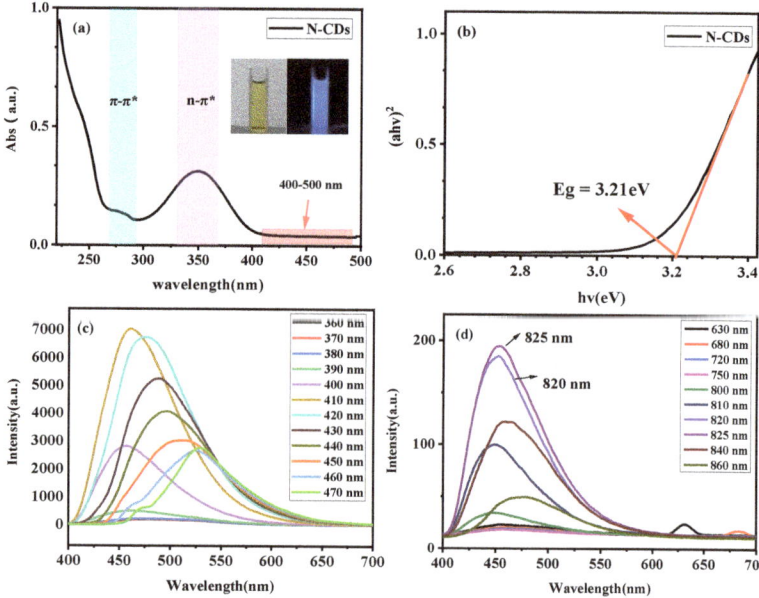

Figure 1. UV–Vis absorption and PL spectra of the DN-CDs. (* represents antibonding molecular orbital) (a) The UV-Vis absorption of DN-CDs. Inset: photograph taken under daylight and 365 nm UV light. (b) The Tauc plot according to the UV-Vis absorption. (The carbon dots are usually indirect bandgaps, so we selected the value of 'n' as 2) (c) Emission spectra of DN-CDs at different excitation wavelengths. (d) Up-conversion emission spectra of DN-CDs at different excitation wavelengths.

Under daylight, the DN-CD solution was slightly yellow and clear; in contrast, it was a bright blue color if exposed to UV irradiation at 365 nm (inset in Figure 1a). Figure 1c displayed the photoluminescence spectra of DN-CDs, showing a maximum emission at 450 nm (λex = 410 nm). Moreover, when the excitation wavelength was increased, the peak value of fluorescence had a small red shift, which indicated that the PL was dependent on the excitation wavelength. This excitation-dependent fluorescence property reflected the size of the DN-CDs and the distribution of different surface states [56]. In particular, Figure 1d also demonstrates the stable up-conversion PL property. When the excitation wavelength changed from 630 nm to 860 nm, the UCPL spectrum showed a basically unchanged emission at 451 nm, indicating a single emission center in the carbon dots. The peak intensities at 820 and 825 nm were significantly higher than that of other wavelengths. The doping with N atoms might lead to an increase in lone pair electrons on oxygen atoms and nitrogen atoms, increasing the electron density in the π—system. Additionally, it would reduce the π^*—energy level of the excitation, so the excitation required a lower energy [35].

3.2. Photoluminescence Stability of DN-CDs

For the application of fluorescent probes in the detection of metal ions, it is important to explore the up-conversion fluorescence (UCPL) stability of DN-CDs. The PL intensity of DN-CDs under different pH conditions is shown in Figure 2a, indicating a strong UCPL intensity in a wide range of pH levels (pH = 4–12). In particular, DN-CDs had a negligible change in the PL emission range or intensity within pH levels of 4–9, indicating excellent stability of the DN-CDs; the UCPL intensity markedly changed under strong acid–base conditions, due to the protonation and deprotonation of the functional groups (-OH/-COOH and -NH$_2$) on the surface of the DN-CDs, which resulted in the destruction of the aromatic conjugate system and complex system [24,57]. This was proven by the Zeta potentials in different pH aqueous solutions in Figure S2. DN-CDs had a negative potential at pH = 4–13, and the negative potential increased sharply under highly alkaline conditions. DN-CDs had a high positive potential at pH = 2, which indicated that the same protonation in a strong acidic environment. Meanwhile, the DN-CDs had a higher negative potential at pH = 11–13, which indicated intense deprotonation. In order to study the optical stability of carbon dots, we tested their up-conversion fluorescence intensity under daily storage conditions and simulated sunlight conditions using a xenon lamp (300 W). As shown in Figure 2b, we could see that the up-conversion fluorescence intensity of the DN-CD solution remained high after 90 days, which indicated that the DN-CDs had long-time UCPL stability; at the same time, the carbon dot solution was still clear and transparent. As shown in Figure 2c, the up-conversion fluorescence intensity of the carbon dot solution was basically unchanged within the first hour of xenon irradiation (300 W), where F_0 is the initial up-conversion fluorescence intensity, and F is the up-conversion fluorescence intensity after xenon irradiation. In summary, the prepared carbon dots (DN-CDs) had strong pH stability and optical stability.

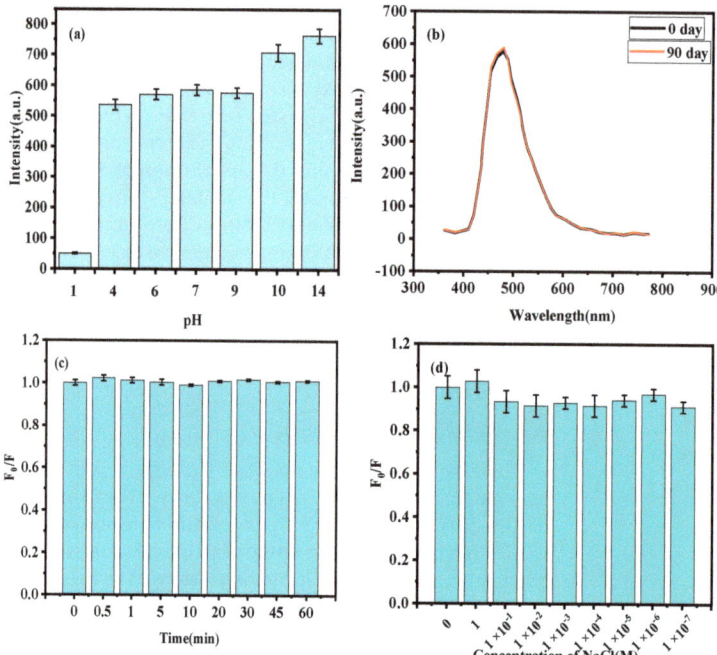

Figure 2. (**a**) The effect of pH on the UCPL intensity of DN-CDs in a range from 1 to 13. (**b**) The effect of storage time on PL intensity of DN-CDs after 90 days. (**c**) The effect of 60 min of UV irradiation on the UCPL intensity (F_0 is the initial UCPL intensity and F is the UCPL intensity after under UV irradiation). (**d**) The effect of salt concentration on the UCPL intensity for DN-CDs at different concentrations of NaCl ('F_0' is the PL intensity of a blank sample and 'F' is the intensity of the experimental sample).

3.3. Morphological and Surface Properties of DN-CDs

The morphological and surface properties of the DN-CDs were investigated by TEM, FT-IR, XRD and XPS; the results are shown in Figures 3 and 4. DN-CDs were nearly spherical with a size distribution between 1.5 and 4.5 nm and the average particle size was about 2.4 nm (Figure 3a,b). Furthermore, the dynamic light scattering (DLS) results in Figure S3 showed that the dynamic diameter of the DN-CDs was about 2.7 nm, which was basically consistent with the TEM. Additionally, we also tested the DLS of DN-CDs in strong acidic and alkaline environments, and found that there was a slight aggregation phenomenon with strong acids, with average size of 3.7 nm. This was also consistent with the fluorescence and Zeta potential changes under strong acidic conditions in Section 3.2. Meanwhile, we found that the crystalline lattice fringe had a space of 0.21 nm, which corresponded to the (100) lattice of graphene in the inset in Figure 3a. Through the Zeta potential of the DN-CDs, it was suggested that the DN-CDs had a negative charge (Figure S3). This was attributed to the partial ionization of the hydroxyl or carboxyl groups in the DN-CDs, which caused the formation of oxygen negative ions. Figure 3c shows the XRD pattern of the DN-CDs with broad peaks at $2\theta = 23.9°$, indicating that the carbon atom in DN-CDs was amorphous. Additionally, this also could be verified in the Raman spectrum of N-CDs in Figure S4. This demonstrated that the fluorescence properties of DN-CDs were mainly determined by the surface groups of the carbon dots.

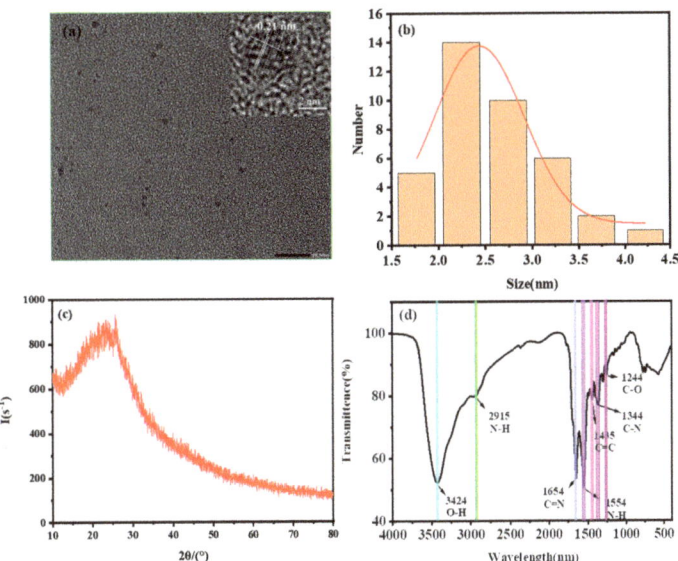

Figure 3. TEM image (**a**), size distribution (**b**), XRD spectra (**c**) and FT-IR spectra (**d**) of DN-CDs (the inset in (**a**) is the HRTEM of the DN-CDs).

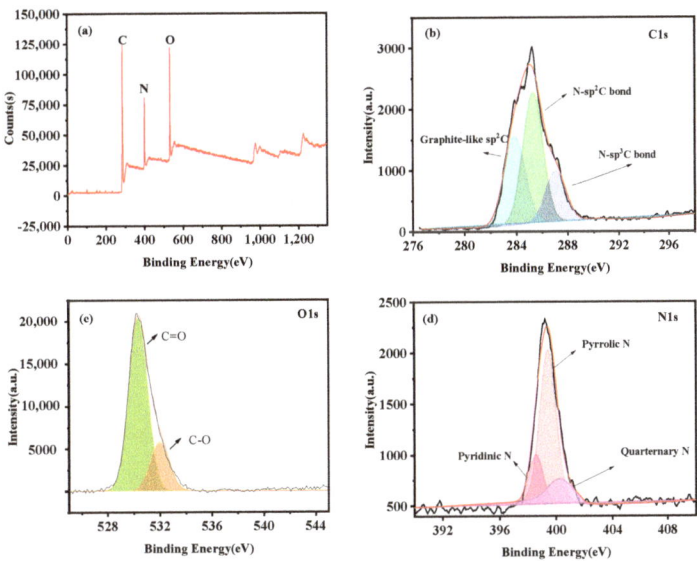

Figure 4. (**a**) The full-scan XPS spectra of the DN-CDs; (**b**) C1s spectra of the DN-CDs; (**c**) O1s spectra of the DN-CDs; (**d**) N1s spectra of the DN-CDs.

FT-IR spectra were obtained to detect the functional groups of the DN-CDs. As shown in Figure 3d, the DN-CDs exhibited absorption peaks at 3423 cm^{-1} and 2915 cm^{-1}, consistent with -OH and N-H stretch vibrations, showing that the DN-CDs had good water-solubility and contained amino groups on their surfaces. Compared with the FT-IR spectra of DETA, the absorption peak of N-H decreased, indicating that the amine group of DETA and the carboxyl group of citric acid had undergone a dehydration reaction.

In addition, the absorption peaks at 1244 cm^{-1} and 1344 cm^{-1} represented the stretch vibrations of C-O and amide C-N, respectively. The absorption peaks at 1654 cm^{-1} and 1554 cm^{-1} were related to the typical stretch vibrations of C=N and N-H. The absorption peak of 1435 cm^{-1} was the typical stretch vibration of C=C due to the establishment of an unsaturated aromatic ring structure during the hydrothermal treatment. Moreover, it was confirmed by FT-IR spectra that some of the carbonyl groups of citric acid were converted into amide groups in the hydrothermal process. These functional groups improved the hydrophilicity of the DN-CDs in an aqueous system, suggesting the potential of DN-CDs as sensors for aqueous samples.

The elemental composition of the DN-CDs was examined through XPS (Figure 4). As shown in Figure 4a, according to the XPS full-scan profile of the DN-CDs, C, N and O exhibited distinct peaks in the spectra at 284.71 eV, 399.83 eV and 530.95 eV corresponding to the binding energies of C1s, N1s and O1s, respectively. Moreover, the contents of C, N, and O elements in the DN-CDs were 38.4%, 44.1%, and 17.5%, respectively, according to the EDS results (Figure S5). Figure 4b shows three peaks at 284.7 eV, 285.9 eV and 287.5 eV related to the binding energies of graphite-like C_{sp2}, N-C_{sp2} and N-C_{sp3} bonds, respectively. The O1s spectra showed two peaks at 530.38 eV and 532.02 eV, attributed to the binding energy of C=O and C-O, respectively (Figure 4c). Additionally, the N1s spectra showed that it could be deconvoluted into three peaks at 398.8 eV, 400.5 eV and 401.2 eV which were consistent with the binding energies of pyridinic N, pyrrolic N and quarternary N, respectively (Figure 4d). The large amount of pyrrolic N indicated that most of the N atoms existed in a π conjugated system. In summary, we confirmed that the N elements were successfully doped onto the structure of the carbon dots.

3.4. DN-CDs as a Photoluminescent Probe for Cu^{2+} Ions

To investigate the potential use of the DN-CDs as a PL probe, we examined the effects of different metal ions on PL intensity (Figure 5). It could be seen from Figure 5a that at the same concentration, Na^+, K^+, Ca^{2+}, Fe^{2+}, Fe^{3+}, Cr^{3+}, Mg^{2+}, Mn^{2+}, Pb^{2+}, Zn^{2+} and Cd^{2+} ions basically did not change the PL intensity of the carbon dots; with the same concentration of Cu^{2+}, the PL intensity was 60% of that of the control, showing an obvious fluorescence quenching phenomenon. In addition, the interference of other metal ions on the detection of Cu^{2+} was examined by adding the mixed solution of various metal ions and Cu^{2+} into the DN-CDs solution. Meanwhile, the effect of the addition of other metal ions on the detection of Cu^{2+} ions could be nearly ignored, which indicated that DN-CDs had excellent selective quenching for Cu^{2+}. Moreover, we also explored the quenching kinetics of the DN-CDs by monitoring the PL intensity as a function of time. Figure 5b showed that the PL intensity of the DN-CD solution gradually weakened and finally stabilized with increasing Cu^{2+} ions concentration. This result indicated that the detection of Cu^{2+} ions in the DN-CD system was a rapid and effective process and the equilibrium was reached within 3 min. Thus, 3 min was chosen as the detection time in the following experiments.

To investigate the sensitivity of the DN-CDs for detecting Cu^{2+}, the PL spectra of the DN-CDs were examined by adding Cu^{2+} solution at various concentrations (2.5–50 μM). Obviously, in Figure 5c, the intensity of PL quenching increased when the concentration of Cu^{2+} increased; therefore, the decrease in PL intensity was caused by the Cu^{2+} ions instead of the associated anions and was quantitatively related to the Cu^{2+} concentration. This indicated that the DN-CDs were capable of quantitatively detecting Cu^{2+} at low levels with high sensitively. Additionally, the PL quenching spectra of Cu^{2+} was analyzed by the Stern–Volmer (Equation (3)):

$$\frac{F_0}{F} = K_{sv}[Q] + 1 \tag{3}$$

where 'F_0' and 'F' are the PL intensities of the DN-CDs in the absence and presence of Cu^{2+}, respectively, 'Q' is the concentration of Cu^{2+} and 'K_{sv}' is the Stern–Volmer quenching constant. This equation could represent the relationship between the relative PL intensity (F_0/F) and Cu^{2+} concentration (Q).

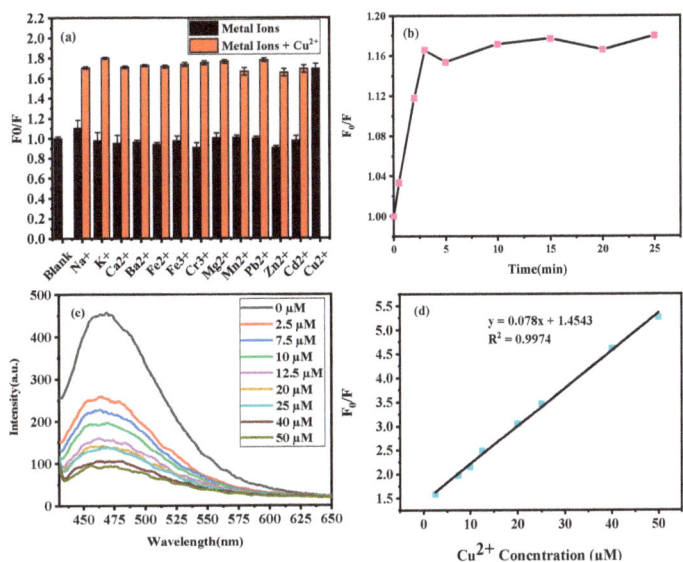

Figure 5. (a) The selectivity of DN-CDs to various metal ions: relative PL intensity of the DN-CDs (20 μg/mL) in the absence (black column) and presence (red column) of other metal ions (10 μM) (λex = 820 nm and λem = 450 nm); (b) the effect of reaction time on the quenching equilibrium of DN-CDs; (c) PL quenching efficiency of DN-CDs at different concentrations of Cu^{2+} ions (2.5–50 μM); (d) the linear relationship between PL intensity and Cu^{2+} ion concentration (F_0 is the PL intensity in the absence of Cu^{2+} ions and F is in the presence of Cu^{2+} ions).

According to Equation (3), we observed strong linearity with a correlation coefficient (R^2) of 0.9974 when the concentration of Cu^{2+} changed from 2.5 to 50 μM. The linear relationship curve is shown in Figure 5d. Based on Equation LOD = $3 \cdot \sigma / S$ (where 'σ' is the standard deviation of the blank PL intensity of DN-CDs and 'S' is the slope of the linear relationship curve), the limit of detection (LOD) of Cu^{2+}, was calculated to be 42 nM. Compared with the threshold of the United States Environmental Protection Agency (USEPA) guidelines, the LOD of the DN-CDs for Cu^{2+} in an aqueous system was much lower. We summarized the PL sensing performances of the other reported sensors for detecting Cu^{2+} in Table 1; it was clear that the LOD of our product was also lower than those of the other sensors.

Table 1. The range of detection and LOD of other PL sensors for Cu^{2+} ions.

Sensor	Range of Detection	LOD	Ref.
BP/NS-CDs	0–50 μM	0.18 μM	[24]
OPD-CDs	0.5–40 μM	0.28 μM	[23]
NS-CDs	1–10 μM	0.29 μM	[58]
FA-MoO$_x$ QDs	0.2–500 μM	29 nM	[59]
HBT-H	0–5 μM	0.308 μM	[60]
Fe$_3$O$_4$@SiO$_2$-PAP	0.1–0.2 mg/L	0.125 μM	[61]

3.5. Mechanism of DN-CD Interaction with Cu^{2+}

3.5.1. Exploration of Quenching Mechanism

Carbon dots have already been used in the field of PL sensors and the PL quenching mechanisms of carbon dots had been discussed in other articles. The quenching mechanisms can be roughly divided into two types, which included static quenching and dynamic quenching. There were also other mechanisms, such as Förster resonance energy transfer

(FRET), photo-induced electron transfer (PET) and inner filter effect (IFE). Static quenching would occur through the interaction between the carbon dots and metal ions, forming a non-fluorescent ground-state complex. The existence of the non-fluorescent complex could be verified by UV-Vis spectra. If there was complexation between the functional groups in the carbon dots and metal ions, the absorption peaks on the spectra would have a corresponding blue shift or red shift or even produce an obvious new absorption peak due to a new chemical bond. Dynamic quenching would be caused by the collision of carbon dots through the mechanisms of charge transfer or energy transfer, so that the excited state of the carbon dots would return to its ground state. In contrast to static quenching, dynamic quenching would only affect the excited state of the carbon dots, so there would be no change in the UV-Vis spectra. However, a significant change in the lifetime of fluorescence would be detected. FRET would occur when the emission spectra of the carbon dots overlapped with the absorption spectra of the metal ions due to long range dipole–dipole interactions between the excited state of carbon dots and the ground state of metal ions. Additionally, if a FRET mechanism existed, it could change the lifetime of fluorescence. PET could be defined that the electron transfer would occur between carbon dots (as electron donor or electron receptor) and metal ions (as electron donor or electron receptor), if there is an energy gap between the lowest unoccupied molecular orbitals (LUMO) and the highest occupied molecular orbitals (HOMO) between the carbon dots and metal ions. Additionally, IFE would result from an overlap of the absorption spectra of the metal ions with the excitation or emission spectra of the carbon dots. This could cause energy transfer between the carbon dots and metal ions. Therefore, in this part, we explored the quenching mechanisms of our DN-CDs in detecting Cu^{2+} as static quenching, FRET, PET or IFE by UV-vis spectra.

As shown in Figure 6a, the absorption peak of the DN-CD solution did not change after adding the Cu^{2+} solution compared with the absorption peak without the Cu^{2+} solution, proving that there was no static quenching behavior in our DN-CD PL sensor system.

To explore other possible quenching mechanisms, we compared the PL spectra of DN-CDs with the absorption spectra of the Cu^{2+} aqueous solution (Figure 6b). This indicated that the significant fluorescence emission peak of DN-CDs at 450 nm basically did not overlap with the absorption peak of Cu^{2+}. This concluded that there was no FRET quenching mechanism between the DN-CDs and Cu^{2+}. At the same time, as shown in Figure 6c, there was no obvious overlap between the absorption spectra and fluorescence emission spectra of DN-CDs. Additionally, we used the up-conversion PL as the excitation wavelength, which was far from the absorption wavelength of Cu^{2+} (Figure S6). This could eliminate the IFE effect of the PL quenching progress, which might reduce the number of photons available for the DN-CDs [62].

The fluorescence lifetime decay of DN-CDs in the absence and presence of Cu^{2+} were measured at λex = 820 nm and λem = 450 nm (Figure 6d). After adding a certain amount of Cu^{2+} solution, the fluorescence lifetime of the DN-CD solution decreased from 10.69 ns to 5.14 ns. This suggested that dynamic quenching occurred in the sensor system. Additionally, we proposed a non-radiative electron-transfer mechanism (PET). We calculated the highest occupied molecular orbital (HOMO) and lowest unoccupied molecular (LUMO) energy levels of the DN-CDs according to the following Equation (4) [63]:

$$E_{LUMO} = -e(E_{Red} + 4.4) \quad E_{HOMO} = -e(E_{Ox} + 4.4) \quad E_{HOMO} = E_{LUMO} - E_g \quad (4)$$

where 'E_{Ox}' and 'E_{Red}' are the onset of oxidation and reduction potential of DN-CDs, respectively, and 'E_g' is the optical energy gap resulting from the UV-Vis adsorption. Additionally, the 'E_{Red}' was −0.69 eV according to the CV plot (Figure S7). The 'E_{LUMO}' was calculated to be −3.71 eV and 'E_{HOMO}' was calculated to be −6.92 eV. Obviously, the band gap could overlap the crystal field stabilization energy of Cu^{2+} (1.98 eV) [64]. As illustrated in Figure 7, the excited electrons of the DN-CDs were prone to transfer to the d orbital of Cu^{2+} [65]. Thus, the fluorescence was suppressed and became a partially non-radiative progress.

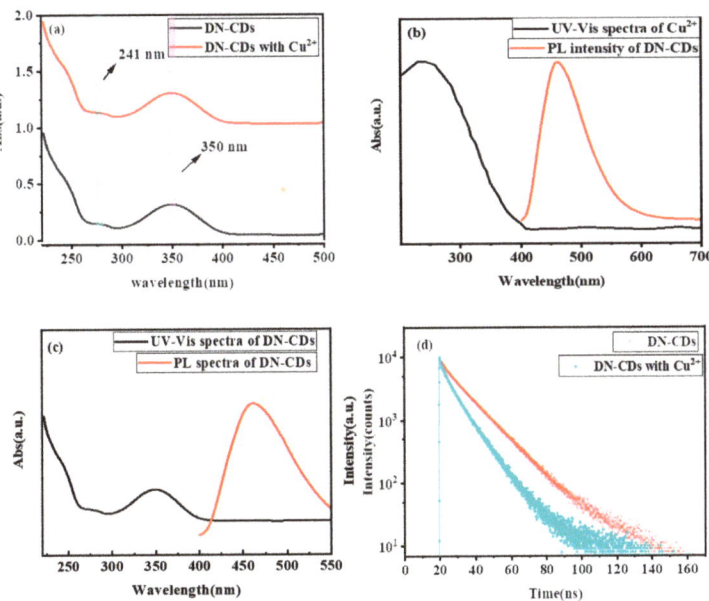

Figure 6. (a) The UV-Vis spectra of DN-CDs in the absence and presence of Cu^{2+} ions; (b) the comparison of the absorption spectra of Cu^{2+} ions and the emission spectra of DN-CDs at λex = 820 nm; (c) the UV-Vis spectra and emission spectra of DN-CDs at λex = 820 nm; (d) the fluorescence lifetime decay of DN-CDs in the absence and presence of Cu^{2+} ions (λex = 410 nm and λem = 450 nm).

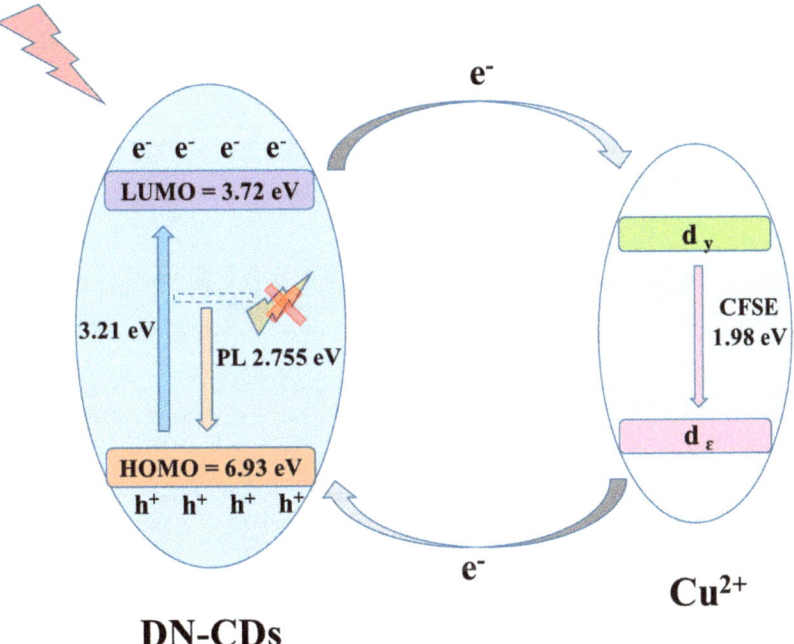

Figure 7. The mechanism of fluorescence quenching detection for Cu^{2+}.

3.5.2. Discussion on the Selectivity of Cu^{2+} and Fe^{3+} Ions

We synthesized two other N-doped carbon dots in our laboratory named N-CDs-1 and N-CDs-2 and we found that they were capable of detecting Fe^{3+} ions. The PL spectra and FT-IR spectra demonstrated that we had successfully synthesized the previously reported carbon dots (Figure S8). The specific and detailed steps of detection of metal ions by these two carbon dots are displayed in the supporting information.

The results of testing the selectivity of N-CDs-1 and N-CDs-2 to metal ions are shown in Figure 8. As seen in Figure 8a, under the same conditions, obvious fluorescence quenching occurred after adding the Fe^{3+} solution to the carbon dot solutions. In contrast, other metal ions had little influence on the PL intensity of N-CDs-1. N-CDs-1 also had an up-conversion PL, so its UCPL sensing system was investigated. The result was consistent with the above one showing that N-CDs-1 with up-conversion PL was also selective for Fe^{3+} ions. Similarly, N-CDs-2 showed selectivity for the detection of Fe^{3+} as shown in Figure 8c. The addition of Fe^{3+} ions caused a strong quenching of the PL intensity, with a quenching degree of more than 80%. Instead, the highest quenching effect of the other ions was only 40% ($CrCl_3$), which might be due to the larger band gap of the carbon dots, which could overlap with the crystal field energy of Cr^{3+} and produce PET behaviors. In summary, N-CDs-1 and N-CDs-2 could detect Fe^{3+} ions with a good linear relationship in the 1–1000 μM and 20–100 μM ranges and the LODs were 100 nM and 3.93 μM, respectively (Figure S9).

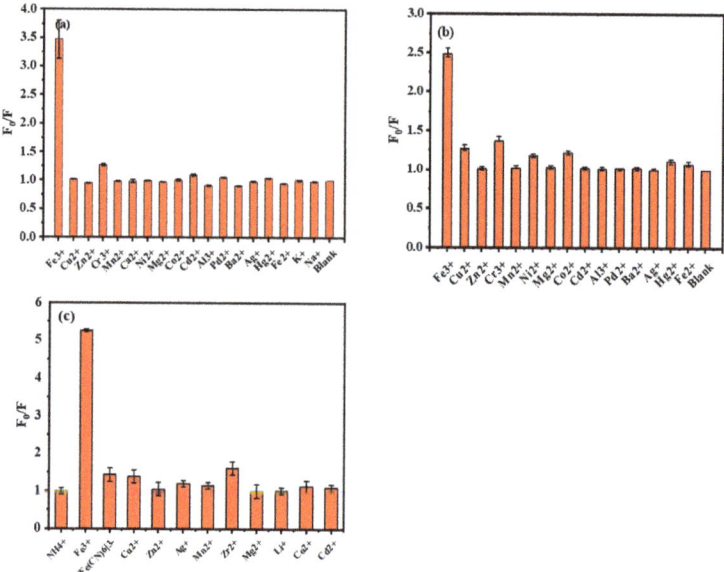

Figure 8. (a) The selectivity of N-CDs-1 to various metal ions at λex = 410 nm, λem = 510 nm and (b) λex = 808 nm, λem = 510 nm; (c) the selectivity of N-CDs-2 to various metal ions at λex = 380 nm, λem = 450 nm.

We were interested in the similarities and differences between the three N-doped carbon dots which could detect Cu^{2+} and Fe^{3+} ions. Through the comparison of the differences between DN-CDs, N-CDs-1 and N-CDs-2, the ratio of N and O atoms were found to be a potential reason for the ability to detecting Cu^{2+} and Fe^{3+} ions. Table S1 summarized the detection parameters for the selective detection of metal ions (Cu^{2+} and Fe^{3+}) with the different N-doped carbon dots [66–79]. It could be found that although N-doped carbon dots contained different electron-rich functional groups, we suggest that the proportion of the N, O atoms might affect the selectivity towards Cu^{2+} and Fe^{3+} ions. The carbon dots with relatively rich N-based groups and more N atoms were more likely to selectively detect Cu^{2+} ions. More simply, the selective detection of Cu^{2+} and Fe^{3+} ions by carbon dots could be adjusted or converted by altering the ratio of N and O atoms. Moreover, the ratio of N/O of DN-CDs, N-CDs-1 and N-CDs-2 were 14.8:15.6, 21.7:12.8 and 27.1:5.2, which are consistent with our conclusion.

Hence, we preliminarily concluded that the selective detection of Cu^{2+} or Fe^{3+} ions by N-doped carbon dots could be estimated according to the ratio of N/O: when the ratio of N/O is about 1:1 or higher, the N-doped carbon dots tend to selectively detect Cu^{2+} ions; when the N/O ratio is about 1:2 or smaller, the N-doped carbon dots prefer to selectively detect Fe^{3+} ions. The Hard-Soft-Acid-Base theory could explain the difference: 'acids' and 'bases' can be classified as 'hard' or 'soft'. 'Hard' refers to particles with a higher charge density and smaller radius, while 'soft' refers to particles with a lower charge density and larger radius. The interactions between a soft acid and soft base are more stable, while the interactions between a hard acid and hard base are more stable. Thus, Cu^{2+} ions are a borderline acid and can more easily interact with N atoms with a lower electronegativity and bigger radius; Fe^{3+} ions are similar to hard acids, so they are more likely to interact with O atoms with a higher electronegativity and smaller radius. Of course, the quenching mechanisms of carbon dots are complex and other situations would also exist, which require more detailed analysis methods.

3.6. Detection of Cu^{2+} Ions in Real Samples

To prove the practical application of the synthesized DN-CDs, we investigated the ability of the DN-CDs to detect Cu^{2+} ions in water samples as shown in Table S2. The relative standard deviations (RSDs) of Cu^{2+} ion detection in the three samples were less than 8% and the recovery rates were in the range of 95.4–105.9%. This verified that the DN-CDs have great potential in Cu^{2+} ion detection.

4. Conclusions

In conclusion, N-doped carbon dots with down-conversion and up-conversion fluorescence were synthesized by a simple one-step hydrothermal method. The prepared carbon dots, DN-CDs, had excellent fluorescence properties, good stability and a high quantum yield QY of 63.87%. They showed a significant fluorescence intensity at 410 nm, and also had up-conversion fluorescence at 820 nm. Additionally, the abundant N, O-based functional groups endowed the DN-CDs with excellent water solubility and excellent sensitivity and selectivity for Cu^{2+} ion detection with a rapid reaction time of 3 min. Therefore, it had better practicality. We also synthesized other N-doped carbon dots with different carbon sources and nitrogen sources to study their ability to detect metal ions. They had the ability to detect Fe^{3+} ions with a detection limit (LOD) of 100 nM and 3.39 μm. Additionally, although the principle of selective metal ion detection is very complex, by comparing the results in this paper and published reports, a method of synthesizing N-doped carbon dots to detect Cu^{2+} or Fe^{3+} ions exclusively through adjusting the ratio of N and O atoms was proposed: (1) when the ratio of N/O was about 1:1 or higher, the N-doped carbon dots tended to selectively detect Cu^{2+} ions; (2) when the N/O ratio was about 1:2 or smaller, the N-doped carbon dots preferred to selectively detect Fe^{3+} ions. This provided a possible direction for the design and synthesis of N-doped carbon dots with the ability to detect Cu^{2+} or Fe^{3+} ions in the future.

Supplementary Materials: The following supporting information can be downloaded at: https://www.mdpi.com/article/10.3390/cryst13050812/s1, Figure S1: The trends of the quantum yield fluorescence of N-doped carbon dots at different ratio of carbon source and nitrogen source (a), reaction temperature (b), reaction time (c); Figure S2: The Zeta potential of DN-CDs under different pH; Figure S3: The DLS of DN-CDs in aqueous solution at (a) pH = 1, (b) pH = 7 and (c) pH = 13; Figure S4: The Raman spectrum of DN-CDs; Figure S5: The EDS of DN-CDs; Figure S6: The up-conversion PL spectra of DN-CDs. (The emission spectrum at maximum excitation wavelength at 820 nm and the excitation spectrum at the maximum emission wavelength of 450 nm); Figure S7: The CV plot of DN-CDs; Figure S8: (a-c) The fluorescence emission spectra at λex = 450 nm and

λex = 600 nm and FT-IR spectra of N-CDs-1; (d-e) The fluorescence emission spectra at λex = 380 nm and FT-IR spectra of N-CDs-2; Figure S9: The PL quenching efficiency of (a) N-CDs-2 at λex = 380 nm and (c) N-CDs-1 at λex = 410 nm and (e) at λex = 808 nm at different concentration of Fe^{3+} ions and the linear relationship between PL intensity and Fe^{3+} ions ('F0' is PL intensity in the absence of Fe^{3+} ions and 'F' is in the presence of Fe^{3+} ions) of (b) N-CDs-2 at λex = 380 nm, (d) N-CDs-1 at λex = 410 nm and (f) at λex = 808 nm; Table S1: The detection parameters for selective detection of metal ions (Cu^{2+} and Fe^{3+}) with different N-doped carbon dots; Table S2: The detection of Cu^{2+} in real samples.

Author Contributions: Conceptualization, Y.X. and Q.Z.; validation, J.H.; investigation, H.M., P.L. and W.Z.; data curation, M.C., Z.M. and Y.D.; writing—original draft preparation, Y.X.; writing—review and editing, Q.Z.; funding acquisition, Q.Z. All authors have read and agreed to the published version of the manuscript.

Funding: This research was funded by an Industry-university-research cooperation of Sichuan University (No. 19H0340).

Data Availability Statement: The authors confirm that the data supporting the reported results can be found within the article. Other relevant information can be obtained from the author.

Conflicts of Interest: The authors declare that they have no known competing financial interests or personal relationships that could have influenced the work reported in this paper.

References

1. Zhang, R.; Liu, B.; Ma, J.; Zhu, R. Preparation and characterization of carboxymethyl cellulose/chitosan/alginic acid hydrogels with adjustable pore structure for adsorption of heavy metal ions. *Eur. Polym. J.* **2022**, *179*, 111577. [CrossRef]
2. Ansari, S.; Karimi, M. Novel developments and trends of analytical methods for drug analysis in biological and environmental samples by molecularly imprinted polymers. *Trends Anal. Chem.* **2017**, *89*, 146–162. [CrossRef]
3. El-Shahawi, M.S.; Hamza, A.; Bashammakh, A.S.; Al-Saggaf, W.T. An overview on the accumulation, distribution, transformations, toxicity and analytical methods for the monitoring of persistent organic pollutants. *Talanta* **2010**, *80*, 1587–1597. [CrossRef] [PubMed]
4. Martín-Pozo, L.; de Alarcón-Gómez, B.; Rodríguez-Gómez, R.; García-Córcoles, M.T.; Çipa, M.; Zafra-Gómez, A. Analytical methods for the determination of emerging contaminants in sewage sludge samples. A review. *Talanta* **2019**, *192*, 508–533. [CrossRef] [PubMed]
5. Kumbhar, H.S.; Gadilohar, B.L.; Shankarling, G.S. A highly selective quinaldine–indole based spiropyran with intramolecular H-bonding for visual detection of Cu(II) ions. *Sens. Actuators B Chem.* **2016**, *222*, 35–42. [CrossRef]
6. Gao, F.; Tu, X.; Yu, Y.; Gao, Y.; Zou, J.; Liu, S.; Qu, F.; Li, M.; Lu, L. Core-shell Cu@C@ZIF-8 composite: A high-performance electrode material for electrochemical sensing of nitrite with high selectivity and sensitivity. *Nanotechnology* **2022**, *33*, 225501. [CrossRef] [PubMed]
7. Chen, H.; Yang, T.; Liu, F.; Li, W. Electrodeposition of gold nanoparticles on Cu-based metal-organic framework for the electrochemical detection of nitrite. *Sens. Actuators B Chem.* **2019**, *286*, 401–407. [CrossRef]
8. Carolan, D.; Doyle, H. Germanium nanocrystals as luminescent probes for rapid, sensitive and label-free detection of Fe^{3+} ions. *Nanoscale* **2015**, *7*, 5488–5494. [CrossRef]
9. Ferreira, S.L.C.; Bezerra, M.A.; Santos, A.S.; dos Santos, W.N.L.; Novaes, C.G.; de Oliveira, O.M.C.; Oliveira, M.L.; Garcia, R.L. Atomic absorption spectrometry—A multi element technique. *Trends Anal. Chem.* **2018**, *100*, 1–6. [CrossRef]
10. Matusch, A.; Depboylu, C.; Palm, C.; Wu, B.; Höglinger, G.U.; Schäfer, M.K.H.; Becker, J.S. Cerebral bioimaging of Cu, Fe, Zn, and Mn in the MPTP mouse model of Parkinson's disease using laser ablation inductively coupled plasma mass spectrometry (LA-ICP-MS). *J. Am. Soc. Mass. Spectrom.* **2010**, *21*, 161–171. [CrossRef]
11. Trung, L.G.; Subedi, S.; Dahal, B.; Truong, P.L.; Gwag, J.S.; Tran, N.T.; Nguyen, M.K. Highly efficient fluorescent probes from chitosan-based amino-functional carbon dots for the selective detection of Cu^{2+} traces. *Mater. Chem. Phys.* **2022**, *291*, 126772. [CrossRef]
12. Xu, H.; Zhou, S.; Xiao, L.; Wang, H.; Li, S.; Yuan, Q. Fabrication of a nitrogen-doped graphene quantum dot from MOF-derived porous carbon and its application for highly selective fluorescence detection of Fe^{3+}. *J. Mater. Chem. C* **2015**, *3*, 291–297. [CrossRef]
13. Yang, C.-X.; Ren, H.-B.; Yan, X.-P. Fluorescent Metal–Organic Framework MIL-53(Al) for Highly Selective and Sensitive Detection of Fe^{3+} in Aqueous Solution. *Anal. Chem.* **2013**, *85*, 7441–7446. [CrossRef]
14. Mohandoss, S.; Palanisamy, S.; You, S.; Lee, Y.R. Synthesis of cyclodextrin functionalized photoluminescent metal nanoclusters for chemoselective Fe^{3+} ion detection in aqueous medium and its applications of paper sensors and cell imaging. *J. Mol. Liq.* **2022**, *356*, 118999. [CrossRef]

15. Xin, J.-Y.; Li, Y.; Liu, F.-Y.; Sun, L.-R.; Wang, Y.; Xia, C.-G. Visible and Ultraviolet Dual-Readout Detection of Cu(II) in Preserved Vegetables Based on Self-Assembly and Peroxidase Simulation Properties of Mb-AuNPs. *Sci. Adv. Mater.* **2022**, *14*, 1410–1418. [CrossRef]
16. Cetinkaya, Y.; Yurt, M.N.Z.; Avni Oktem, H.; Yilmaz, M.D. A Monostyryl Boradiazaindacene (BODIPY)-based lanthanide-free colorimetric and fluorogenic probe for sequential sensing of copper (II) ions and dipicolinic acid as a biomarker of bacterial endospores. *J. Hazard. Mater.* **2019**, *377*, 299–304. [CrossRef]
17. Zhu, S.; Meng, Q.; Wang, L.; Zhang, J.; Song, Y.; Jin, H.; Zhang, K.; Sun, H.; Wang, H.; Yang, B. Highly photoluminescent carbon dots for multicolor patterning, sensors, and bioimaging. *Angew. Chem. Int. Ed. Engl.* **2013**, *52*, 3953–3957. [CrossRef]
18. Han, Z.; Nan, D.; Yang, H.; Sun, Q.; Pan, S.; Liu, H.; Hu, X. Carbon quantum dots based ratiometric fluorescence probe for sensitive and selective detection of Cu^{2+} and glutathione. *Sens. Actuators B Chem.* **2019**, *298*, 126842. [CrossRef]
19. Mitra, S.; Chakraborty, A.J.; Tareq, A.M.; Emran, T.B.; Nainu, F.; Khusro, A.; Idris, A.M.; Khandaker, M.U.; Osman, H.; Alhumaydhi, F.A.; et al. Impact of heavy metals on the environment and human health: Novel therapeutic insights to counter the toxicity. *J. King. Saud. Univ. Sci.* **2022**, *34*, 101865. [CrossRef]
20. Zhou, W.; Mo, F.; Sun, Z.; Luo, J.; Fan, J.; Zhu, H.; Zhu, Z.; Huang, J.; Zhang, X. Bright red-emitting P, Br co-doped carbon dots as "OFF-ON" fluorescent probe for Cu^{2+} and L-cysteine detection. *J. Alloys Compd.* **2022**, *897*, 162731. [CrossRef]
21. Kumari, A.; Kumar, A.; Sahu, S.K.; Kumar, S. Synthesis of green fluorescent carbon quantum dots using waste polyolefins residue for Cu^{2+} ion sensing and live cell imaging. *Sens. Actuators B Chem.* **2018**, *254*, 197–205. [CrossRef]
22. Li, H.; Xu, T.; Zhang, Z.; Chen, J.; She, M.; Ji, Y.; Zheng, B.; Yang, Z.; Zhang, S.; Li, J. Photostable and printable fluorescence carbon quantum dots for advanced message encryption and specific reversible multiple sensing of Cu^{2+} and cysteine. *Chem. Eng. J.* **2023**, *453*, 139722. [CrossRef]
23. Lv, W.; Lin, M.; Li, R.; Zhang, Q.; Liu, H.; Wang, J.; Huang, C. Aggregation-induced emission enhancement of yellow photoluminescent carbon dots for highly selective detection of environmental and intracellular copper(II) ions. *Chin. Chem. Lett.* **2019**, *30*, 1410–1414. [CrossRef]
24. Sonaimuthu, M.; Ganesan, S.; Anand, S.; Kumar, A.J.; Palanisamy, S.; You, S.; Velsankar, K.; Sudhahar, S.; Lo, H.-M.; Lee, Y.R. Multiple heteroatom dopant carbon dots as a novel photoluminescent probe for the sensitive detection of Cu^{2+} and Fe^{3+} ions in living cells and environmental sample analysis. *Environ. Res.* **2023**, *219*, 115106. [CrossRef]
25. Jiang, D.; Zheng, M.; Yan, X.; Huang, B.; Huang, H.; Gong, T.; Liu, K.; Liu, J. A "turn-on" ESIPT fluorescence probe of 2-(aminocarbonyl)phenylboronic acid for the selective detection of Cu(ii). *RSC Adv.* **2022**, *12*, 31186–31191. [CrossRef] [PubMed]
26. Mahajan, P.G.; Dige, N.C.; Vanjare, B.D.; Eo, S.-H.; Kim, S.J.; Lee, K.H. A nano sensor for sensitive and selective detection of Cu^{2+} based on fluorescein: Cell imaging and drinking water analysis. *Spectrochim. Acta A* **2019**, *216*, 105–116. [CrossRef]
27. Issa, M.A.; Abidin, Z.Z.; Sobri, S.; Rashid, S.A.; Mahdi, M.A.; Ibrahim, N.A. Fluorescent recognition of Fe^{3+} in acidic environment by enhanced-quantum yield N-doped carbon dots: Optimization of variables using central composite design. *Sci. Rep.* **2020**, *10*, 11710. [CrossRef] [PubMed]
28. Nagaraj, M.; Ramalingam, S.; Murugan, C.; Aldawood, S.; Jin, J.-O.; Choi, I.; Kim, M. Detection of Fe^{3+} ions in aqueous environment using fluorescent carbon quantum dots synthesized from endosperm of Borassus flabellifer. *Environ. Res.* **2022**, *212*, 113273. [CrossRef]
29. Wang, Y.; Chang, Q.; Hu, S. Carbon dots with concentration-tunable multicolored photoluminescence for simultaneous detection of Fe^{3+} and Cu^{2+} ions. *Sens. Actuators B Chem.* **2017**, *253*, 928–933. [CrossRef]
30. Yuan, Y.H.; Liu, Z.X.; Li, R.S.; Zou, H.Y.; Lin, M.; Liu, H.; Huang, C.Z. Synthesis of nitrogen-doping carbon dots with different photoluminescence properties by controlling the surface states. *Nanoscale* **2016**, *8*, 6770–6776. [CrossRef]
31. Zhang, H.; Chen, Y.; Liang, M.; Xu, L.; Qi, S.; Chen, H.; Chen, X. Solid-phase synthesis of highly fluorescent nitrogen-doped carbon dots for sensitive and selective probing ferric ions in living cells. *Anal. Chem.* **2014**, *86*, 9846–9852. [CrossRef]
32. Mejía Ávila, J.; Rangel Ayala, M.; Kumar, Y.; Pérez-Tijerina, E.; Robles, M.A.R.; Agarwal, V. Avocado seeds derived carbon dots for highly sensitive Cu (II)/Cr (VI) detection and copper (II) removal via flocculation. *Chem. Eng. J.* **2022**, *446*, 137171. [CrossRef]
33. Rahmani, Z.; Ghaemy, M. One-step hydrothermal-assisted synthesis of highly fluorescent N-doped carbon dots from gum tragacanth: Luminescent stability and sensitive probe for Au^{3+} ions. *Opt. Mater.* **2019**, *97*, 109356. [CrossRef]
34. Luo, T.; Bu, L.; Peng, S.; Zhang, Y.; Zhou, Z.; Li, G.; Huang, J. One-step microwave-assisted preparation of oxygen-rich multifunctional carbon quantum dots and their application for Cu2+-curcumin detection. *Talanta* **2019**, *205*, 120117. [CrossRef] [PubMed]
35. Zhao, Q.; Li, X.; Wang, X.; Zang, Z.; Liu, H.; Li, Y.; Yu, X.; Yang, X.; Lu, Z.; Zhang, X. Surface amino group modulation of carbon dots with blue, green and red emission as Cu^{2+} ion reversible detector. *Appl. Surf. Sci.* **2022**, *598*, 153892. [CrossRef]
36. Xu, J.; Wang, C.; Li, H.; Zhao, W. Synthesis of green-emitting carbon quantum dots with double carbon sources and their application as a fluorescent probe for selective detection of Cu^{2+} ions. *RSC Adv.* **2020**, *10*, 2536–2544. [CrossRef]
37. Jiao, T.; Wen, H.; Dong, W.; Wen, W.; Li, Z. Metal ion (Fe^{3+} and Cu^{2+}) catalyzed synthesis of high quantum yield carbon dots. *Mater. Lett.* **2022**, *324*, 132575. [CrossRef]
38. Siahcheshm, P.; Heiden, P. High quantum yield carbon quantum dots as selective fluorescent turn-off probes for dual detection of Fe^{2+}/Fe^{3+} ions. *J. Photochem. Photobiol. A* **2023**, *435*, 114284. [CrossRef]
39. Atabaev, T.S. Doped carbon dots for sensing and bioimaging applications: A minireview. *Nanomaterials* **2018**, *8*, 342. [CrossRef]

40. Gao, X.; Du, C.; Zhuang, Z.; Chen, W. Carbon quantum dot-based nanoprobes for metal ion detection. *J. Mater. Chem. C* **2016**, *4*, 6927–6945. [CrossRef]
41. Molaei, M.J. Principles, mechanisms, and application of carbon quantum dots in sensors: A review. *Anal. Methods* **2020**, *12*, 1266–1287. [CrossRef]
42. Mohandoss, S.; Ganesan, S.; Palanisamy, S.; You, S.; Velsankar, K.; Sudhahar, S.; Lo, H.-M.; Lee, Y.R. Nitrogen, sulfur, and phosphorus Co-doped carbon dots-based ratiometric chemosensor for highly selective sequential detection of Al^{3+} and Fe^{3+} ions in logic gate, cell imaging, and real sample analysis. *Chemosphere* **2023**, *313*, 137444. [CrossRef] [PubMed]
43. Tian, L.; Li, Z.; Wang, P.; Zhai, X.; Wang, X.; Li, T. Carbon quantum dots for advanced electrocatalysis. *J. Energy Chem.* **2021**, *55*, 279–294. [CrossRef]
44. Salehtabar, F.; Ghaemy, M. Preparation of strongly photoluminescent nanocomposite from DGEBA epoxy resin and highly fluorescent nitrogen-doped carbon dots. *Polym. Bull.* **2023**, *80*, 3247–3264. [CrossRef]
45. Liu, R.; Haruna, S.A.; Ali, S.; Xu, J.; Ouyang, Q.; Li, H.; Chen, Q. An Up-conversion signal probe-MnO2 nanosheet sensor for rapid and sensitive detection of tetracycline in food. *Spectrochim. Acta A* **2022**, *270*, 120855. [CrossRef]
46. Fu, A.C.; Hu, Y.; Zhao, Z.-H.; Su, R.; Song, Y.; Zhu, D. Functionalized paper microzone plate for colorimetry and up-conversion fluorescence dual-mode detection of telomerase based on elongation and capturing amplification. *Sens. Actuators B Chem.* **2018**, *259*, 642–649. [CrossRef]
47. Liang, J.-M.; Zhang, F.; Zhu, Y.-L.; Deng, X.-Y.; Chen, X.-P.; Zhou, Q.-J.; Tan, K.-J. One-pot hydrothermal synthesis of Si-doped carbon quantum dots with up-conversion fluorescence as fluorescent probes for dual-readout detection of berberine hydrochloride. *Spectrochim. Acta A* **2022**, *275*, 121139. [CrossRef]
48. Fu, J.; Zhou, S.; Tang, S.; Wu, X.; Zhao, P.; Tang, K.; Chen, Y.; Yang, Z.; Zhang, Z. Imparting down/up-conversion dual channels fluorescence to luminescence metal-organic frameworks by carbon dots-induced for fluorescence sensing. *Talanta* **2022**, *242*, 123283. [CrossRef]
49. Deng, Y.; Chen, M.; Chen, G.; Zou, W.; Zhao, Y.; Zhang, H.; Zhao, Q. Visible–Ultraviolet Upconversion Carbon Quantum Dots for Enhancement of the Photocatalytic Activity of Titanium Dioxide. *ACS Omega* **2021**, *6*, 4247–4254. [CrossRef]
50. Mao, Z.; Li, H.; Gan, N.; Suo, Z.; Zhang, H.; Zhao, Q. Contribution of nicotinamide as an intracyclic N dopant to the structure and properties of carbon dots synthesized using three α-hydroxy acids as C sources. *Nanotechnology* **2022**, *33*, 215705. [CrossRef]
51. Fang, C.; Lihua, Z.; Hong, W. Preparation of Carbon Dots and Determination of Their Fluorescence Quantum Yield. *Univ. Chem.* **2019**, *34*, 67–72.
52. Ali, M.; Anjum, A.S.; Bibi, A.; Wageh, S.; Sun, K.C.; Jeong, S.H. Gradient heating-induced bi-phase synthesis of carbon quantum dots (CQDs) on graphene-coated carbon cloth for efficient photoelectrocatalysis. *Carbon* **2022**, *196*, 649–662. [CrossRef]
53. Liang, C.; Xie, X.; Shi, Q.; Feng, J.; Zhang, D.; Huang, X. Nitrogen/sulfur-doped dual-emission carbon dots with tunable fluorescence for ratiometric sensing of ferric ions and cell membrane imaging. *Appl. Surf. Sci.* **2022**, *572*, 151447. [CrossRef]
54. Chen, X.; Song, Z.; Yuan, B.; Li, X.; Li, S.; Thang Nguyen, T.; Guo, M.; Guo, Z. Fluorescent carbon dots crosslinked cellulose Nanofibril/Chitosan interpenetrating hydrogel system for sensitive detection and efficient adsorption of Cu (II) and Cr (VI). *Chem. Eng. J.* **2022**, *430*, 133154. [CrossRef]
55. Preeyanghaa, M.; Vinesh, V.; Sabarikirishwaran, P.; Rajkamal, A.; Ashokkumar, M.; Neppolian, B. Investigating the role of ultrasound in improving the photocatalytic ability of CQD decorated boron-doped g-C3N4 for tetracycline degradation and first-principles study of nitrogen-vacancy formation. *Carbon* **2022**, *192*, 405–417. [CrossRef]
56. Dong, Y.; Pang, H.; Yang, H.B.; Guo, C.; Shao, J.; Chi, Y.; Li, C.M.; Yu, T. Carbon-Based Dots Co-doped with Nitrogen and Sulfur for High Quantum Yield and Excitation-Independent Emission. *Angew. Chem. Int. Ed.* **2013**, *52*, 7800–7804. [CrossRef]
57. Gao, Y.; Zhang, H.; Jiao, Y.; Lu, W.; Liu, Y.; Han, H.; Gong, X.; Shuang, S.; Dong, C. Strategy for Activating Room-Temperature Phosphorescence of Carbon Dots in Aqueous Environments. *Chem. Mater.* **2019**, *31*, 7979–7986. [CrossRef]
58. Wang, H.; Wang, P.; Niu, L.; Liu, C.; Xiao, Y.; Tang, Y.; Chen, Y. Carbazole-thiophene based fluorescent probe for selective detection of Cu^{2+} and its live cell imaging. *Spectrochim. Acta A* **2022**, *278*, 121257. [CrossRef]
59. Kateshiya, M.R.; Malek, N.I.; Kailasa, S.K. Folic acid functionalized molybdenum oxide quantum dots for the detection of Cu^{2+} ion and alkaline phosphatase via fluorescence turn off–on mechanism. *Spectrochim. Acta A* **2022**, *268*, 120659. [CrossRef]
60. Zhao, L.-X.; Chen, K.-Y.; Xie, K.-B.; Hu, J.-J.; Deng, M.-Y.; Zou, Y.-L.; Gao, S.; Fu, Y.; Ye, F. A benzothiazole-based "on-off" fluorescence probe for the specific detection of Cu^{2+} and its application in solution and living cells. *Dye. Pigm.* **2023**, *210*, 110943. [CrossRef]
61. Zhang, Y.; Zhu, T.; Xu, Y.; Yang, Y.; Sheng, D.; Ma, Q. Quaternized salicylaldehyde Schiff base side-chain polymer-grafted magnetic Fe3O4 nanoparticles for the removal and detection of Cu^{2+} ions in water. *Appl. Surf. Sci.* **2023**, *611*, 155632. [CrossRef]
62. John, B.K.; John, N.; Korah, B.K.; Thara, C.; Abraham, T.; Mathew, B. Nitrogen-doped carbon quantum dots as a highly selective fluorescent and electrochemical sensor for tetracycline. *J. Photochem. Photobiol. A* **2022**, *432*, 114060. [CrossRef]
63. Sun, S.; Bao, W.; Yang, F.; Yan, X.; Sun, Y.; Zhang, G.; Yang, W.; Li, Y. Electrochemical synthesis of FeNx doped carbon quantum dots for sensitive detection of Cu^{2+} ion. *Green. Energy Environ.* **2021**, *8*, 141–150. [CrossRef]
64. Bossu, F.P.; Chellappa, K.L.; Margerum, D.W. Ligand effects on the thermodynamic stabilization of copper(III)-peptide complexes. *J. Am. Chem. Soc.* **1977**, *99*, 2195–2203. [CrossRef] [PubMed]

65. Wang, J.; Sheng Li, R.; Zhi Zhang, H.; Wang, N.; Zhang, Z.; Huang, C.Z. Highly fluorescent carbon dots as selective and visual probes for sensing copper ions in living cells via an electron transfer process. *Biosens. Bioelectron.* **2017**, *97*, 157–163. [CrossRef] [PubMed]
66. Wu, H.; Pang, L.-F.; Fu, M.-J.; Guo, X.-F.; Wang, H. Boron and nitrogen co-doped carbon dots as fluorescence sensor for Fe^{3+} with improved selectivity. *J. Pharm. Biomed. Anal.* **2020**, *180*, 113052. [CrossRef]
67. Zhang, Z.; Chen, X.; Wang, J. Bright blue emissions N-doped carbon dots from a single precursor and their application in the trace detection of Fe^{3+} and F^-. *Inorg. Chim. Acta* **2021**, *515*, 120087. [CrossRef]
68. Zhou, Y.; Chen, G.; Ma, C.; Gu, J.; Yang, T.; Li, L.; Gao, H.; Xiong, Y.; Wu, Y.; Zhu, C.; et al. Nitrogen-doped carbon dots with bright fluorescence for highly sensitive detection of Fe^{3+} in environmental waters. *Spectrochim. Acta A Mol. Biomol. Spectrosc.* **2023**, *293*, 122414. [CrossRef]
69. Zhang, W.J.; Liu, S.G.; Han, L.; Luo, H.Q.; Li, N.B. A ratiometric fluorescent and colorimetric dual-signal sensing platform based on N-doped carbon dots for selective and sensitive detection of copper (II) and pyrophosphate ion. *Sens. Actuators B Chem.* **2019**, *283*, 215–221. [CrossRef]
70. Yang, L.; Zeng, J.; Quan, T.; Liu, S.; Deng, L.; Kang, X.; Xia, Z.; Gao, D. Liquid-liquid extraction and purification of oil red O derived nitrogen-doped highly photoluminescent carbon dots and their application as multi-functional sensing platform for Cu^{2+} and tetracycline antibiotics. *Microchem. J.* **2021**, *168*, 106391. [CrossRef]
71. Yang, F.; Zhou, P.; Duan, C. Solid-phase synthesis of red dual-emissive nitrogen-doped carbon dots for the detection of Cu^{2+} and glutathione. *Microchem. J.* **2021**, *169*, 106534. [CrossRef]
72. Praneerad, J.; Thongsai, N.; Supchocksoonthorn, P.; Kladsomboon, S.; Paoprasert, P. Multipurpose sensing applications of biocompatible radish-derived carbon dots as Cu^{2+} and acetic acid vapor sensors. *Spectrochim. Acta A* **2019**, *211*, 59–70. [CrossRef] [PubMed]
73. Li, Z.; Zhou, Q.; Li, S.; Liu, M.; Li, Y.; Chen, C. Carbon dots fabricated by solid-phase carbonization using p-toluidine and l-cysteine for sensitive detection of copper. *Chemosphere* **2022**, *308*, 136298. [CrossRef]
74. Krishnaiah, P.; Atchudan, R.; Perumal, S.; Salama, E.-S.; Lee, Y.R.; Jeon, B.-H. Utilization of waste biomass of Poa pratensis for green synthesis of n-doped carbon dots and its application in detection of Mn^{2+} and Fe^{3+}. *Chemosphere* **2022**, *286*, 131764. [CrossRef]
75. Gao, B.; Chen, D.; Gu, B.; Wang, T.; Wang, Z.; Xie, F.; Yang, Y.; Guo, Q.; Wang, G. Facile and highly effective synthesis of nitrogen-doped graphene quantum dots as a fluorescent sensing probe for Cu^{2+} detection. *Curr. Appl. Phys.* **2020**, *20*, 538–544. [CrossRef]
76. Liu, L.; Qin, K.; Yin, S.; Zheng, X.; Li, H.; Yan, H.; Song, P.; Ji, X.; Zhang, Q.; Wei, Y.; et al. Bifunctional carbon dots derived from an anaerobic bacterium of porphyromonas gingivalis for selective detection of Fe and bioimaging. *Photochem. Photobiol.* **2021**, *97*, 574–581. [CrossRef]
77. Fu, Y.; Zhao, S.; Wu, S.; Huang, L.; Xu, T.; Xing, X.; Lan, M.; Song, X. A carbon dots-based fluorescent probe for turn-on sensing of ampicillin. *Dye. Pigment.* **2020**, *172*, 107846. [CrossRef]
78. Qian, Z.; Shan, X.; Chai, L.; Ma, J.; Chen, J.; Feng, H. Si-doped carbon quantum dots: A facile and general preparation strategy, bioimaging application, and multifunctional sensor. *ACS Appl. Mater. Interfaces* **2014**, *6*, 6797–6805. [CrossRef] [PubMed]
79. Zhang, L.C.; Yang, Y.M.; Liang, L.; Jiang, Y.J.; Li, C.M.; Li, Y.F.; Zhan, L.; Zou, H.Y.; Huang, C.Z. Lighting up of carbon dots for copper (II) detection using an aggregation-induced enhanced strategy. *Analyst* **2022**, *147*, 417–422. [CrossRef]

Disclaimer/Publisher's Note: The statements, opinions and data contained in all publications are solely those of the individual author(s) and contributor(s) and not of MDPI and/or the editor(s). MDPI and/or the editor(s) disclaim responsibility for any injury to people or property resulting from any ideas, methods, instructions or products referred to in the content.

Article

High-Performance Nanoplasmonic Enhanced Indium Oxide—UV Photodetectors

Eric Y. Li [1], Andrew F. Zhou [2,*] and Peter X. Feng [1,*]

[1] Department of Physics, University of Puerto Rico, San Juan, PR 00936, USA
[2] Department of Chemistry, Biochemistry, Physics, and Engineering, Indiana University of Pennsylvania, Indiana, PA 15705, USA
* Correspondence: fzhou@iup.edu (A.F.Z.); peter.feng@upr.edu (P.X.F.)

Abstract: In this paper, high-performance UV photodetectors have been demonstrated based on indium oxide (In_2O_3) thin films of approximately 1.5–2 µm thick, synthesized by a simple and quick plasma sputtering deposition approach. After the deposition, the thin-film surface was treated with 4–5 nm-sized platinum (Pt) nanoparticles. Then, titanium metal electrodes were deposited onto the sample surface to form a metal–semiconductor–metal (MSM) photodetector of 50 mm^2 in size. Raman scattering spectroscopy and scanning electron microscope (SEM) were used to study the crystal structure of the synthesized In_2O_3 film. The nanoplasmonic enhanced In_2O_3-based UV photodetectors were characterized by various UV wavelengths at different radiation intensities and temperatures. A high responsivity of up to 18 A/W was obtained at 300 nm wavelength when operating at 180 °C. In addition, the fabricated prototypes show a thermally stable baseline and excellent repeatability to a wide range of UV lights with low illumination intensity when operating at such a high temperature.

Keywords: UV photodetectors; surface functionalization; indium oxide; nanoplasmonic

Citation: Li, E.Y.; Zhou, A.F.; Feng, P.X. High-Performance Nanoplasmonic Enhanced Indium Oxide—UV Photodetectors. *Crystals* 2023, 13, 689. https://doi.org/10.3390/cryst13040689

Academic Editor: Alessandro Chiasera

Received: 23 March 2023
Revised: 12 April 2023
Accepted: 14 April 2023
Published: 17 April 2023

Copyright: © 2023 by the authors. Licensee MDPI, Basel, Switzerland. This article is an open access article distributed under the terms and conditions of the Creative Commons Attribution (CC BY) license (https://creativecommons.org/licenses/by/4.0/).

1. Introduction

Indium oxide (In_2O_3) has received much attention in multidisciplinary device applications such as flat-panel displays, light-emitting diodes (LEDs), and solar cells. Moreover, in recent years, indium oxide nanomaterials with different morphologies (nanoparticles, nanoflowers, nanosheets, nanowires, etc.) [1,2] have been found to be promising candidates for gas sensors [3–5]. On the other hand, indium oxide is a desirable material for developing smart, flexible, and wearable optoelectronics and multifunctional sensors in the field of environmental indicators [6], owing to its wide band gap (with reported values ranging from 2.7 to 3.7 eV) [7], optical transparency, thermal and chemical resistance, flexibility and stretchability, ease of manufacturing and low fabrication cost. Hence, UV photodetectors (PDs) based on In_2O_3 nanostructures, such as nanowires [8–10], quantum dots [11], nanosheets [12], and nanoparticles [13], have been explored, which demonstrates that In_2O_3 is a candidate for developing UV PDs. However, there are neither reports on the UV photodetector based on In_2O_3 thin films, nor studies on PD performance comparison at different temperatures.

Recently, surface functionalization with noble metal nanoparticles (NPs, e.g., Au and Ag) was effectively used to improve ZnO [14], GaN [15] and UNCD [16] UV photodetector performances. The effects of different metals such as Pt, Al and Ag as well as nanoparticle size have been discussed in our previous publications [16–19]. The nanoplasmonic resonance in metal nanoparticles causes the strong absorption and scattering of incident light. As a result, the photodetector showed high responsivity, which is defined as the ratio of generated photocurrent and incident optical power. Although the surfaced functionalization of noble metal nanoparticles was demonstrated in indium oxide-based gas sensors [20], to date, no papers have been published on surface functionalized In_2O_3 UV photodetectors with noble metal nanoparticles.

In the present work, we demonstrated a high-performance In_2O_3 UV photodetector with Pt nanoparticle surface functionalization. Pt nanoparticles were chosen due to their localized plasmon resonance featured in the UV region [21]. Very high responsivities have been obtained both at room temperature and an elevated temperature, which is much better than that of UV photodetectors with MSM (metal–semiconductor–metal) structures based on SiC [22], diamond [23], and oxides [24] that can operate at high temperatures. The fabricated prototypes show a thermally stable baseline and excellent repeatability to a wide range of UV lights with low illumination intensity.

2. Materials and Methods

Compared to metal–organic chemical vapor deposition (MOCVD) and molecular beam epitaxy (MBE) techniques, sputtering has features of a high deposition rate, low cost, acceptable quality of thin films, and capability to deposit on large-area substrates. Owing to these advantages and their flexibility in the control of composition and microstructure, magnetron sputtering is widely employed for the synthesis of various thin films. The In_2O_3 thin films used in this study were grown on Si (100) substrates under deposition durations of 15 min using a homemade RF reactive magnetron sputtering system [25] with a power of 200 W. The substrates were cleaned in acetone and methanol with ultrasonic vibration and then dried with blowing dry nitrogen gas before deposition. The target for the growth of In_2O_3 films was sintered In_2O_3 (99.9%), 2-inch in diameter. The distance between the target and substrate was approximately 7–8 cm. All the films were deposited at room temperature and the obtained thickness of the sample was approximately 1.5–2.0 μm measured using a stylus surface profiler (DektakXT, BRUKER, Billerica, MA). After deposition, the samples were annealed at 400, 600 and 800 °C for 2 h, respectively. The base pressure of the chamber was on the order of 10^{-6} Torr and the working pressure with argon gas was maintained at approximately 8–10 mTorr. To enhance the responsivity of the In_2O_3 thin–film material to UV radiation, the samples were functionalized using platinum (Pt) nanoparticles before photodetector fabrication. The surface treatments with Pt nanoparticles lasted for 1–2 s by using a 200 W RF magnetron sputtering deposition technique.

3. Results and Discussions

SEM was utilized to generate high-resolution photographs of surface morphology, along with a Raman spectrometer enabling a deeper insight into structural information. Figure 1a–c show SEM images of three In_2O_3 samples after annealing at 400, 600 and 800 °C, respectively, for 2 h. The surfaces of the obtained samples are flat but many particles on the surface were clearly visible. Following an increase in the temperature on annealing, the sizes of these particles increased from the sub-micrometer range to approximately 1 micrometer. A tentative interpretation is that melting and dissociation of partial In_2O_3 thin films into droplets on the surfaces of the samples during annealing at very high temperatures results in a rough/coarse surface.

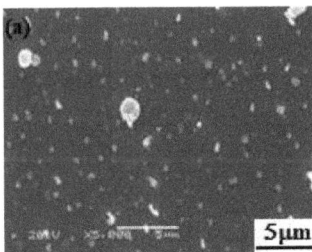
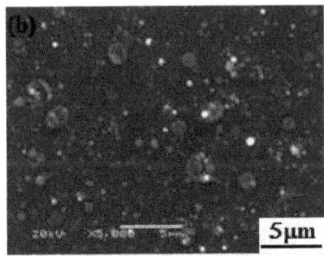
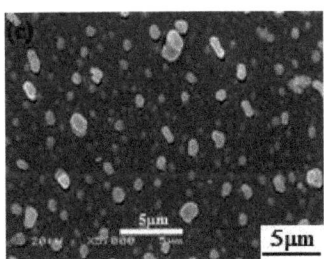

Figure 1. SEM images of In_2O_3 samples annealed at (**a**) 400 °C, (**b**) 600 °C, and (**c**) 800 °C, respectively, for 2 h.

Generally, the coarse surface of sample-based devices would be seriously affected by the surrounding environment such as humidity or dust. Therefore, in the following part, the focus would be on the first sample (as shown in Figure 1a) annealed at 400 °C with better surface quality. As an n-type semiconductor, different preparation methods and conditions influence the crystalline structures of In_2O_3 synthesized, such as cubic or rhombohedral, although both have the same band gap energy. The body-centered cubic (bcc) In_2O_3 with lattice parameter a = 1.011 nm is the most stable structure that is preferred for electronic [26] applications due to its high conductivity.

The In_2O_3 films prepared for this study were characterized by room-temperature Raman spectroscopy. An Ar^+ ion laser (Model 95, Lexcel Laser Inc., Fremont, CA) with an excitation wavelength of 514 nm was used along with a triple-grating monochromator to measure the Raman spectra. The ~2 mW laser beam, after passing through an 80× Olympus microscope objective, was focused onto the sample surface with a spot size of 3~4 μm in diameter, corresponding to a laser power density of approximately 2×10^4 W/cm^2. It was noticed that at this power density, no obvious surface morphology change in the In_2O_3 samples was observed because of the annealing effect caused by the laser beam with an accumulation time of 30 s required by the Raman measurement. Figure 2 shows Raman spectra of In_2O_3 film deposited on Si(100) substrate before and after the surface treatment with platinum nanoparticles. As displayed in Figure 2, the strongest peak at approximately 521 cm^{-1} is from the Si substrate. The spectra from In_2O_3 are dominated by the optical phonon lines at 132, 307, 366, 495, and 627 cm^{-1}, corresponding to bcc In_2O_3. The narrow spectral lines at 132 and 307 cm^{-1} confirmed the good crystalline quality of the bcc In_2O_3 films.

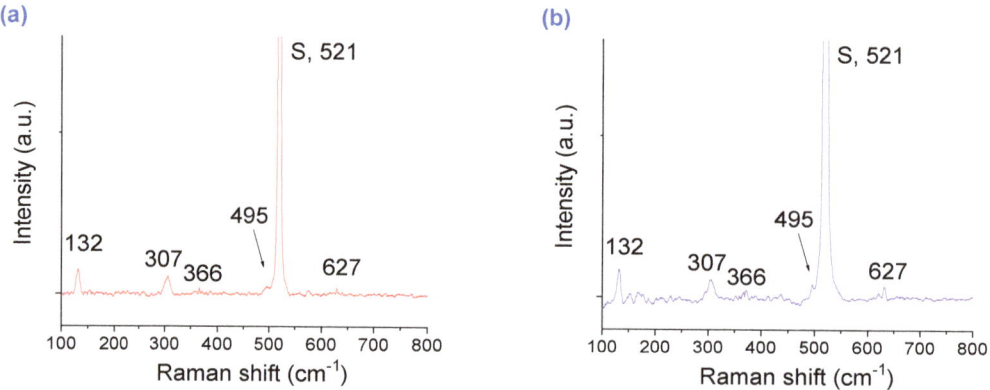

Figure 2. Raman scattering spectra of In_2O_3 thin film prepared on Si substrate (**a**) before and (**b**) after Pt treatment. The large peak at 521 cm^{-1} from the Si substrate is labeled as "S".

The fast fabrication process of the In_2O_3 thin-film-based photodetector is shown in Figure 3a. After the In_2O_3 thin film was deposited on Si(100) substrate, the sample surface was decorated with Pt nanoparticles. Then, a mask made of a metal wire of 400 μm in diameter was placed over the top surface of the In_2O_3 sample, followed by a 2-min direct coating of Ti metal electrodes of a thickness of 0.5–1 μm. The fabricated In_2O_3 UV photodetector was finally annealed at 200 °C for 60 min in the same vacuum chamber. Figure 3b shows the edge between the Ti electrode (grey) and In_2O_3 sensing material (black). The MSM geometry of the photodetector used in this investigation was similar to our previous boron nitride-based and SiC-based UV photodetectors [22,27] except for the In_2O_3 sensing layer. The gap formed by this simple and quick fabrication process was 400 μm between a pair of electrical electrodes, leading to a totally exposed surface area of 15 mm^2, although smaller gaps can be obtained with a more tedious lithography process.

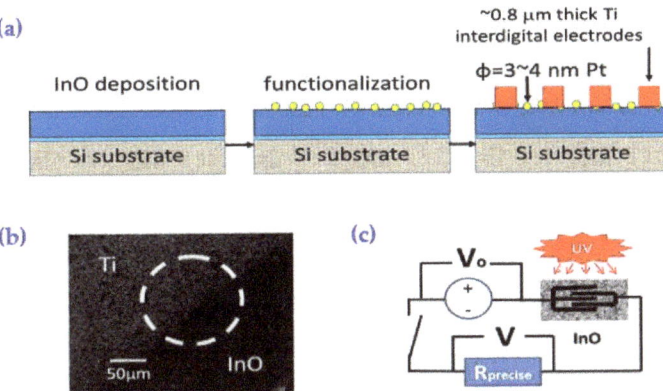

Figure 3. (a) Schematic diagram of the fabrication of the In_2O_3-based UV photodetector. (b) SEM image of the edge between the Ti electrode (grey) and sensing material (black) of the prototype. (c) Experimental setup for measuring electric properties and light responsivity of the UV photodetector.

Once the process flow of the MSM UV photodetector was completed, the prototypic UV photodetector was tested for its I–V graph and the photon-induced electrical current I_{ph} when the sensing material was exposed to UV radiation. As shown in Figure 3c, the prototype is connected in series with a precise resistor $R_{precise}$, a switch, and an Agilent voltage source power supply V_o. The voltage variations during the characterization across the precise resistor were monitored with a HEWLETT 34,401 electrical multimeter controlled by the LabVIEW program. To control the operating temperature of the tested prototype, a thermocouple and tungsten filament were deployed. The error of the measurements was approximately 5%.

The UV sensing mechanism of the In_2O_3 photodetector can be understood as a photoelectric phenomenon. As an n-type semiconductor, the conductivity of the In_2O_3 photodetector is largely due to the depletion region formed when there is no UV illumination because oxygen molecules trap free electrons to be ionized and chemisorbed on the surface. With UV radiation of photon energy slightly equal to or greater than the energy band gap of In_2O_3, the photoelectrically generated hole carriers recombine negatively charged adsorbed oxygen ions on the surface and discharge oxygen molecules that reduce the depletion region and increase the conductivity of the In_2O_3 sensing layer. Further illumination of UV light will generate more electron–hole pairs in the conduction and valence bands, which decreases the resistance significantly.

Figure 4a exhibits the four I–V graphs, i.e., the measured dark current as a function of the applied bias voltage when the prototype operates at room temperature (RT, 23 °C), 40 °C, 80 °C, and 120 °C, respectively, when the bias increases from zero to a maximum of 11 V. The linear relationship indicates a good ohmic contact formed between the electrode and the semiconductor except for a slightly curved I–V graph at room temperature. At room temperature, the unpaired electrons are collected at the anode under bias, which leads to a decrease in the width of the depletion layer. Therefore, an increase in conductivity occurs with an increase in the bias voltage. As a comparison, at room temperature, the dark current is approximately 28 µA at 4 V bias and 87 µA at 11 V.

The increase in dark current at higher temperatures is directly attributed to material properties. An increase in the operating temperature up to 40 and 80 °C results in an increase in the dark current to 42 µA and 80 µA at 4 V bias, respectively. With the increase in operating temperature, a lower resistance takes place with the departure of the chemisorbed oxygen molecules from the surface. As shown in Figure 4a, when biased at 4 V, the dark currents are 28, 43, and 80 µA for room temperature, 40 °C and 80 °C operation, respectively. When the temperature is high enough and the sample surface is already maintained in a chemisorbed oxygen molecule free or rather a dynamic equilibrium state, as indicated by the I–V graphs at 80 °C and 120 °C in Figure 4a, further increase in the operating temperature does not boost the current. Similarly, as revealed in Figure 4b, the resistance of the device decreases with the increase in operating temperature. After the operating temperature reaches 80 °C, the resistance remains the same and no longer declines with the temperature rise.

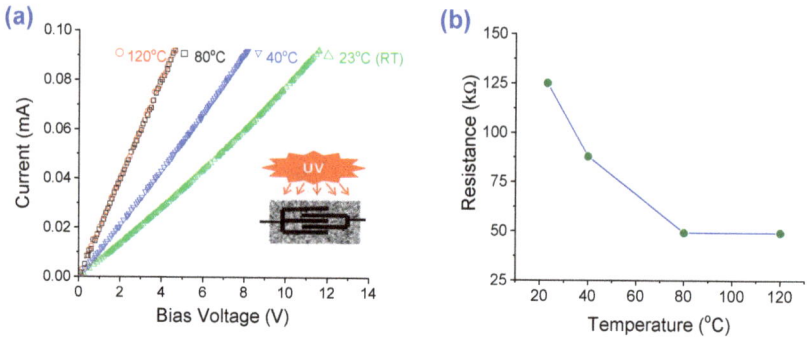

Figure 4. (a) Electrical current as a function of bias voltage at four different temperatures: room temperature (RT, 23 °C), 40 °C, 80 °C, and 120 °C. Inset: the schematic of the photodetector. (b) The temperature effect on the electrical resistance property of the In_2O_3 prototype.

To verify the authenticity of the photocurrent, the fabricated prototype was tested with UV light illuminations at different wavelengths for a period of 4 min and 50% duty, as shown in Figure 5. The device was operating at room temperature when the incident UV light intensity was 2 mW/cm^2 on the detector surface, which was controlled by varying the distance between the UV light source and the detector surface. As shown in Figure 5, the photocurrent rises and falls as the light is turned on and off when biased at 3.5 V. Besides, the device also displays good stability during the illumination cycle test. Under 250 nm, 300 nm, and 350 nm UV light illumination, the induced photocurrent is 0.39 mA, 0.48 mA, and 0.78 mA, respectively. At a longer wavelength, although the photocurrent is greater, the In_2O_3 prototype takes a longer time to respond. As indicated in Figure 5a,b, the measured photocurrent builds up quicker under 250 nm and 300 nm UV light, compared to 350 nm UV light illumination. On the other hand, noise under 250 nm UV light is greater than that under 300 nm and 350 nm UV light, since noise is greater at a short wavelength when the light intensity is fixed.

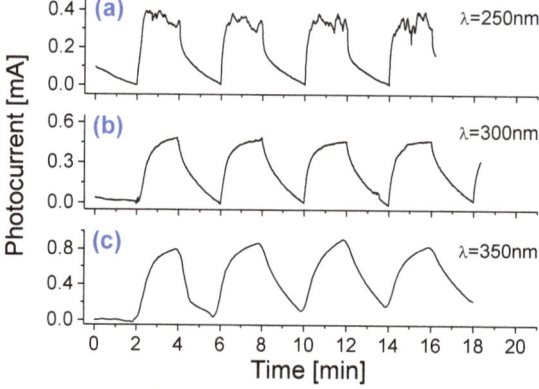

Figure 5. Photocurrent responding to 2 mW/cm^2 UV signals of a period of 4 min and a 50% duty cycle at (**a**) 250 nm, (**b**) 300 nm, and (**c**) 350 nm, respectively, when operating at room temperature and 3.5 V bias voltage.

Figure 6 shows the typical dependence of the photocurrent on UV light intensity under a bias of 3.5 V at room temperature. When the detector was exposed with a period of 4 min between the "switch-on" and "switch-off" to a 250 nm UV light at an intensity of 0.013 mW/cm², a well-defined light-induced photocurrent was obtainable with sharp edges. When the UV source was switched off, the photocurrent responded sharply down and then gradually decayed to zero. As expected, with the enhancement of illumination intensity, more electron-hole pairs are excited that contribute to the enriched photocurrent. As shown in Figure 6a–c, when exposed with the light intensities of 0.013, 0.17, and 2 mW/cm², the photocurrent measured are 0.10, 0.16 and 0.38 mA, respectively.

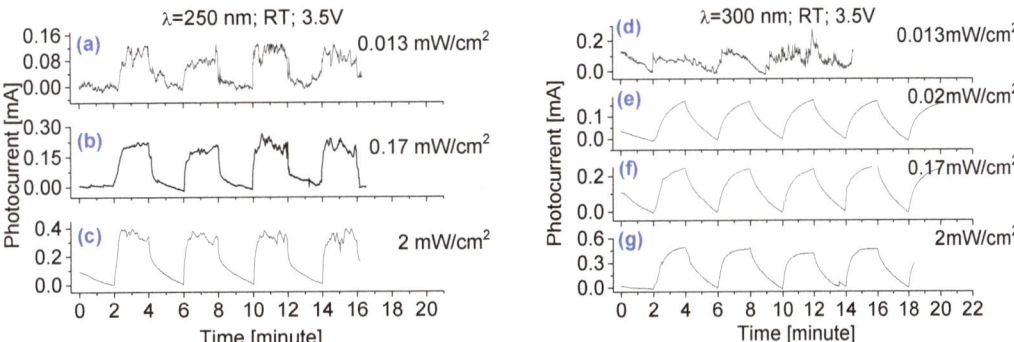

Figure 6. The measured photocurrents when illuminated with 250 nm of (**a**) 0.013 mW/cm², (**b**) 0.17 mW/cm², (**c**) 2 mW/cm², and when illuminated with 300 nm of (**d**) 0.013 mW/cm², (**e**) 0.02 mW/cm², (**f**) 0.17 mW/cm² and (**g**) 2 mW/cm², respectively. The device was operated at room temperature and biased at 3.5 V.

The photodetector's responsivity, R_λ, is defined as

$$R_\lambda = \frac{I_{ph}}{p_{light}}, \qquad (1)$$

where I_{ph} is the maximal photocurrent from the detector, and P_{light} is the total incident light power on the surface of the detector. It can be estimated based on the experimental data. Since the exposed surface area of the detector is 15 mm², 250 nm light power on the surface of the detector at 13 μW/cm² intensity is approximately 1.95 μW, leading to a responsivity of 51 A/W at the bias of 3.5 V. When the intensities were increased to 0.17 mW/cm², and 2 mW/cm², the responsivity dropped to 6.3 A/W and 1.3 A/W, respectively. The reduction in responsivity under greater UV light intensity could be caused by the saturation of the sensing layer.

The external quantum efficiency (EQE), η, can be expressed by

$$\eta = R_\lambda \frac{hC}{\lambda q} \qquad (2)$$

where h is Planck's constant, c the speed of light, q the electron charge, and λ the incident wavelength. A responsivity of 51 A/W at 250 nm corresponds to a quantum efficiency of 25.296% when operating at room temperature under an illumination intensity of 13 μW/cm² and a bias voltage of 3.5 V. We can also estimate the special detectivity D^* at this wavelength according to the following equation:

$$D^* = R_\lambda \sqrt{\frac{A}{2qI_d}}, \qquad (3)$$

where I_d = 0.025 mA is the dark current at room temperature (Figure 4a). The calculated special detectivity is 7.0×10^{12} Jones at a power density of 13 µW/cm².

To find the photodetector's response at different UV wavelengths, 300 nm UV light was also used with different intensities ranging from 0.013, 0.02, 0.17 to 2 mW/cm², as shown in Figure 6d–g, corresponding to photocurrents of 0.1, 0.17, 0.24 and 0.46 mA, respectively. Higher photocurrents were obtained, compared to 250 nm UV light at the same intensities. The rise time and recovery time, on the other hand, were slightly longer than those under 250 nm wavelength UV light. Because of the MSM geometry, charge carriers generated close to the surface and top region can contribute to a fast rise time. While charge carriers generated further away from the surface must take extra time to diffuse to the electrodes, leading to a longer rise time. We believe that at 300 nm wavelength, the photocurrent of the prototype was dominated by the photo-generated charge carriers inside the layer rather than that from the surface, which travels a longer average distance, hence a longer rise time, to generate the current. Our observations also indicate that the absorption of 250 nm could be greater than that of 300 nm UV light. However, its lower responsivity at relatively greater illumination light intensity could be caused by the possible saturation due to its higher absorption. A trade-off should be made between the photocurrent and response time as the In_2O_3 layer thickness is optimized. The response of a device to different light illumination may also rely on the bandgap width or cross-section parameter of the active layer.

Since the temperature effect of In_2O_3 UV photodetectors has not been studied by other groups, it is interesting to see how an In_2O_3 UV photodetector behaves at different operating temperatures. Therefore, additional experiments have also been carried out and the photocurrents responding to 250 nm and 300 nm UV illumination when operated at room temperature, 180 °C and 300 °C are displayed in Figure 7. Under 250 nm UV light, with the increase in operating temperature from room temperature to 180 °C, the yielded photocurrent increases from 0.2 mA to 0.6 mA. However, the contribution of thermal noise to the obtained photocurrent increases too, as shown in Figure 7a–c. When the operating temperature was further raised to 300 °C, the photocurrent was completely buried under noise.

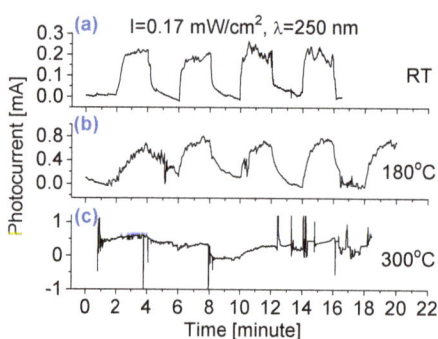
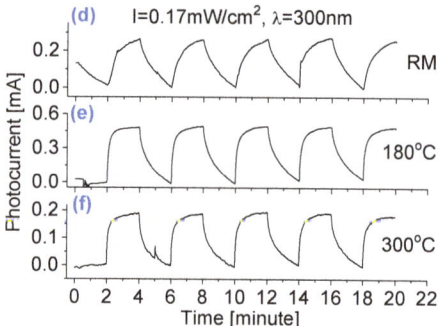

Figure 7. Photocurrents under UV light of 250 nm and 300 nm at an intensity of 0.17 mW/cm² with a period of 4 min at 50% duty cycle when operated at (**a,d**) room temperature (RT), (**b,e**) 180 °C, and (**c,f**) 300 °C. The bias voltage is 3.5 V.

As shown in Figure 7d–f, under 300 nm UV light of an intensity of 0.17 mW/cm², following the increase in temperature from 23 °C (RT) to 180 °C and then to 300 °C, the photocurrent was first increased from 0.28 mA to 0.47 mA and then decreased to 0.19 mA, giving a responsivity of 18 A/W at 180 °C and 7.3 A/W at 300 °C. The trend observed is similar to what was discovered with the In_2O_3-based oxygen sensors [28,29]. It is believed that with the temperature increase, apart from the resistance decrease due to the departure of adsorbed oxygen on the In_2O_3 surface, it is possible that the charge-carrier concentration increases, and the Debye length decreases and thus the sensitivity decreases [16].

The device also showed a faster response time and recovery time, and other good features such as excellent stability and repeatability, and a highly stable baseline when operating at both 180 and 300 °C. Upon receiving the UV light, the photocurrent reached its peak quickly and then dropped to zero gradually after the light source was switched off. The rise time, τ_r, which is defined as the interval from 10% to 90% of I_{max}, and fall time, τ_f, which is defined as the interval from 90% to 10% of I_{max}, are $\tau_r = 0.57$ s and $\tau_f = 1.4$ s, respectively, when the device is irradiated with UV pulses. The short rise time is attributed to the Pt nanoparticle decoration and the low capacitance of the MSM geometry, although its responsivity is limited due to the small effective absorbing area caused by the metallization of the electrodes.

In fact, the noble metal nanoparticles deposited on a metal oxide semiconductor surface act as electron-hole separators to promote the interfacial charge-transfer kinetics between the metal and semiconductor as reported in [14]. For example, the Ag-nanoparticle-decorated ZnO nanowire-based UV photodetector effectively eliminated the common persistent photoconductivity effect, leading to a significant decrease in the rise time as well as the dark current. However, the relatively long fall time is possible owing to the well-known persistent photoconductivity (PPC) effect in metal oxide nanostructure-based UV photodetectors, governed by the presence of numerous trapping states at crystal defects and the depth of the trap that prevent the photogenerated minority carriers from recombination and prolong the recovery time. As a comparison, Table 1 outlines the reported performance parameters of UV photodetectors operating at elevated temperatures.

Table 1. Comparison of the key device performance of the reported UV photodetectors.

Material	Peak λ (nm)	Responsivity (A/W)	Bias Voltage (V)	Response Time	Temperature Tested (°C)	Ref
SiC	250	0.1	10	18 s	180	[22]
Diamond	350	2.0	0	<1 s	RT	[23]
TiO$_2$	220	0.2	30	20 μs	RT	[24]
In$_2$O$_3$ quantum dots	290	70	5	-	RT	[11]
In$_2$O$_3$ nanosheets	254	0.172	-	-	RT	[12]
In$_2$O$_3$ nanoparticles	340	11	20	500 s	RT	[13]
In$_2$O$_3$ film	300	18	3.5	0.57 s	180	This work

4. Conclusions

In$_2$O$_3$ thin films were synthesized on Si substrates using the RF sputtering technique, followed by the surface decoration of Pt nanoparticles. The Raman spectroscopy measurements confirmed that the thin films consisted of crystalline cubic In$_2$O$_3$. Finally, an MSM UV photodetector was fabricated using a simple and low-cost approach. The synthesis of a thin-film layer used for the UV photodetector eliminated the tight control needed for a nanostructure counterpart. Furthermore, there was no requirement for an expansive photolithography process in the device fabrication. A high responsivity to a wide range of UV light from 250 to 350 nm was demonstrated for the fabricated photodetector, as well as good stability and repeatability. When operated with a 3.5 V bias at room temperature, a responsivity of 50 A/W was achieved at 250 nm wavelength. Following a slight increase in the operating temperature, the fabricated photodetector runs well with improved responsivity, stability, and repeatability, as well as a faster rise time and fall time. However, a further increase in operating temperature would eventually lead to high thermal noise that dominates the output. The obtained experimental data indicated that at under 300 nm radiation, the best performance was achieved at 180 °C although the device was still working properly at temperatures up to 300 °C. This characteristic is much better than that of UV photodetectors based on other materials that were reported to operate at high temperatures but were generally limited to less than 150 °C. This experimental investigation suggests that the simple approach described provides quick fabrication of high-performance UV photodetectors, which could possibly handle a high operating temperature of up to 300 °C.

Author Contributions: Designing and performing the experiments, as well as analyzing the data and writing this manuscript, E.Y.L., A.F.Z. and P.X.F. All authors have read and agreed to the published version of the manuscript.

Funding: NSF-CREST Center for Innovation, Research and Education in Environmental Nanotechnology (CIRE2N), grant HRD-1736093.

Data Availability Statement: Not applicable.

Acknowledgments: This work was financially supported by the NSF-CREST Center for Innovation, Research and Education in Environmental Nanotechnology (CIRE2N), grant HRD-1736093. A.F.Z. acknowledges the receipt of an IUP USRC research grant award.

Conflicts of Interest: The authors declare no conflict of interest.

References

1. Gu, F.; Li, C.; Han, D.; Wang, Z. Manipulating the defect structure (VO) of In_2O_3 nanoparticles for enhancement of formaldehyde detection. *ACS Appl. Mater. Interfaces* **2018**, *10*, 933–942. [CrossRef]
2. Zhao, S.; Shen, Y.; Zhou, P.; Hao, F.; Xu, X.; Gao, S.; Wei, D.; Ao, Y.; Shen, Y. Enhanced NO_2 sensing performance of ZnO nanowires functionalized with ultra-fine In2O3 nanoparticles. *Sens. Actuators B Chem.* **2020**, *308*, 127729. [CrossRef]
3. Wang, C.; Chen, D.; Jiao, X.; Chen, C. Lotus-root-like In_2O_3 nanostructures: Fabrication, characterization, and photoluminescence properties. *J. Phys. Chem. C* **2007**, *111*, 13398–13403. [CrossRef]
4. Staerz, A.; Weimar, U.; Barsan, N. Current state of knowledge on the metal oxide based gas sensing mechanism. *Sens. Actuators B Chem.* **2022**, *358*, 131531. [CrossRef]
5. Liu, W.; Xie, Y.; Chen, T.; Lu, Q.; Rehman, S.U.; Zhu, L. Rationally designed mesoporous In_2O_3 nanofibers functionalized Pt catalysts for high-performance acetone gas sensors. *Sens. Actuators B Chem.* **2019**, *298*, 126871. [CrossRef]
6. Yoon, Y.; Truong, P.L.; Lee, D.; Ko, S.H. Metal-oxide nanomaterials synthesis and applications in flexible and wearable sensors. *ACS Nanosci. Au* **2021**, *2*, 64–92. [CrossRef]
7. Bierwagen, O. Indium oxide—A transparent, wide-band gap semiconductor for (opto)electronic applications. *Semicond. Sci. Technol.* **2015**, *30*, 024001. [CrossRef]
8. Zhang, D.; Li, C.; Han, S.; Liu, X.; Tang, T.; Jin, W.; Zhou, C. Ultraviolet photodetection properties of indium oxide nanowires. *Appl. Phys. A* **2003**, *77*, 163–166. [CrossRef]
9. Meng, M.; Wu, X.; Ji, X.; Gan, Z.; Liu, L.; Shen, J.; Chu, P.K. Ultrahigh quantum efficiency photodetector and ultrafast reversible surface wettability transition of square In_2O_3 nanowires. *Nano Res.* **2017**, *10*, 2772–2781. [CrossRef]
10. Meng, M.; Yang, L.; Wu, X.; Gan, Z.; Pan, W.; Liu, K.; Li, C.; Qin, N.; Li, J. Boosted photoelectrochemical performance of In_2O_3 nanowires via modulating oxygen vacancies on crystal facets. *J. Alloys Compd.* **2020**, *845*, 156311. [CrossRef]
11. Rajamani, S.; Arora, K.; Konakov, A.; Belov, A.; Korolev, D.; Nikolskaya, A.; Mikhaylov, A.; Surodin, S.; Kryukov, R.; Nikolitchev, D.; et al. Deep UV narrow-band photodetector based on ion beam synthesized indium oxide quantum dots in Al_2O_3 matrix. *Nanotechnology* **2018**, *29*, 305603. [CrossRef] [PubMed]
12. Zhang, M.; Yu, H.; Li, H.; Jiang, Y.; Qu, L.; Wang, Y.; Gao, F.; Feng, W. Ultrathin In_2O_3 Nanosheets toward High Responsivity and Rejection Ratio Visible-Blind UV Photodetection. *Small* **2023**, *19*, 2205623. [CrossRef] [PubMed]
13. Shao, D.; Qin, L.; Sawyer, S. Near ultraviolet photodetector fabricated from polyvinyl-alcohol coated In_2O_3 nanoparticles. *Appl. Surf. Sci.* **2012**, *261*, 123–127. [CrossRef]
14. Tzeng, S.K.; Hon, M.H.; Leu, C. Improving the performance of a zinc oxide nanowire ultraviolet photodetector by adding silver nanoparticles. *J. Electrochem. Soc.* **2012**, *159*, H440. [CrossRef]
15. Li, D.; Sun, X.; Song, H.; Li, Z.; Chen, Y.; Jiang, H.; Miao, G. Realization of a high-performance GaN UV detector by nanoplasmonic enhancement. *Adv. Mater.* **2012**, *24*, 845–849. [CrossRef]
16. Zhou, A.F.; Velázquez, R.; Wang, X.; Feng, P.X. Nanoplasmonic 1D diamond UV photodetectors with high performance. *ACS Appl. Mater. Interfaces* **2019**, *11*, 38068–38074. [CrossRef]
17. Chu, J.; Peng, X.; Sajjad, M.; Yang, B.; Feng, P.X. Nanostructures and sensing properties of ZnO prepared using normal and oblique angle deposition techniques. *Thin Solid Film.* **2012**, *520*, 3493–3498. [CrossRef]
18. Zhou, A.F.; Wang, X.; Pacheco, E.; Feng, P.X. Ultrananocrystalline Diamond Nanowires: Fabrication, Characterization, and Sensor Applications. *Materials* **2021**, *14*, 661. [CrossRef]
19. Aldalbahi, A.; Velázquez, R.; Zhou, A.F.; Rahaman, M.; Feng, P.X. Bandgap-Tuned 2D Boron Nitride/Tungsten Nitride Nanocomposites for Development of High-Performance Deep Ultraviolet Selective Photodetectors. *Nanomaterials* **2020**, *10*, 1433. [CrossRef]
20. Kim, S.S.; Park, J.Y.; Choi, S.W.; Kim, H.S.; Na, H.G.; Yang, J.C.; Kim, H.W. Significant enhancement of the sensing characteristics of In_2O_3 nanowires by functionalization with Pt nanoparticles. *Nanotechnology* **2010**, *21*, 415502. [CrossRef]

21. Zhang, X.; Liu, Q.; Liu, B.; Yang, W.; Li, J.; Niu, P.; Jiang, X. Giant UV photoresponse of a GaN nanowire photodetector through effective Pt nanoparticle coupling. *J. Mater. Chem. C* **2017**, *5*, 4319–4326. [CrossRef]
22. Aldalbahi, A.; Li, E.; Rivera, M.; Velazquez, R.; Altalhi, T.; Peng, X.; Feng, P.X. A new approach for fabrications of SiC based photodetectors. *Sci. Rep.* **2016**, *6*, 1–10. [CrossRef]
23. Pacheco, E.; Zhou, B.; Aldalbahi, A.; Zhou, A.F.; Feng, P.X. Zero-biased and visible-blind UV photodetectors based on nitrogen-doped ultrananocrystalline diamond nanowires. *Ceram. Int.* **2022**, *48*, 3757–3761. [CrossRef]
24. Liu, Z.; Li, F.; Li, S.; Hu, C.; Wang, W.; Wang, F.; Lin, F.; Wang, H. Fabrication of UV photodetector on TiO2/diamond film. *Sci. Rep.* **2015**, *5*, 14420. [CrossRef] [PubMed]
25. Feng, P.X.; Aldalbahi, A. A compact design of a characterization station for far UV photodetectors. *Rev. Sci. Instrum.* **2018**, *89*, 015001. [CrossRef] [PubMed]
26. Eranna, G. *Metal Oxide Nanostructures as Gas Sensing Devices*; CRC Press: Boca Raton, FL, USA, 2011; pp. 1–336.
27. Rivera, M.; Velázquez, R.; Aldalbahi, A.; Zhou, A.F.; Feng, P.X. UV photodetector based on energy bandgap shifted hexagonal boron nitride nanosheets for high-temperature environments. *J. Phys. D Appl. Phys.* **2018**, *51*, 045102. [CrossRef]
28. Neri, G.; Bonavita, A.; Micali, G.; Rizzo, G.; Pinna, N.; Niederberger, M. In_2O_3 and $Pt-In_2O_3$ nanopowders for low temperature oxygen sensors. *Sens. Actuators B Chem.* **2007**, *127*, 455–462. [CrossRef]
29. Rumyantseva, M.N.; Makeeva, E.A.; Badalyan, S.M.; Zhukova, A.A.; Gaskov, A.M. Nanocrystalline SnO_2 and In_2O_3 as materials for gas sensors: The relationship between microstructure and oxygen chemisorption. *Thin Solid Film.* **2009**, *518*, 1283–1288. [CrossRef]

Disclaimer/Publisher's Note: The statements, opinions and data contained in all publications are solely those of the individual author(s) and contributor(s) and not of MDPI and/or the editor(s). MDPI and/or the editor(s) disclaim responsibility for any injury to people or property resulting from any ideas, methods, instructions or products referred to in the content.

Article

Colorimetric Plasmonic Hydrogen Gas Sensor Based on One-Dimensional Nano-Gratings

Majid Zarei [1], Seyedeh M. Hamidi [1,*] and K. -W. -A. Chee [2,3,4,5,*]

1. Magneto-plasmonic Laboratory, Laser and Plasma Research Institute, Shahid Beheshti University, Tehran 1983969411, Iran
2. National Education Center for Semiconductor Technology, Kyungpook National University, Daegu 41566, Republic of Korea
3. Institute of Semiconductor Fusion Technology, Kyungpook National University, Daegu 41566, Republic of Korea
4. Department of Electronics and Electrical Engineering, Kyungpook National University, Daegu 41566, Republic of Korea
5. School of Electronics Engineering, College of IT Engineering, Kyungpook National University, Daegu 41566, Republic of Korea
* Correspondence: m_hamidi@sbu.ac.ir (S.M.H.); aghjuee@knu.ac.kr (K.-W.-A.C.)

Abstract: Plasmonic hydrogen gas sensors have become widely used in recent years due to their low cost, reliability, safety, and measurement accuracy. In this paper, we designed, optimized, and fabricated a palladium (Pd)-coated nano-grating-based plasmonic hydrogen gas sensor; and investigated using the finite-difference time-domain method and experimental spectral reflectance measurements, the calibrated effects of hydrogen gas exposure on the mechano-optical properties of the Pd sensing layer. The nanostructures were fabricated using DC sputter deposition onto a one-dimensional nano-grating optimized with a thin-film gold buffer to extend the optical response dynamic range and performance stability; the color change sensitivity of the Pd surface layer was demonstrated for hydrogen gas concentrations as low as 0.5 vol.%, up to 4 vol.%, based on the resonance wavelength shift within the visible band corresponding to the reversible phase transformation. Visual color change detection of even the smallest hydrogen concentrations indicated the high sensitivity of the gas sensor. Our technique has potential for application to high-accuracy portable plasmonic sensors compatible with biochemical sensing with smartphones.

Keywords: lattice structure; phase change; plasmonic sensing; nano-gratings; thin films; surface plasmon resonance; polarization

Citation: Zarei, M.; Hamidi, S.M.; Chee, K.-W.-A. Colorimetric Plasmonic Hydrogen Gas Sensor Based on One-Dimensional Nano-Gratings. Crystals 2023, 14, 363. https://doi.org/10.3390/cryst 13020363

Academic Editors: Andrew F. Zhou and Peter X. Feng

Received: 18 January 2023
Revised: 11 February 2023
Accepted: 14 February 2023
Published: 20 February 2023

Copyright: © 2024 by the authors. Licensee MDPI, Basel, Switzerland. This article is an open access article distributed under the terms and conditions of the Creative Commons Attribution (CC BY) license (https://creativecommons.org/licenses/by/4.0/).

1. Introduction

Over the last few decades, the use of hydrogen gas has proliferated due to its numerous applications in industry, medicine, and everyday life as a chemical reactant and an energy carrier [1,2]. Hydrogen is the most abundant chemical element in the universe [3,4], and is considered to be an excellent alternative to fossil fuels as a clean and renewable fuel [5–7]. Hydrogen gas is currently being used as a fuel in the aerospace sector [4]. Several chemical compounds, including ammonia, are synthesized using hydrogen; and some drugs, such as hydrogen peroxide, are manufactured using hydrogen [4]. Other applications of hydrogen include the oil and gas industry, petrochemicals, the food industry, fertilizer manufacturing, welding, and metallurgy [8]. The increased industrial use of hydrogen in combination with its unusual properties, such as very low density, lightness, and high flammability (with 4 vol.% as the explosive lower limit in air), call for the need to monitor and carefully control its concentration [9,10]. Furthermore, hydrogen is a completely odorless, colorless, and tasteless gas, and so precise and sensitive sensors are required for detection and measurement in routine and industrial applications [4,5,11].

Thus, in the past few years, several attempts have been made to develop inexpensive and precise hydrogen gas sensors, employing various electrical, optical, mechanical, thermochemical, acoustic, and catalytic strategies [4,7,12]. Particularly, optical, or colorimetric sensing can be devised with a material that changes color in a visually discernible manner corresponding to the change in the lattice structure upon reaction with a target gas. No electrical power is required, and the sensor can operate at room temperature, with a high resistance to electromagnetic interference. As all-optical gas detectors do not generate electrical sparks, they do not pose a risk of explosion [2,13]. Furthermore, optical sensors provide higher accuracy alongside a lower limit of detection, shorter response times, as well as good stability and recyclability (sensing mechanisms based on reversible chemical reactions) [6,10,14]. Furthermore, in the presence of hydrogen, optical gas sensors based on surface plasmon resonance (SPR) are more sensitive thanks to the localized field [5]. The plasmonic effect due to collective oscillations of conduction electrons excited by an external electromagnetic radiation serves as the mechanism in SPR sensors, which in turn depends on the morphology of the metal-dielectric interface. The optical response can be induced through localized surface plasmon resonance (LSPR) or propagating surface plasmon polaritons (SPP). In fact, SPR sensors have attracted much attention from researchers in recent years because of their ease of construction, affordability, and portability [15–17]. More specifically, nano-grating structures have attracted the most attention due to their cost-effectiveness, high sensitivity, and tunable resonance wavelength, and consequently, they have been widely adopted in scientific research and industry [18–20]. The coupling of LSPR-enhanced electric fields significantly enhances the sensitivity performance, and, indeed, subwavelength interactions of nano-structures supporting the SPP and LSPR modes can refine the spectral transmission characteristics.

Usually, hydrogen sensing palladium (Pd) or its alloy plays a major role as a catalyst layer in plasmonic sensors [21,22] due to its high absorption capability (almost three orders of magnitude greater than its volume), high selectivity, as well as the adjustable LSPR spectrum [11,23,24]. Upon hydrogen adsorption, the metal Pd transforms into Pd hydride (PdH_x), so that the electrical conductance and optical properties significantly change, especially the dielectric constant in the visible region [25–27]. The optical response can be dichotomized into three distinct phases, namely the α-phase, mixed phase, and the fully formed metal-hydride β-phase. These phases correspond to important response regions for SPR hydrogen gas sensing. The spectral extinction/transmission/reflection magnitudes, peak/dip position, and full width at half maximum of the peak/dip can characterize the reversible chemical reactions of the hydrogen with the metal sensing layer. While a lot of work has been carried out on hydrogen gas sensing so far, in which plasmonic hydrogen sensor schemes have also played an important role, a very limited number of these works have been about hydrogen gas sensor design based on the colorimetric method and that use practical plasmonic structures that have been experimentally demonstrated. To address this technology gap, we introduced a novel, easy-to-fabricate, low-cost and portable sensor platform that works based on color changes of the sensing layer (Pd) surface. Fabrication of the sensors tend to involve spin coating, RF magnetron sputtering, photochemical deposition, and solution-stirring methods. Conversely, for successful commercial exploitation, a facile and economical fabrication procedure is required.

Colorimetric sensors utilizing surface plasmon resonance offer a straightforward yet sensitive method of gas detection that has a simpler structure and better stability than other colorimetric methods [12,28,29]. In addition, the structural and optical parameters can control the plasmonic color. Common optical hydrogen gas sensors use optical fibers including the distributed Bragg grating and tapered single-mode fiber. By applying coatings at the optical fiber end comprising a silver layer, a thin film of chemochromic material, and a catalytic layer of Pd, all-optical guided-wave hydrogen gas detectors can be realized [30]. Serhatlioglu et al. [31] numerically studied the effect of varying the dielectric layer thickness in the metal-insulator-metal interference structure on the spectral absorption characteristics upon hydrogen exposure that changes the surface color. In another recent work, Duan and

Liu [32] made scanning plasmonic color displays possible using aluminum nanoparticles as plasmonic pixels that are switched on/off by hydrogen atom adsorption/desorption on a magnesium screen. Nugroho et al. [33] inversely designed using a particle swarm optimization algorithm, a plasmonic metasurface based on a periodic array of Pd nanoparticles that led to ultrasensitive optical hydrogen detection. Luong et al. [34] developed an optical hydrogen gas sensor platform based on Pd bilayer plasmonic nano-lattices that exhibited an extended optical response range and enhanced sensitivity for hydrogen gas detection, as well as an order of magnitude higher optical response speed in the low hydrogen pressure regime. Hosseini and Ranjar [35] demonstrated a linear blueshift of the LSPR for optical hydrogen sensing using sputter-deposited Pd film on flame-synthesized nanostructured MoO_3 films.

In this study, we present a colorimetric plasmonic hydrogen sensor based on a one-dimensional nano-grating and thin-film Pd. Since the nano-gratings used in this sensor design were extracted from a digital optical disk data storage format (Digital Video Disk or DVD), the sensor inherits attractive technology features of being lightweight, inexpensive, and very accessible. To enhance the structural quality of the sensor and to improve the gas sensing performance, we also explored the advantages of having sputter-deposited gold (Au) as a noble metal buffer layer. It is important to correlate the hydrogenation/dehydrogenation process with property modifications. When the surface dissociated hydrogen atoms diffuse into the Pd lattice, phase transitions lead to color changes according to the hydrogen concentration. Hence, the concentration of hydrogen gas in the surrounding environment can be determined by measuring the relative intensity contrast and resonance wavelength shift from the reflectance spectrum. Sensors developed in this way are low-cost, ocular-safe, portable, and highly sensitive. When paired with smart systems such as smartphones utilizing machine learning and digital technologies for environmental monitoring and mobile health, powerful, cost-effective analytical devices can be realized offering direct and convenient diagnostic solutions.

2. Materials and Methods

A one-dimensional nano-grating structure extracted from a DVD was used as a substrate (see Figure 1a). In contrast to prisms or optical waveguides, these nano-gratings are small, lightweight, compact, readily available, and inexpensive, allowing them to be used in portable plasmonic systems. First, the protective layer of the optical disk must be removed to extract the gratings, as shown in Figure 1a. We removed the pigments from the surface using ethanol without damage to the grating structure. Thereafter, we washed the sample in distilled water followed by drying. Pd (40-nm thick) and Au (35-nm thick) layers were then deposited onto the nano-grating substrate using the DC sputter method. To protect the sensing Pd layer from damaging ambient gasses and deactivation, the sample was placed in a vacuum chamber. Figure 1b,c illustrates, respectively, the optical configuration in the vacuum chamber and the overall experimental setup. The cylinder-shaped steel chamber had a window through which the sample was connected from the back by an index-matched gel. An index-matching medium is essential to allow efficient radiative coupling. A mass flow controller (MFC) was used to control the gas flow of hydrogen and argon into the chamber so that the gas concentrations could be precisely regulated and monitored.

For this study, the optical apparatus was based on a versatile micro-reflectivity setup to investigate the optical properties of hydrogen adsorbent surfaces. Unpolarized light from a broadband halogen lamp (Thorlabs, Newton, NJ, USA, OSL2 fiber optic illuminator) was polarized with a Glan-Taylor prism (GT10-A) to produce TM polarization. The polarized light was then focused onto the sample using an objective lens (Thorlabs), and the reflected light was collected and coupled into a UV-VIS spectrometer (Ocean Optics, Ostfildern, Germany, NANOCALC-XR) using an achromatic lens (Thorlabs). The intensity reading of the broadband light source below 420 nm and above 730 nm was almost zero (see Figure 1b).

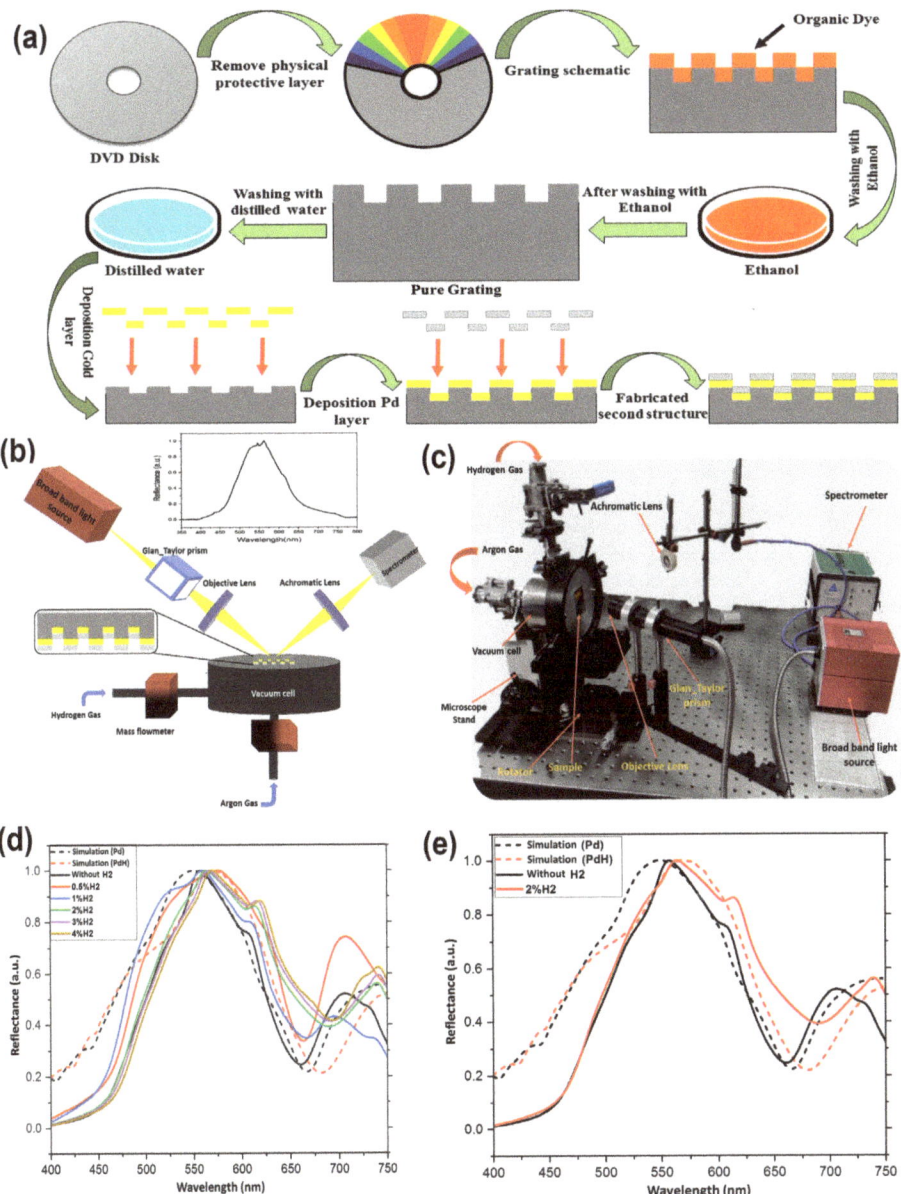

Figure 1. Fabrication, experimental configuration, and measurements. (**a**) Fabrication process of the plasmonic hydrogen gas sensor (with Au buffer layer). (**b**) Schematic of the optical setup in the vacuum chamber for reflectance measurements. The inset shows the spectrum of the broadband light source measured using the same spectrometer. (**c**) Photograph of experimental setup including indicators of the hardware components used. (**d**) FDTD simulated and experimental reflectance spectra from the nano-grating/Pd structure prototype without or with hydrogen exposure at different volumetric concentrations (0 to 4 vol.%), and (**e**) experimental and theoretical comparison of the reflectance spectra in the presence or absence of hydrogen, from the DVD-Pd nanostructure film. The incidence angle was 44°.

The sensing Pd crystal has a face-centered cubic (FCC) lattice structure. When the hydrogen atoms are adsorbed on the surface, they occupy the interstitial (octahedral) sites of the FCC lattice [36]. The high selectivity toward hydrogen is because of the higher hydrogen permeability in the Pd lattice compared to that of other gasses. In general, hydrogen penetration into the Pd lattice leads to the formation of metal hydride (PdH$_x$) in different phases and correspondingly alters its structure and properties. At low hydrogen gas concentrations, the α-phase is formed of the Pd lattice and there is no significant change to the lattice structure [11,37]. As the hydrogen gas concentration increases further, patches of the metal hydride can be formed as the so-called β-phase in the Pd lattice. Hence, the α-phase and β-phase will be simultaneously present in the FCC lattice. The metal hydride's behavior is dependent on the temperature [24]. With increasing penetration of hydrogen, the Pd lattice transforms from the mixed α- and β-phases into the pure β-phase and the Fermi level shifts due to a significant volume expansion. Furthermore, the dielectric response function of the Pd hydride will be completely modified from that of pure Pd, with significant changes to the real and imaginary parts of the Pd hydride's dielectric response function [21,25]. This phase transition in the sensing layer is therefore responsible for modifications of the electrical and optical properties, such that the modulating surface color can provide opportunities for remote sensing.

We utilized the Lumerical software and the finite-difference time-domain (FDTD) method to compute and investigate the optical characteristics of the sensor structures employed. We established the geometric specifications of the nano-grating, defining one period as a unit cell with three-dimensional periodic boundary conditions. A perfectly matched layer (PML) boundary condition was applied on the structure perimeter. Using a frequency-domain field and power monitor in the simulation environment, we computed the reflection spectra from the structures. To achieve acceptable convergence characteristics, we used a mesh size of 2 nm × 2 nm × 2 nm. We modeled the thin BK7 borosilicate glass of the vacuum chamber window as a substrate in the simulation domain. A Pd metal layer directly fabricated on a one-dimensional nano-grating forms the basis of the initial structure. Therefore, we determined the optimal Pd layer thickness to obtain the most suitable visible resonance wavelength, and then collected the spectral reflectance for the range of angles of incidence from 40° to 60°. Then, by using reputable sources for the refractive index and extinction coefficient for the Pd hydride [38], we compared its reflectance spectrum with that of pure Pd (see Figure 1d). To excite the surface plasmons, we used a one-dimensional nano-grating that equates the incident light wave vector (K_{light}) with the plasmon wave vector (K_{SPP}). Using atomic force microscopy (AFM) imaging, we obtained measurements of the periodicity, height, and the widths of the nano-grating. The nano-grating geometry will have an influence on the sensor performance, and one of the most important parameters is the periodicity. The relationship with nano-grating geometry when the surface plasmon polaritons are excited at a specified wavelength λ can be expressed as [39]:

$$\frac{2\pi}{\lambda}\sin\theta + m\frac{2\pi}{P} = \pm\frac{2\pi}{\lambda}\sqrt{\frac{\varepsilon_m n_d^2}{\varepsilon_m + n_d^2}} \qquad (1)$$

where $m = 0, \pm 1, \pm 2, \pm 3, \ldots$ is the diffraction order, P is the grating period and ε_m is the dielectric constant of the metal. θ is the light incidence angle on the grating and n_d is the refractive index of the dielectric medium coating the grating. The resonance wavelength redshifts with periodicity. The height and width of the nano-grating are the other parameters influencing the resonance wavelength and shape of the reflectance spectrum, in turn governing the sensor performance. When the height or width of the nano-grating increases, the resonance wavelength redshifts or blueshifts, respectively [39,40]. For the range of hydrogen concentrations studied, the optical simulations results, accounting for the properties of the Pd sensing layer (and Au buffer layer), show that in terms of the periodicity and height of the nano-grating, the structure parameters in the optical disk compare well with that required for placement of the resonance wavelength in the optical regime. We therefore find that the nano-grating geometry extracted from the DVD is most suitable to establish our colorimetric scale, thus combining a high sensitivity performance with the unique benefits of the ease of availability of substrate materials, as well as the low-cost and low-complexity fabrication of the sensor, and portability.

After the sample preparation, we placed the sample inside the vacuum chamber, which operated with a turbopump until a 10^{-6} mbar pressure was reached. This allowed us to analyze the effect of varying quantities of hydrogen adsorption on the structure more precisely. By using MFCs, a steady stream of argon and hydrogen gas was introduced into the chamber at the targeted volumetric concentrations. The sample was affixed to a microscope stand and attached to the center of a rotator onto which an optical system was mounted (concentric with a microscope stand). For the aforementioned angles, we recorded the spectral reflectance at room temperature with and without hydrogen exposure. We then pumped the sample chamber to a high vacuum and flushed the sample with pure argon gas after each loading of a hydrogen concentration for 5 cycles to ensure minimization of the hysteresis effect. Adsorption/desorption hysteresis and phase transformation properties can otherwise be very common in metal-hydride configurations. The experiments were repeated at least three times under the same conditions to confirm good reproducibility of the measurements. That the readings were essentially identical also ensured the repeatability and reliability of the results. By evaluating and comparing the theoretical and experimental reflection spectra, we determined 44° as the ideal angle for measurement due to the resonance depth and its placement inside the visible range (Figure 1d,e). For this structure, Figure 1e demonstrates close agreement of the simulated and experimental reflectance spectra and plasmonic resonance wavelength characteristics. Notably, the feature reflectance dip was caused by the activation of the surface plasmon at this wavelength since the light strikes the nano-grating from behind and there is no waveguide mode in this case [41].

3. Results and Discussion

Our goal was to determine the amount of hydrogen gas present in the environment by calibrating the number of color changes (RGB) in the sensing layer (Pd). Additionally, we do not alloy the sensing metal with other metals or use a protective layer because that is known to diminish the color variation of the sensing layer and more complex fabrication processes are mandatory for alloy nanostructures. The resonance wavelength for the as-fabricated nano-grating/Pd structure was approximately 660 nm without hydrogen exposure. As shown in Figure 1d,e, peak broadening and redshift in the resonance wavelength were observed upon hydrogen gas exposure. In principle, the amount of hydrogen adsorbed on the Pd sensing material leads to a corresponding proportionate change in the physical and optical properties, thereby affecting the reflectance spectra and the surface color. When hydrogen adsorbs on the Pd layer, the metal hydride (PdH_x) forms. The conversion of Pd into PdH_x upon hydrogen gas exposure is accompanied by a refractive index change in the sensing layer. Hence, the redshift in the resonance wavelength, where the absorption is maximum, indicates the refractive index change of the sensing material. For the hydrogen concentrations ranging from 0.5 to 4 vol.%, the resonance wavelength was ideally located within the visible range (see Figure 1d–e), so that the prototyped platform is feasible for colorimetric sensing applications. However, owing to the cyclic adsorption and desorption of hydrogen gas and the consequent expansion and contraction of the sensing metal lattice, damage is caused to the surface of the Pd sensing material and cracking can occur when a single layer is used, thereby degrading the nanostructure's mechanical stability. Furthermore, due to rapid saturation by gas adsorption, sensors based on a single metal layer operate well only within a narrow range of gas pressures [10,18]. To resolve the abovementioned challenges, Au was deposited as a buffer layer between the nano-grating and the Pd layer to enhance surface adhesion, lower pressure on the Pd sensing layer, and further stabilize the mechanical integrity. The addition of the Au layer may also enhance the measurement of rise and fall times [10,37,42].

In our simulation results (Figure 2a), we determined the appropriate thicknesses of the Pd and Au layers and identified the optimal resonance wavelength from the reflection spectrum for incidence angles between 40° to 60°. The resonance wavelength was approximately 670 nm without hydrogen exposure. When loading the appropriate amounts of hydrogen gas concentration, the experimental spectral reflectance showed similar characteristics to that in the simulations (Figure 2a,b). As can be seen, the inclusion of the Au layer caused a redshift in the resonance wavelength, while the resonance wavelength remained in the visible range at the volumetric concentrations under study, allowing the device to be suited for colorimetric sensing operations.

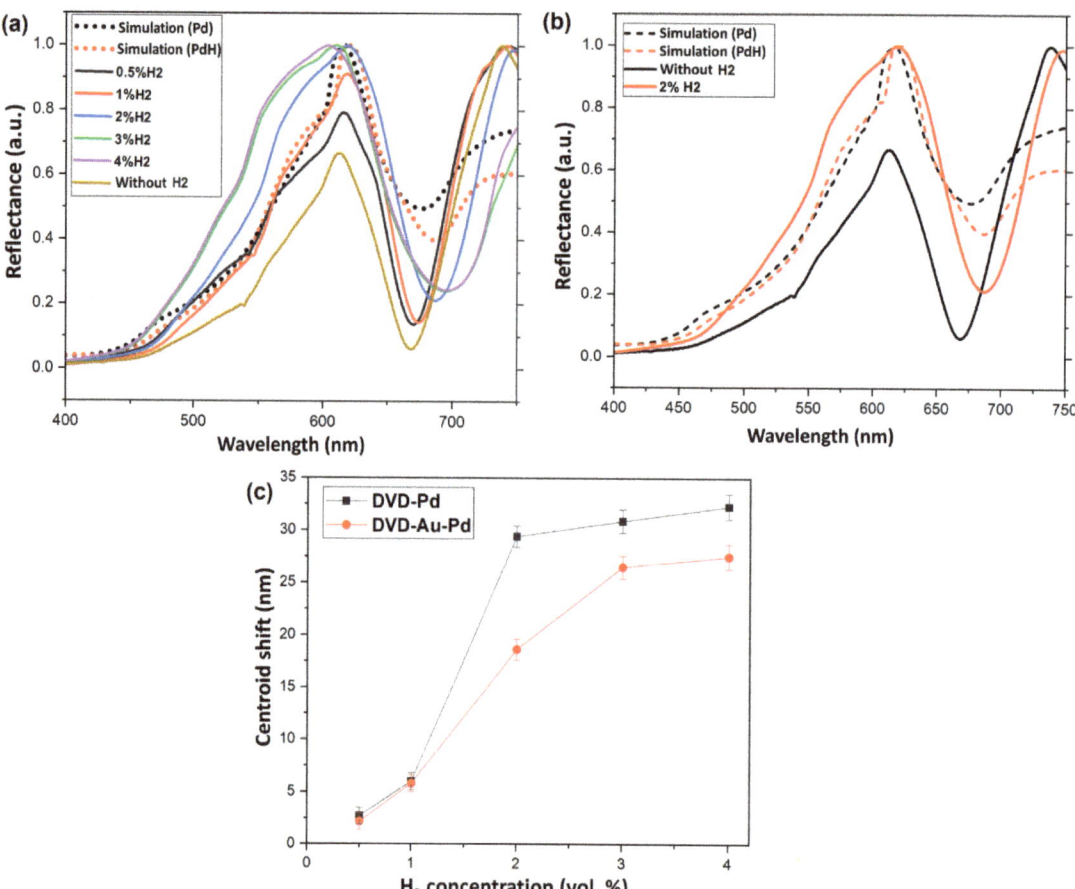

Figure 2. Optical measurements from the nano-grating/Au/Pd structure prototype. (**a**) FDTD simulations and experimental reflectance spectra without or with hydrogen gas exposure at different concentrations (0 to 4 vol.%). (**b**) Theoretical and experimental reflectance spectra in the presence or absence of hydrogen gas. (**c**) Centroid wavelength shift resulting from hydrogen adsorption with respect to volumetric concentration, with a measurement relative error estimated at ± 0.8 nm. Data from the nano-grating/Pd structure are also included for comparison. The incidence angle was 44°.

The spectral shift of the resonance wavelength with hydrogen concentration is shown in Figure 2c for the nano-grating/Pd and nano-grating/Au/Pd structures. We find that when compared to the nano-grating/Au/Pd structure, the nano-grating/Pd structure without the Au buffer exhibits a larger wavelength shift when exposed to hydrogen. Moreover, due to modifications of the Pd lattice constant in either structure by hydrogen adsorption, the phase transition from the α-phase to the mixed α- and β-phases is nonlinear, exhibiting a large, abrupt transformation in phase above a threshold volumetric concentration.

As shown in Figure 3a, the secondary electron image confirms the periodicity and uniformity of the nano-gratings. The sensitivity performance is governed by the structural configurations of the prototyped nano-grating/Pd and nano-grating/Au/Pd designs (Figure 3b). Compared to the nano-grating/Pd structure, the Pd phase transition for the nano-grating/Au/Pd structure is slower with respect to the amount of hydrogen adsorption on the surface sensing layer, resulting in an improved dynamic range and performance stability (see Figure 2c). Whereas the sensing response for the nano-grating/Pd structure rapidly saturates after the transition from the α- to β-phase of the sensing material, the saturation of the sensitivity performance is deferred to higher hydrogen concentrations for the nano-grating/Au/Pd structure due to the slower phase transition, thus extending the dynamic range. Sensors generally perform in a limited pressure range; these sensors usually exhibit a large response around the hydride formation pressure. Without the Au buffer layer, the expansion and contraction of the Pd lattice will be relatively large and rapid, thus directly affecting the stability of the Pd sensing layer. Conversely, adding the Au buffer layer between the Pd and nano-grating layer not only enhances the surface adhesion between the sensing and substrate material, but also reduces the pressure in the Pd lattice; this is because the Au layer acts as a diffusion barrier, preventing hydrogen atoms from diffusing too deeply into the Pd layer and causing excessive lattice expansion. The slower and milder Pd lattice expansions/contractions with respect to the hydrogen adsorption/desorption underlie an enhanced mechanical stability of the sensing layer.

Figure 3c exhibits the visual color changes of the Pd layer according to the CIE 1931 color space diagram for the various volumetric concentrations of hydrogen. The phase transition occurred between 1 to 2 vol.% and 2 to 3 vol.% for the nano-grating/Pd and nano-grating/Au/Pd nanostructure films, respectively. We can utilize these gasochromic results to operate an efficient system platform for domestic and industrial sensing applications. This platform may include a smartphone that can accurately and directly measure the color changes of the sensing layer surface in a continuous fashion through real-time image capture and processing, adopting high-performance industrial machine vision standards involving specialized image processing software. As a result, we will have a compact and accurate colorimetric plasmonic hydrogen sensing platform suitable for any form factor. While our focus has been to investigate the color change sensitivity of the Pd surface layer in response to the different gas concentrations up to 4 vol.%, which is the flammability limit at normal atmospheric pressure in oxygen, our work has so far demonstrated sensitivity to hydrogen concentrations as low as 0.5 vol.%. The corresponding color changes extending through to this low gas concentration are completely noticeable, and thus we have numerically and experimentally demonstrated a more sensitive and accurate plasmonic sensor than that reported in other papers that also use the colorimetric method [43,44]. Color changes on the sensing layer are virtually insignificant at lower hydrogen concentrations, and they either border on or exceed the threshold for human visual detection. Nevertheless, we can still measure the amount of resonance wavelength shift (although very small) for hydrogen concentrations as low as 0.2 vol.% using a spectrometer. However, it is usually not practical or necessary to detect such trace amounts of hydrogen in routine domestic and industrial screening processes.

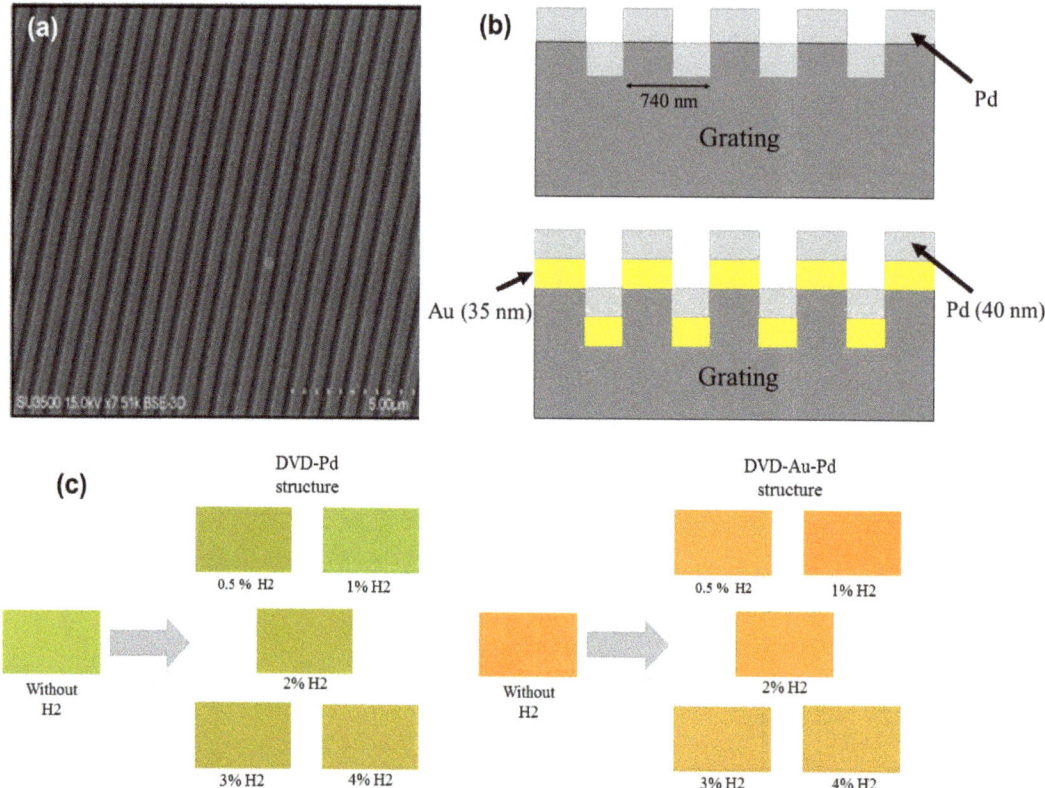

Figure 3. Physical and color structure of the prototyped sensor. (**a**) Scanning electron microscope image of nano-grating. (**b**) Schematic cross-sections of the nano-grating/Pd and nano-grating/Au/Pd structures. (**c**) MATLAB simulations of color changes of the Pd layer due to hydrogen adsorption at different concentrations (vol.%) according to the CIE 1931 color space diagram for the (left): DVD/Pd structure and (right): DVD/Au/Pd structure.

4. Conclusions

The requirement to fabricate a hydrogen gas sensor is crucial. Plasmonic sensors have received more attention than electrical, mechanical, thermochemical, acoustic and chemical sensors because of their manifold advantages, such as being more sensitive, reliable and safer without electrical contacts. Here, we designed, optimized, fabricated and demonstrated a plasmonic structure consisting of a one-dimensional nano-grating for excitation of the surface plasmons, using Pd as the sensing metal. For sensitivity performance optimization, an Au buffer layer was incorporated between the Pd and the nano-grating. By fine-tuning the thickness of the active nanomaterial layers, the optical response to hydrogen adsorption at full concentration was located within the visible region. The simulations and experiments demonstrated nearly similar behaviors in the reflectance spectra and plasmonic resonance wavelengths, with or without ambient hydrogen gas. We calculated the optical response based on the color changes of the plasmonic structures and demonstrated the feasibility of the gasochromic and optical hydrogen gas sensor based on the metal-coated one-dimensional nano-grating. This work paves a route for portable and accurate colorimetric plasmonic hydrogen sensors fabricated using a facile thin-film DC sputter deposition approach. Nevertheless, challenges will include the need to record high-quality and high-resolution images in combination with powerful and advanced programming logic devices for industrial-scale utility. Detection accuracy will rely on advanced computational methods and hardware, especially for detection at ultra-low hydrogen concentrations (<0.5 vol.%).

Author Contributions: Conceptualization, S.M.H. and K.-W.-A.C.; Methodology, M.Z., S.M.H. and K.-W.-A.C.; Software, M.Z.; Validation, M.Z., S.M.H. and K.-W.-A.C.; Formal analysis, M.Z., S.M.H. and K.-W.-A.C.; Investigation, M.Z., S.M.H. and K.-W.-A.C.; Data curation, M.Z.; Writing—original draft, M.Z.; Writing—review & editing, S.M.H. and K.-W.-A.C.; Visualization, M.Z., S.M.H. and K.-W.-A.C.; Supervision, S.M.H. and K.-W.-A.C.; Project administration, S.M.H. and K.-W.-A.C. All authors have read and agreed to the published version of the manuscript.

Funding: No funding to declare.

Institutional Review Board Statement: Not applicable.

Informed Consent Statement: Not applicable.

Data Availability Statement: The data presented in this study are available on request from the corresponding author.

Conflicts of Interest: The authors declare no conflict of interest.

References

1. Marbán, G.; Valdés-Solís, T. Towards the hydrogen economy? *Int. J. Hydrog. Energy* **2007**, *32*, 1625–1637. [CrossRef]
2. Yip, H.K.; Zhu, X.; Zhuo, X.; Jiang, R.; Yang, Z.; Wang, J. Gold Nanobipyramid-Enhanced Hydrogen Sensing with Plasmon Red Shifts Reaching ≈140 nm at 2 vol% Hydrogen Concentration. *Adv. Opt. Mater.* **2017**, *5*, 1700740. [CrossRef]
3. Okolie, J.A.; Patra, B.R.; Mukherjee, A.; Nanda, S.; Dalai, A.K.; Kozinski, J.A. Futuristic applications of hydrogen in energy, biorefining, aerospace, pharmaceuticals and metallurgy. *Int. J. Hydrog. Energy* **2021**, *46*, 8885–8905. [CrossRef]
4. Wadell, C.; Syrenova, S.; Langhammer, C. Plasmonic hydrogen sensing with nanostructured metal hydrides. *ACS Nano* **2014**, *8*, 11925–11940. [CrossRef] [PubMed]
5. Cerea, A.; Garoli, D.; Zilio, P.; Dipalo, M.; Calandrini, E.; Jacassi, A.; Caprettini, V.; Carrara, A.; Pelizzo, M.G.; De Angelis, F. Modified three-dimensional nanoantennas for infrared hydrogen detection. *Microelectron. Eng.* **2016**, *162*, 105–109. [CrossRef]
6. Watkins, W.L.; Borensztein, Y. Ultrasensitive and fast single wavelength plasmonic hydrogen sensing with anisotropic nanostructured Pd films. *Sens. Actuators B Chem.* **2018**, *273*, 527–535. [CrossRef]
7. Hassan, K.; Iftekhar Uddin, A.S.M.; Chung, G.S. Fast-response hydrogen sensors based on discrete Pt/Pd bimetallic ultra-thin films. *Sens. Actuators B Chem.* **2016**, *234*, 435–445. [CrossRef]
8. Ramachandran, R.; Menon, R.K. An overview of industrial uses of hydrogen. *Int. J. Hydrog. Energy* **1998**, *23*, 593–598. [CrossRef]
9. Mazloomi, K.; Gomes, C. Hydrogen as an energy carrier: Prospects and challenges. *Renew. Sustain. Energy Rev.* **2012**, *16*, 3024–3033. [CrossRef]
10. Wadell, C.; Nugroho, F.A.A.; Lidström, E.; Iandolo, B.; Wagner, J.B.; Langhammer, C. Hysteresis-free nanoplasmonic pd-au alloy hydrogen sensors. *Nano Lett.* **2015**, *15*, 3563–3570. [CrossRef]
11. Hu, Y.; Lei, J.; Wang, Z.; Yang, S.; Luo, X.; Zhang, G.; Chen, W.; Gu, H. Rapid response hydrogen sensor based on nanoporous Pd thin films. *Int. J. Hydrog. Energy* **2016**, *41*, 10986–10990. [CrossRef]
12. Ngene, P.; Radeva, T.; Slaman, M.; Westerwaal, R.J.; Schreuders, H.; Dam, B. Seeing hydrogen in colors: Low-cost and highly sensitive eye readable hydrogen detectors. *Adv. Funct. Mater.* **2014**, *24*, 2374–2382. [CrossRef]
13. Houlihan, N.M.; Karker, N.; Potyrailo, R.A.; Carpenter, M.A. High Sensitivity Plasmonic Sensing of Hydrogen over a Broad Dynamic Range Using Catalytic Au-CeO$_2$ Thin Film Nanocomposites. *ACS Sens.* **2018**, *3*, 2684–2692. [CrossRef] [PubMed]
14. Darmadi, I.; Nugroho, F.A.A.; Kadkhodazadeh, S.; Wagner, J.B.; Langhammer, C. Rationally Designed PdAuCu Ternary Alloy Nanoparticles for Intrinsically Deactivation-Resistant Ultrafast Plasmonic Hydrogen Sensing. *ACS Sens.* **2019**, *4*, 1424–1432. [CrossRef]
15. Tittl, A.; Giessen, H.; Liu, N. Plasmonic gas and chemical sensing. *Nanophotonics* **2014**, *3*, 157–180. [CrossRef]
16. Zaman, M.A.; Hesselink, L. Plasmonic Response of Nano-C-apertures: Polarization Dependent Field Enhancement and Circuit Model. *Plasmonics* **2023**, *18*, 155–164. [CrossRef]
17. Maleki, M.; Mehran, M.; Mokhtari, A. Design of a near-infrared plasmonic gas sensor based on graphene nanogratings. *J. Opt. Soc. Am. B* **2020**, *37*, 3478–3486. [CrossRef]
18. Zhu, J.; Wang, X.; Wu, Y.; Su, Y.; Jia, T.; Yang, H.; Zhang, L.; Qi, Y.; Wen, X. Plasmonic Refractive Index Sensors Based on One- and Two-Dimensional Gold Grating on a Gold Film. *Photonic Sens.* **2020**, *10*, 375–386. [CrossRef]
19. Iqbal, T.; Afsheen, S. One Dimensional Plasmonic Grating: High Sensitive Biosensor. *Plasmonics* **2017**, *12*, 19–25. [CrossRef]
20. Afsheen, S.; Iqbal, T.; Aftab, M.; Bashir, A.; Tehseen, A.; Khan, M.Y.; Ijaz, M. Modeling of 1D Au plasmonic grating as efficient gas sensor. *Mater. Res. Express* **2019**, *6*, 126203. [CrossRef]

21. Shegai, T.; Johansson, P.; Langhammer, C.; Käll, M. Directional scattering and hydrogen sensing by bimetallic Pd-Au nanoantennas. *Nano Lett.* **2012**, *12*, 2464–2469. [CrossRef] [PubMed]
22. Sterl, F.; Strohfeldt, N.; Both, S.; Herkert, E.; Weiss, T.; Giessen, H. Design Principles for Sensitivity Optimization in Plasmonic Hydrogen Sensors. *ACS Sens.* **2020**, *5*, 917–927. [CrossRef] [PubMed]
23. Hamidi, S.M.; Ramezani, R.; Bananej, A. Hydrogen gas sensor based on long-range surface plasmons in lossy palladium film placed on photonic crystal stack. *Opt. Mater.* **2016**, *53*, 201–208. [CrossRef]
24. Langhammer, C.; Zorić, I.; Kasemo, B.; Clemens, B.M. Hydrogen storage in Pd nanodisks characterized with a novel nanoplasmonic sensing scheme. *Nano Lett.* **2007**, *7*, 3122–3127. [CrossRef]
25. Tittl, A.; Mai, P.; Taubert, R.; Dregely, D.; Liu, N.; Giessen, H. Palladium-based plasmonic perfect absorber in the visible wavelength range and its application to hydrogen sensing. *Nano Lett.* **2011**, *11*, 4366–4369. [CrossRef] [PubMed]
26. Yan, H.; Zhao, X.; Zhang, C.; Li, Q.Z.; Cao, J.; Han, D.F.; Hao, H.; Wang, M. A fast response hydrogen sensor with Pd metallic grating onto a fiber's end-face. *Opt. Commun.* **2016**, *359*, 157–161. [CrossRef]
27. Avila, J.I.; Matelon, R.J.; Trabol, R.; Favre, M.; Lederman, D.; Volkmann, U.G.; Cabrera, A.L. Optical properties of Pd thin films exposed to hydrogen studied by transmittance and reflectance spectroscopy. *J. Appl. Phys.* **2010**, *107*, 023504. [CrossRef]
28. Neubrech, F.; Duan, X.; Liu, N. Dynamic plasmonic color generation enabled by functional materials. *Sci. Adv.* **2020**, *6*, eabc2709. [CrossRef]
29. Xu, M.; Bunes, B.R.; Zang, L. Paper-based vapor detection of hydrogen peroxide: Colorimetric sensing with tunable interface. *ACS Appl. Mater. Interfaces* **2011**, *3*, 642–647. [CrossRef]
30. Benson, D.K.; Tracy, C.E.; Hishmeh, G.A.; Ciszek, P.E.; Lee, S.-H.; Haberman, D.P. Low-cost fiber optic hydrogen gas detector using guided-wave surface-plasmon resonance in chemochromic thin films. In Proceedings of the Photonics East Symposium, Boston, MA, USA, 1–5 November 1998; Volume 3535, pp. 185–202.
31. Serhatlioglu, M.; Ayas, S.; Biyikli, N.; Dana, A.; Solmaz, M.E. Perfectly absorbing ultra thin interference coatings for hydrogen sensing. *Opt. Lett.* **2016**, *41*, 1724. [CrossRef]
32. Duan, X.; Liu, N. Scanning plasmonic color display. *ACS Nano* **2018**, *12*, 8817–8823. [CrossRef] [PubMed]
33. Nugroho, F.A.A.; Bai, P.; Darmadi, I.; Castellanos, G.W.; Fritzsche, J.; Langhammer, C.; Gómez Rivas, J.; Baldi, A. Inverse designed plasmonic metasurface with parts per billion optical hydrogen detection. *Nat. Commun.* **2022**, *13*, 5737. [CrossRef]
34. Luong, H.M.; Pham, M.T.; Madhogaria, R.P.; Phan, M.-H.; Larsen, G.K.; Nguyen, T.D. Bilayer plasmonic nano-lattices for tunable hydrogen sensing platform. *Nano Energy* **2020**, *71*, 104558. [CrossRef]
35. Hosseini, M.A.; Ranjbar, M. Optical hydrogen sensing by MoO3 films deposited by a facile flame synthesis method. *Appl. Surf. Sci.* **2023**, *618*, 156641. [CrossRef]
36. Bellini, S.; Sun, Y.; Gallucci, F.; Caravella, A. Thermodynamic aspects in non-ideal metal membranes for hydrogen purification. *Membranes* **2018**, *8*, 82. [CrossRef]
37. Strohfeldt, N.; Tittl, A.; Giessen, H. Long-term stability of capped and buffered palladium-nickel thin films and nanostructures for plasmonic hydrogen sensing applications. *Opt. Mater. Express* **2013**, *3*, 194. [CrossRef]
38. Palm, K.J.; Murray, J.B.; Narayan, T.C.; Munday, J.N. Dynamic Optical Properties of Metal Hydrides. *ACS Photonics* **2018**, *5*, 4677–4686. [CrossRef]
39. Long, S.; Cao, J.; Wang, Y.; Gao, S.; Xu, N.; Gao, J.; Wan, W. Grating coupled SPR sensors using off the shelf compact discs and sensitivity dependence on grating period. *Sens. Actuators Rep.* **2020**, *2*, 100016. [CrossRef]
40. Karabchevsky, A.; Krasnykov, O.; Auslender, M.; Hadad, B.; Goldner, A.; Abdulhalim, I. Theoretical and experimental investigation of enhanced transmission through periodic metal nanoslits for sensing in water environment. *Plasmonics* **2009**, *4*, 281–292. [CrossRef]
41. Gao, H.; Zheng, Z.; Dong, J.; Feng, J.; Zhou, J. Multi-frequency optical unidirectional transmission based on one-way guided mode resonance in an extremely simple dielectric grating. *Opt. Commun.* **2015**, *355*, 137–142. [CrossRef]
42. Strohfeldt, N.; Zhao, J.; Tittl, A.; Giessen, H. Sensitivity engineering in direct contact palladium-gold nano-sandwich hydrogen sensors [Invited]. *Opt. Mater. Express* **2015**, *5*, 2525. [CrossRef]
43. Kalanur, S.S.; Lee, Y.-A.; Seo, H. Eye-readable gasochromic and optical hydrogen gas sensor based on CuS–Pd. *RSC Adv.* **2015**, *5*, 9028–9034. [CrossRef]
44. She, X.; Shen, Y.; Wang, J.; Jin, C. Pd films on soft substrates: A visual, high-contrast and low-cost optical hydrogen sensor. *Light Sci. Appl.* **2019**, *8*, 4. [CrossRef] [PubMed]

Disclaimer/Publisher's Note: The statements, opinions and data contained in all publications are solely those of the individual author(s) and contributor(s) and not of MDPI and/or the editor(s). MDPI and/or the editor(s) disclaim responsibility for any injury to people or property resulting from any ideas, methods, instructions or products referred to in the content.

Article

Sensing and Detection Capabilities of One-Dimensional Defective Photonic Crystal Suitable for Malaria Infection Diagnosis from Preliminary to Advanced Stage: Theoretical Study

Sujit Kumar Saini and Suneet Kumar Awasthi *

Department of Physics and Material Science and Engineering, Jaypee Institute of Information Technology, Noida 201304, India
* Correspondence: suneet_electronic@yahoo.com

Abstract: In the present research work we have examined the biosensing capabilities of one-dimensional photonic crystals with defects for the detection and sensing of malaria infection in humans by investigating blood samples containing red blood cells. This theoretical scheme utilizes a transfer matrix formulation in addition to MATLAB software under normal incidence conditions. The purpose of considering normal incidence is to rule out the difficulties associated with oblique incidence. We have examined the performance of various structures of cavity layer thicknesses 1000 nm, 2200 nm, 3000 nm and 5000 nm. The comparison between the performances of various structures of different cavity thickness helps us to select the structure of particular cavity thicknesses giving optimum biosensing performance. Thus, the proper selection of cavity thickness is one of the most necessary requirements because it also decides how much volume of the blood sample has to be poured into the cavity to produce results of high accuracy. Moreover, the sensing and detection capabilities of the proposed design have been evaluated by examining the sensitivity, figure of merit and quality factor values of the design, corresponding to optimum cavity thickness.

Keywords: photonic crystals; biosensors; transfer matrix method

1. Introduction

The pioneering research work on photonic crystals (PCs) by two scientists, Yablonovitch and John in 1987, has revolutionized the research field of optical engineering and technology [1,2]. PCs have commendable control of the propagation of light passing through them. The periodic modulation of refractive indices of the constituent materials of PC results in the formation of photonic band gap (PBG) due to Bragg scattering of incident waves from the interfaces between the various material layers of the structure [3,4]. PBG restricts the propagation of light of specific frequencies from the structure and allows the propagation of light of other frequencies to pass through. PCs can be classified into three categories, depending upon the modulation of the refractive index of the constituent materials in x, y and z directions as one-dimensional (1D), two-dimensional (2D) and three dimensional (3D) PCs. The ease of fabrication techniques associated with 1D PCs motivated the photonic engineers to explore the biosensing capabilities of 1D photonic structures with a defect. In recent years, the rapid, advanced and accurate biosensing capabilities of 1D defective photonic crystal (DPC) have attracted the attention of photonic technocrats to design and develop photonic biosensors due to their importance in the field of applied sciences, such as for security, medical, defense, food detection, environment, and aerospace worldwide [5]. Actually, the creation of an empty space known as a cavity region inside photonic structures is responsible for the break in periodicity which results in the existence of a sharp tunneling peak inside the PBG of the structure. The optical properties of the tunneling peaks (also called the defect mode) are strongly dependent upon both the refractive index and the thickness of the cavity region. This property of the defect mode is very useful in designing various biosensors consisting of 1D DPCs [6–8]. For example, Zaky et al. suggested a plasma cell sensing device based on 1D DPC for the detection and sensing of convalescent plasma whose refractive index variation is restricted between 1.3246 and 1.3634 [9]. Another photonic design capable of detecting glucose concentration levels has been investigated by Asmaa et al. Their design works on the principle of Fano-resonance, which is excited across the interface between PC and metallic capping mounted on top of the structure [10]. In contrast to the conventional biosensing technologies based on plasmonics and photonic crystal fibers, 1D DPC based biosensors are highly sensitive sensing mechanisms due to the ultra-high localization of light inside the cavity region. Additionally, 1D DPC based biosensing lowers down the volume requirement of the sample under investigation [11]. Moreover, 1D DPC based sensors are compact in size and easily accommodated in a complex environment [12]. Moreover, the compatibility between 1D photonic structures and integrated photonic circuits encourages their extensive role in the fields such as force–strain, temperature, liquid, pressure, displacement, gas and biomedical engineering [13–15].

Nowadays, blood examination is an essential tool for identifying hematological disorders which are responsible for a series of non-communicable diseases such as diabetes, coronary artery, cancerous and respiratory [16]. As per the report of the World Economical Forum, published in September 2011, these diseases were the root cause of around 36 million mortalities across the world [17]. Therefore, examining the human blood sample is one of the cheapest, most necessary and easiest ways to carry out regular and periodic health monitoring. The blood sample examination helps in identifying the diseases and become a foundation for proper treatment. Human blood is made up of a large number of bio-constituents which are approximately more than 4000 in number [18]. Actually, nowadays, blood optics play an important role in biophotonic sensing and clinical therapy applications [19]. The absorption and scattering characteristics of light interacting with the blood sample depend on the refractive index of the erythrocytes present in the blood sample, which is strongly dependent upon the hemoglobin concentration of erythrocytes [20]. Blood is a highly functional bodily fluid whose refractive index is complex in general [21]. More than half of human blood is made up of blood plasma, which contains various proteins such as red and white blood cells, enzymes, albumin, hormones, glucose, minerals, etc. [22] The supply of oxygen from lungs to different body parts is being accomplished by hemoglobin, also known as a main protein present in red blood cells (RBCs). On the other hand, white blood cells (WBCs) which are also known as leukocytes, strengthen our body to fight against various infections [23]. The dielectric properties of human blood have great relevance in various medical applications such as early stage detection of cancer cells in the human body and several other diseases. For example, the dielectric blood coagulometry helps us to analyze the whole spectra of human blood to understand the biological, physical and chemical properties comprehensively [24].

Malaria is one of the fatal diseases caused by protozoan parasites of the genus plasmodium [25,26]. Untreated or misdiagnosed malaria may become a root cause of death globally. According to the World Health Organization (WHO), around 405,000 casualties out of 228 million malaria cases were reported in 2018 worldwide [27]. If someone is bitten by female anopheles mosquito protozoan, parasites enter into the red blood cells of the human body through the liver [28]. The presence of protozoan parasites in the RBCs results in structural and biological change in RBCs. This modification degrades hemoglobin, which is the main constituent of RBCs. This degradation of hemoglobin becomes nutrition for protozoan parasites. These parasites digest hemoglobin of the human body as a free ferrous heme, which is quickly transformed into ferric heme and are highly toxic. This transformation results in the change in the homogeneous structure of RBCs. Malaria diagnosis must be speedy, reliable and very accurate for their eradication via timely treatment [29,30]. At present, various conventional approaches are being used in malaria diagnosis, and all of these conventional approaches have limitations due to their laboratory requirements and/or the complexity involved in investigation. Some other limitations associated with the conventional approaches are sample size requirement, sensitivity, result accuracy, time-consumption and difficulties associated with early-stage detection of malaria, depending upon the stage of infection [31]. On the other hand, 1D photonic biosensors can satisfactorily address all the above issues pertaining to timely and early-stage detection of malaria infection in humans. Moreover, investigations conducted by 1D photonic biosensors are rapid and cost effective, which brings the medical expenses within the reach of poorer people. For example, Somaia et al. explored the biosensing application of 1D PC by studying the propagation of a p polarized wave through 1D PC. They have shown a 714% improvement in the sensitivity of the structure as compared to the waveguide based conventional sensors [32]. Both Mahdi et al. and Taya have exploited the defect mode properties of 1D ternary photonic structure for minute refractometric sensing application loaded with the various analytes having refractive index variation between 1.00 to 1.06 and 1.33 to 1.35, respectively [33,34]. In addition, Banerjee has suggested how a 1D ternary photonic structure can be used as an enhanced sensitivity gas sensor [35]. The surface plasmon resonance driven photonic crystal fiber based biosensing structures are suggested by research groups of Qingli et al. and Zhiwen et al. [36,37]. Tongyu et al. suggested the PC cavity coupled photonic sensor for simultaneously sensing refractive index and temperature by using an electromagnetically induced transparency effect [38]. Recently, Zina et al. suggested how 1D PC consisting of cold magnetic plasma and quartz materials according to the Copper mean sequence can be used for the detection of the magnetic field direction by studying the external magnetic field dependent movement of ultra large PBG of the structure [39]. Parandin et al. suggested a 2D photonic biosensor made up of circular nano-rings between the waveguides for the detection of various blood components [40,41]. Liu et al. demonstrated how a 2D PC based cavity structure can be used as a quality sensor for the detection of ethanol [42]. Olyaee et al. designed a pressure sensor composed of a 2D PC of ultra-high sensitivity and resolution, by performing finite difference time domain simulation [43]. Moreover, Claudia has suggested how porous silicon material based photonic biosensing structures can be used as high performance sensors [44].

The present work is focused on the biosensing properties of 1D DPC for the diagnosis of various stages of malaria infection present in human body. The organization of the present manuscript is as follows. Section 2 deals with the structural design of the proposed work. Theoretical formulation is discussed in Section 3. The results of this work are given in Section 4. Section 5 deals with conclusions of the proposed work.

2. Structural Design

Figure 1 represents the structural design of the present blood sensor composed of 1D DPC for the detection of various stages of malaria infection. The present biosensing structure $(AB)^N C(AB)^N /GS$ can easily be fabricated by creating a defect layer C of air at the middle of the 1D PC composed of alternating layers A and B of materials: silicon (Si) and lanthanum flint (LAFN7), respectively. The alphabet N represent the period number of the structure. The ion-bean sputtering technique can be used for the fabrication of the proposed biosensing structure composed of Si and LANF7 on glass substrate for the detection of malaria infection through red blood cell (RBC) samples containing Cell A, Cell B, Cell C, Cell D and Cell E separately [45–47].

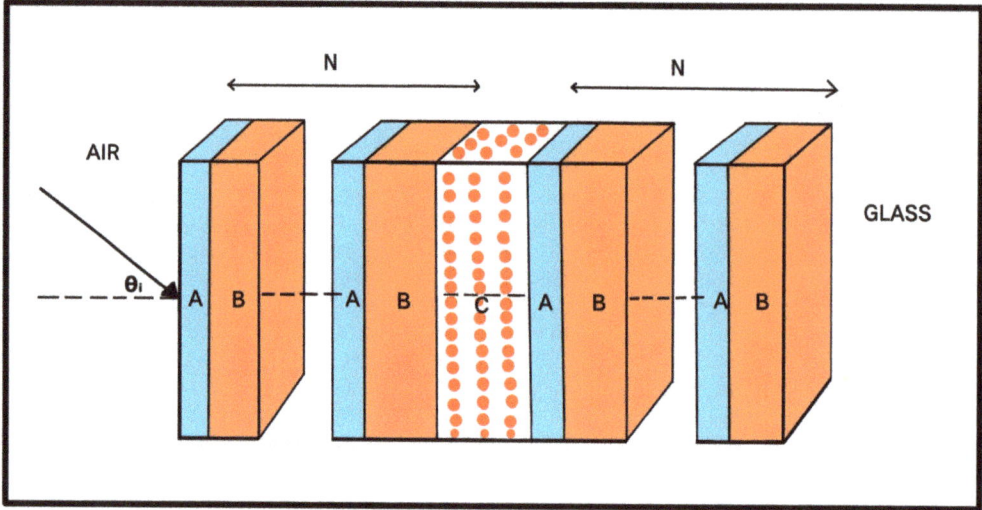

Figure 1. Schematic view of the proposed blood sensor for malaria detection and sensing composed of 1D photonic crystal with single defect.

3. Theoretical Formulation

In order to obtain the simulation results through MATLAB software, we have used a transfer matrix method [48,49]. This is one of the most suitable techniques for the computation of simulation results of the proposed 1D photonic biosensing structure. According to this method, the amplitudes of electric and magnetic fields associated with incident and transmitted electromagnetic radiation at either ends of the structure, i.e., incident and transmitted ends, are connected via transfer matrix as

$$Z = (z_1 z_2)^N z_3 (z_1 z_2)^N = \begin{pmatrix} Z_{11} & Z_{12} \\ Z_{21} & Z_{22} \end{pmatrix} \quad (1)$$

Here, Z_{11}, Z_{12}, Z_{21} and Z_{22} are representing the elements of resultant transfer matrix Z. The z_1, z_2 and z_3 are being used for representing the characteristic matrix of layers A, B and C, respectively [50].

The coefficient of transmission t of the proposed biosensing structure [air/(Si/LAFN7)^N/cavity/(Si/LAFN7)^N/GS] is defined as

$$t = \frac{2p_0}{(Z_{11} + Z_{12} p_s) p_0 + (Z_{21} + Z_{22} p_s)} \quad (2)$$

Here, $p_0 = n_0 \cos(\alpha_0)$ and $p_s = n_s \cos(\alpha_s)$ are corresponding to input and exit ends of the structure, respectively, for s-polarized wave. For p-polarized wave, $p_0 = \cos(\alpha_0)/n_0$ and $p_s = \cos(\alpha_s)/n_s$. Additionally, α_0 and α_s are representing angles of incidence and emergence in incident and exit media, respectively.

Finally, the transmittance $\Im$ of the proposed biosensing structure is

$$\Im = \frac{s_s}{s_0} |t|^2 \times 100 \quad (3)$$

4. Results and Discussions

The transfer matrix method as discussed above has been applied over the proposed 1D defective photonic structure (AB)^{N}C(AB)^N/G as presented in Figure 1. We have used MATLAB software to obtain the transmittance of the proposed biosensor under normal incidence conditions. The purpose of considering the normal incidence is to overlook the challenges associated with the oblique incidence, along with the requirement of transverse electric and transverse magnetic modes of incident light. The entire simulations have been carried out in the visible region of the electromagnetic spectrum, extending from 600 nm to 700 nm. The materials silicon (Si) and lanthanum flint (LAFN7) have been used to fabricate the layers A and B of the proposed 1D multilayer stack of refractive indices, $n_{Si} = 3.5$ and $n_{LAFN7} = 1.7$, respectively, on the glass substrate of refractive index $n_s = 1.57$. The purpose of selecting Si and LAFN7 materials in our design is to ensure a large refractive index contrast between the high and low refractive index layers of the proposed structure, which is one of the essential requirements for getting wider as well as deeper photonic band gap (PBG). The depth of the PBG may also be increased by increasing the period number. However, instead of increasing the period number of the design, we have preferred to ensure large refractive index contrast to obtain wider PBG. The wider PBG also increases the possibility of having a large number of resonant transmission peaks whose central wavelengths are restricted inside the PBG of the structure. Moreover, larger PBG also improves the number of blood samples to be investigated by our design, depending upon their refractive index variation. In the present work the refractive index variation between the blood samples is from 1.371 to 1.408 depending upon the stages of malaria infection (Table 1). In this simulation work, the thicknesses of layers A and B are taken as $d_A = 70$ nm and $d_B = 400$ nm. The period number N has been fixed to 10. The defect layer of thickness $dd = 300$ nm has been created at the middle of the proposed biosensor by disturbing the periodicity of the design, as shown in Figure 1.

Table 1. Refractive index values of various cells depending upon the hemoglobin concentration within RBCs [28].

Stage of Infection	RBC Component	Refractive Index	Hemoglobin Concentration (g/dL)
Healthy	Cell A	1.408	30.9
Ring	Cell B	1.396	25.59
Trophozoite	Cell C	1.381	19.78
Schizont	Cell D	1.372	16.28
Schizont	Cell E	1.371	15.9

4.1. Description of Malaria Samples Used

In this study, we have investigated four samples of malaria-infected red blood cells (RBCs) as B, C, D and E cells with respect to the sample containing healthy RBCs, referred as cell A. Here, cells B, C, D and E correspond to different stages of malaria infections with respect to cell A, which represents the healthy stage. Table 1 gives the refractive index of values of samples containing healthy and malaria infected RBCs obtained by Agnero et al. [51]. They suggested an optical method based on the transportation of the intensity equation which differentiates between malaria infected and healthy RBCs by combining the topography, three dimensional reconstruction of refractive index and deconvolution of RBCs. Actually, RBCs are a mixture of 32% of hemoglobin surrounded by 3% membrane and 65% water [52]. RBCs can be considered as an aqueous solution in which hemoglobin is dissolved. Both Kevin and Tycko et al. suggested that the change in the hemoglobin concentration within RBCs results in the significant change in the refractive index of cells as shown in Table 1 [53,54]. The refractive index and hemoglobin concentration within RBCs are the two essential parameters which are usually used to identify whether or not RBCs belong to a healthy or malaria infected person.

In healthy RBCs, hemoglobin is one of the major of components of cells. These healthy cells are physically identified by their biconcave shape, whose edges are thicker than the middle. The main function of RBCs is to maintain flow of oxygen and carbon dioxide inside the human body. Hence, if RBCs are healthy, it means the flow of O_2 and CO_2 inside body is perfect. For healthy RBCs, the range of hemoglobin concentration of cell A should be between 28 g/dL and 36 g/dL, which corresponds to refractive index values between 1.402 and 1.409, respectively [28]. If someone is bitten by the female Anopheles mosquito, parasites enter into the body and reach the RBCs through the liver. The presence of parasites in RBCs initiates the biochemical and structural changes of host cells due to which homogeneous structure of cell is lost. Moreover, the presence of parasites into the cells also decreases both the hemoglobin concentration and refractive index value of that cell. Therefore, the presence of parasites within the various cells is ensured by the region having a low refractive index [28–30]. This is the first stage of malaria infection and is called the ring stage. In this stage, the shape of the RBC remains biconcave and the infected cell is named as cell B. After the ring stage, the malaria infection reaches the trophozoite stage. In this stage, parasites are mature enough and have a more intense metabolism because host cells C lost their biconcavity. Finally, infection reaches to its prominent stage, called the schizont stage. In this stage, the growth of the parasites reaches to an advanced level and the corresponding infection is called cell D. By knowing the refractive index and concentration of the cell in the RBCs, one can easily identify the schizont stage of malaria infection by means of an optical route. Generally, the refractive index and hemoglobin concentration of quasi-identical cells D and E are different even though both are representing the same stage of infection, as shown in Table 1 [25–30].

We have also performed the linear curve fitting, as shown in Figure 2, over the data given in Table 1, to extract an expression which gives the hemoglobin concentration (C_{Hb}) inside RBCs corresponding to the refractive index (n_{RBC}) of the samples depending upon the distribution of cell. It can be clearly seen from Figure 2 that the increase in the refractive index of the cell is due to the increase in the hemoglobin concentration within RBC samples. The red line in Figure 2 is representing a liner curve fitting equation obtained from simulated data. The change in hemoglobin concentration within RBCs can easily be obtained by putting the value of n_{RBC} in the curve fitting equation given below:

$$C_{Hb} = 402.35 n_{RBC} - 535.8 \left(R^2 = 0.9992 \right) \tag{4}$$

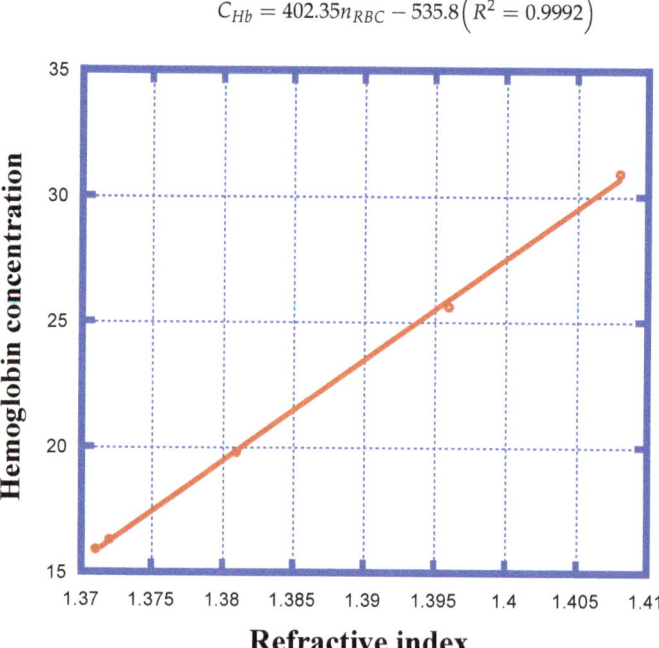

Figure 2. The diagram showing refractive index of RBC components containing cells A, B, C, D and E dependent upon the hemoglobin concentration of blood.

Here, R^2 represents the square of the correlation coefficient which determines the accuracy between the simulated and curve fitting data. The higher value of R^2 is always accepted to validate the results.

4.2. Initialization of Biosensing Application of the Proposed Design Loaded with Water Sample

The empty space of the defect layer is infiltrated by a pure water sample of refractive index 1.333 to initiate the biosensing application of the design. The infiltration of the water sample into the cavity of the proposed biosensor results in the confinement of light into the cavity region. This confinement of light appears as a defect mode of unit transmission inside the photonic band-gap of the structure located at 620.9 nm, as shown in Figure 3.

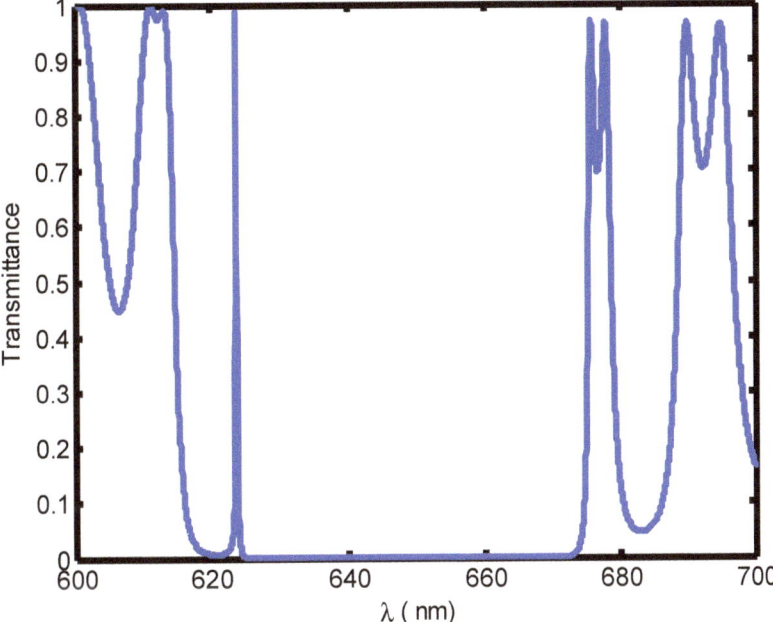

Figure 3. Transmission spectra of proposed biosensor loaded with water sample corresponding to cavity of thickness dd = 300 nm at normal incidence.

For analyzing the performance of the proposed design one may use the approach suggested by us. Firstly, both the ends of the biosensor are connected with single mode fiber (SMF) through precision positioning equipment to avoid errors during measurements. The light from the polychromatic source is launched into the structure via the input end of the design through SMF. The output terminal of the proposed design is connected with the optical spectrum analyzer (OSA) through SMF for the projection of the biosensing results into the monitor via computer. The qualitative setup for analyzing the performance of the proposed biosensing is shown in Figure 4, below, as per our understanding, though the findings of the proposed work are based on theoretical simulation which has been carried out with the help of the transfer matrix method in addition to MATLAB software.

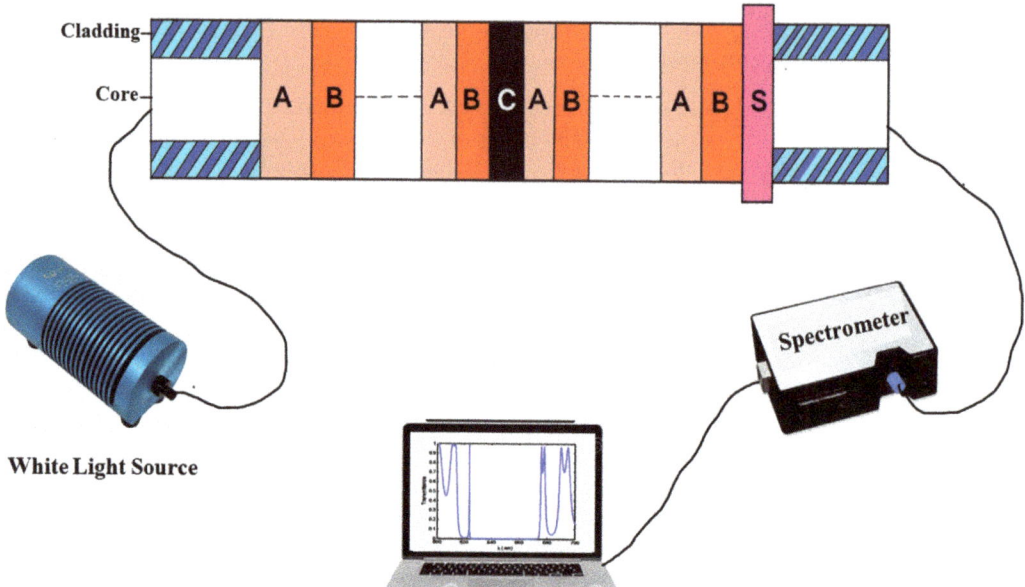

Figure 4. Experimental setup is required for the measurement of transmission response of the biosensor. Here, letters A, B and C represent silicon, lanthanum flint and air layers of the structure fabricated on the glass substrate S.

4.3. Evaluation of Biosensor Performance Loaded with Different Blood Samples

In this section, we are highlighting the biosensing capabilities of the proposed design loaded with hemoglobin blood samples containing Cell A, Cell B, Cell C, Cell D and Cell E, one at a time for the diagnosis of malaria infection. Figure 5, below, shows the transmission spectra of the proposed biosensor loaded with different five RBC samples under examination. The defect mode peaks of unit transmission shown in blue, black, red, yellow and purple solid line colors are corresponding to RBC samples containing Cell A, Cell B, Cell C, Cell D and Cell E, respectively, under investigation. After recording the central wavelength of each defect mode inside PBG with the help of the setup described above, we have calculated the sensitivity of the design of cavity thickness $dd = 1000$ nm with the help of the following equation [6–11]:

$$S = \frac{d\lambda}{dn} (\text{nm/RIU}) \quad (5)$$

Here, $d\lambda$ is representing the change in the position of central wavelength of the defect mode associated with the particular sample with respect to the water sample, and dn is the corresponding difference between the refractive index of that sample with water.

The proposed biosensor could achieve a maximum sensitivity value of 148.1 nm/RIU, corresponding to defect layer thickness $dd = 1000$ nm. A high value of sensitivity is always desirable for the designing of any high performance photonic biosensor, so we have given our efforts to improve the sensitivity further. For this purpose, we have randomly chosen some higher values of cavity thickness such as $dd = 2200$ nm, 3000 nm and 5000 nm, also keeping all other parameters of the design fixed as discussed above. The transmission spectra of the proposed biosensing structures corresponding to defect layer thicknesses $dd = 2200$ nm, 3000 nm and 5000 nm are plotted in Figures 6–8, respectively.

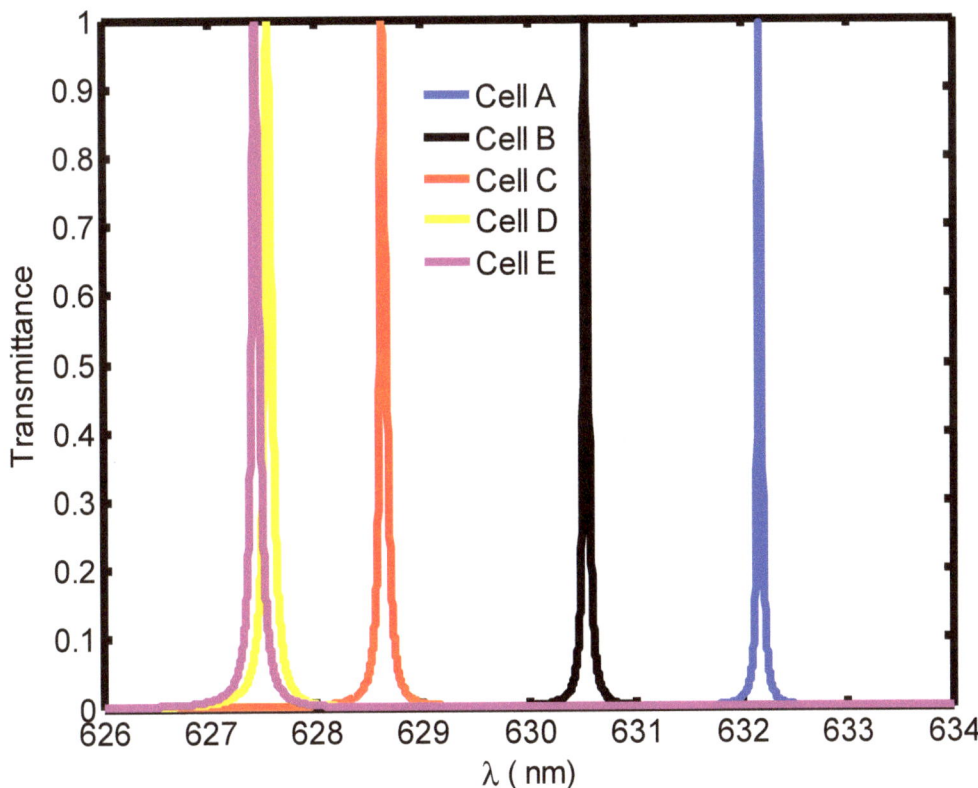

Figure 5. Transmission spectra showing five defect modes in solid blue, black, brown, yellow and purple line colors corresponding to RBC sample containing A, B, C, D and E blood cells separately, one at a time. The thickness of cavity layer is $dd = 1000$ nm under normal incidence.

The comparison of Figures 5–8 shows that as the defect layer thickness increases, the defect modes corresponding to all five samples show red shifting. This shifting is between wavelength range 626 nm to 633 nm corresponding to $dd = 3000$ nm and 658 nm to 675 nm corresponding to $dd = 5000$ nm, respectively. Further increase in the defect layer thickness results in the movement of defect modes beyond 665 nm, i.e., outside the PBG extending from 620 nm to 665 nm (Figure 3). There is one more common observation, that, corresponding to defect layer thickness 2200 nm, 3000 nm and 5000 nm, the intensity of all defect modes associated with the five samples is slightly reduced. However, this reduction does not affect the performance of the design, due the fact that the reduced intensity of defect modes is significantly higher in comparison to the threshold limit of the OSA, which is used for the detection of defect modes under the influence of different RBC samples. The numeric values of the sensitivity of the proposed designs corresponding to cavity thickness $dd = 1000$ nm, 2200 nm, 3000 nm and 5000 nm have been summarized in Table 2 below.

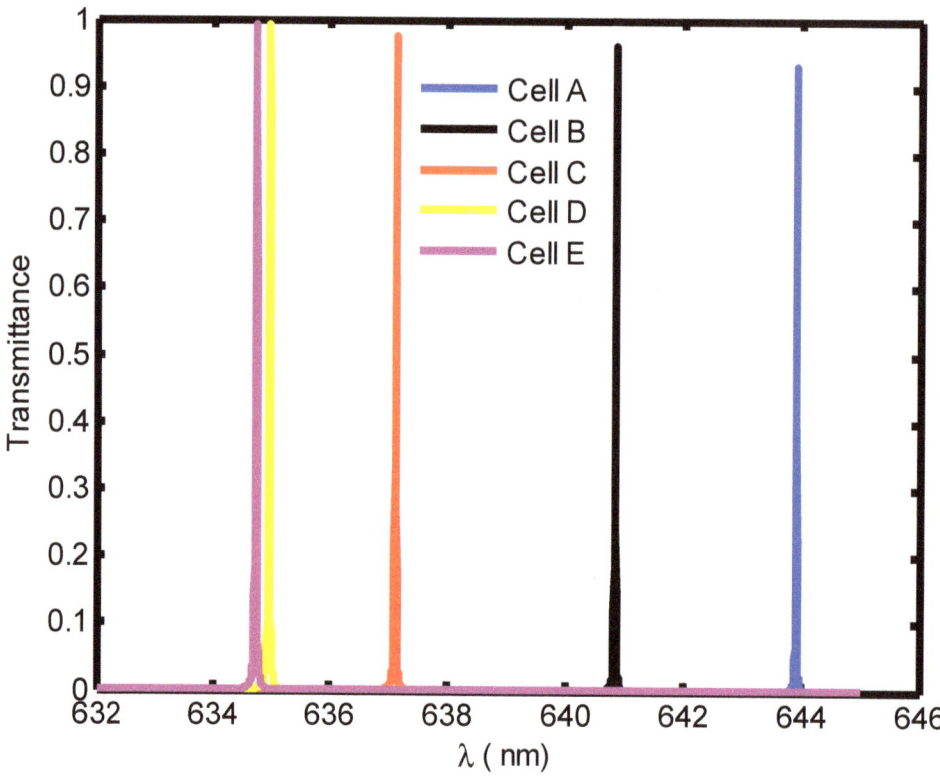

Figure 6. Transmission spectra showing five defect modes in solid blue, black, brown, yellow and purple line colors corresponding to RBC sample containing A, B, C, D and E blood cells separately, one at a time. The thickness of cavity layer is dd = 2200 nm under normal incidence.

Table 2. Sensitivity calculations of proposed biosensors corresponding to different defect layer thicknesses.

Defect Layer Thickness (nm)	Sensitivity (nm/RIU)
dd = 1000	141.6
dd = 2200	248.1
dd = 3000	303
dd = 5000	327.7

The data presented in Table 2 have been visualized by plotting Figure 9, which shows the dependence of sensitivity on the thickness of the defect layer region. Figure 9 shows that, as the thickness of the defect layer increases from 1000 nm to 3000 nm, the sensitivity increases linearly and reaches to 303 nm/RIU. Further increase in the thickness of the defect layer results in a relatively small change in the sensitivity, as shown in Figure 9. The maximum sensitivity of 327.7 nm/RIU is reached, corresponding to a defect layer thickness of 5000 nm. Thus, a defect layer thickness of 5000 nm can be considered as an optimum value of thickness under which our design becomes highly sensitive. Additionally, corresponding to the optimum value of defect layer thickness, our design is capable of detecting very minute changes in the refractive index of RBC samples containing Cell B to Cell E with respect to Cell A.

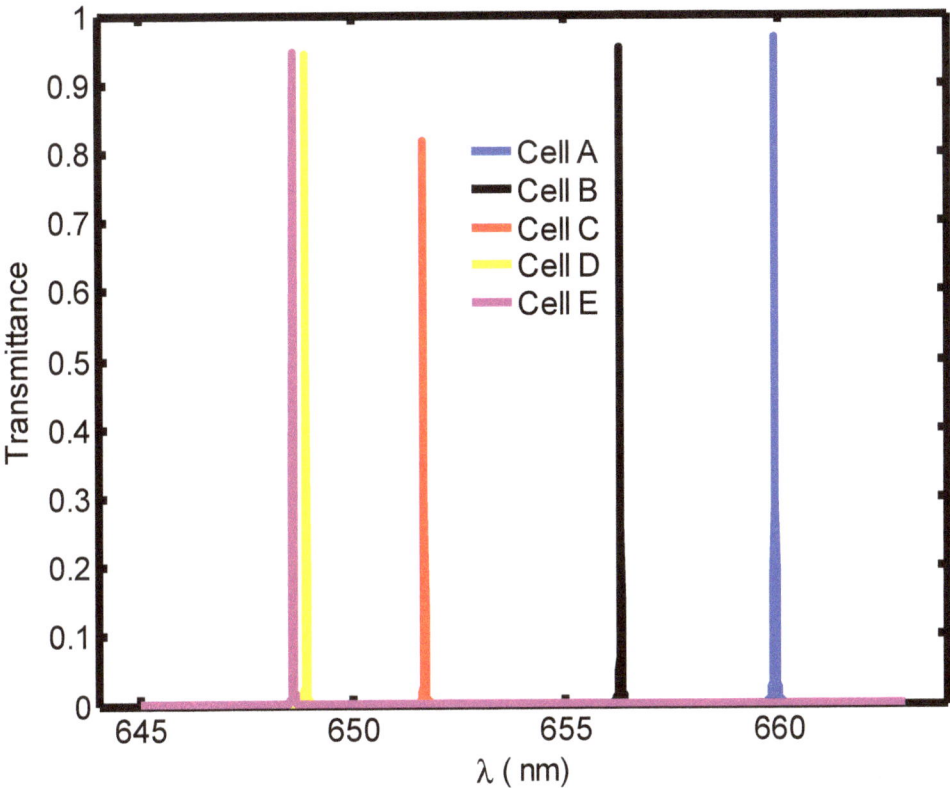

Figure 7. Transmission spectra showing five defect modes in solid blue, black, brown, yellow and purple line colors corresponding to RBC sample containing A, B, C, D and E blood cells separately, one at a time. The thickness of cavity layer is dd = 3000 nm under normal incidence.

4.4. Evaluation of the Performance of Proposed Biosensors Corresponding to Optimum Cavity Thickness under Normal Incidence

Apart from sensitivity, we have also examined to figure of merit (FoM) and quality factor (QF) values of proposed malaria sensors in true sense. These two parameters are also important while evaluating the working efficiency of any photonic biosensor. Mathematically, we can define FoM and QF with the help of following expressions as [6–11]

$$QF = \frac{\lambda_{peak}}{\lambda_{FWHM}} \tag{6}$$

$$FoM = \frac{S}{\lambda_{FWHM}} \tag{7}$$

To conclude our work, we have evaluated the S, FoM and QF of the proposed design under optimum cavity thickness of 5000 nm. The numeric values of S, FoM and QF of the proposed design, loaded with RBC samples containing Cell B, Cell C, Cell D and Cell E with respect to Cell A, are listed in Table 3 below.

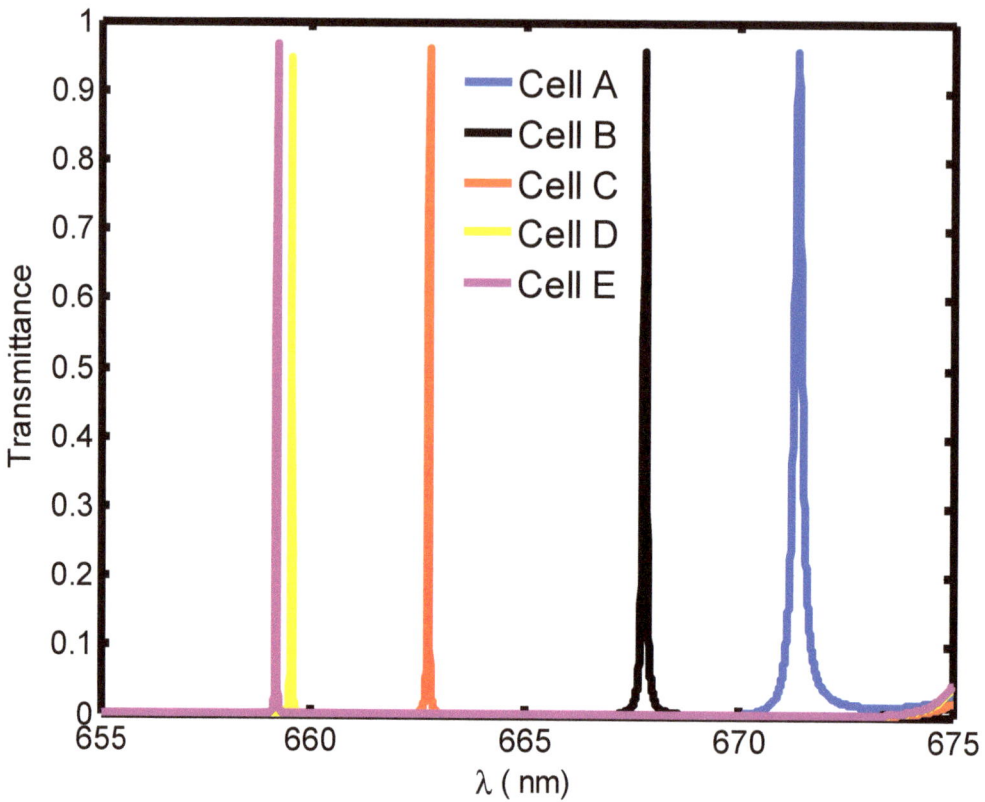

Figure 8. Transmission spectra showing five defect modes in solid blue, black, brown, yellow and purple line colors corresponding to RBC sample containing A, B, C, D and E blood cells separately, one at a time. The thickness of cavity layer is dd = 5000 nm under normal incidence.

Table 3. Performance evaluation table showing the numeric values of sensitivity, full width half maximum, figure of merit and quality factor of the proposed biosensor, corresponding to different RBC components under optimum condition.

Blood Component	Refractive Index	λ_{Peak} (nm)	S (nm/RIU)	λ_{FWHM} (nm)	FoM	QF
Cell A	1.408	671.4	—	0.11	—	6103.63
Cell B	1.396	667.8	300	0.036	8333.3	18550
Cell C	1.381	662.7	322.2	0.0135	23,866.7	49,088.8
Cell D	1.372	659.6	327.7	0.01	32,770	65960
Cell E	1.371	659.2	310.5	0.0095	32,684.2	69,389.47

It can easily be observed from the data in Table 3 that the sensitivity of the proposed biosensor varies between a maximum of 327.7 nm/RIU to a minimum of 300 nm/RIU when the cavity is infiltrated with RBC samples containing cell D and cell B, respectively. On the other hand, FoM and QF values vary between 2.3×10^4 to 8.33×10^3 and 6.93×10^4 to 1.85×10^4, respectively, depending upon the nature of malaria samples with respect to the water sample. Under the light of the above facts, we have come to the conclusion that our proposed design can be efficiently used for the detection of malaria infection from the preliminary stage (ring stage) to advanced stage (schizont stage).

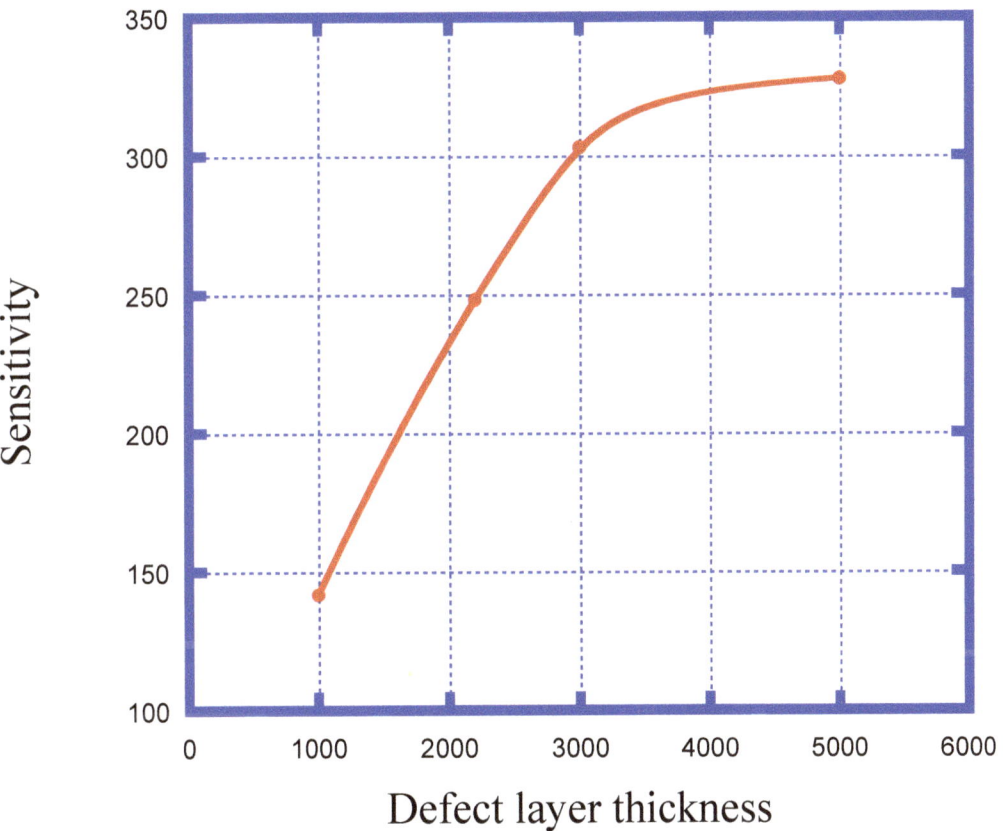

Figure 9. Variation of sensitivity with respect to defect layer thickness.

Finally, we have given our efforts to compare the findings of a proposed blood sensor for malaria detection with the similar kind of work based on various blood sensing applications. This comparison has been presented in Table 4, which highlights the blood sensing applications of various biosensors based on the principle of the refractive index sensing mechanism. This comparison shows that the proposed biosensor is suitable for sensing and detecting malaria infection from preliminary to advanced stages effectively. The dependence of our design on photonic biosensing technology makes it suitable for obtaining rapid, accurate and timely reports to ensure proper diagnosis, treatment and cure.

Table 4. Comparison of sensitivity, figure of merit and quality factor values of proposed 1D photonic blood sensor for malaria sensing and detection with similar kinds of work of other researchers at normal incidence (NR = Not reported).

Year	Structure Details	Type of Analyte	S	FoM	QF	Reference
2019	1D PC with graphene coated cavity walls	Blood plasma	51.49 nm/RIU	NR	NR	[55]
2019	2D PC waveguide structure	10 different blood components	473.38	7324.2	NR	[56]
2019	D shaped PC fiber	Blood glucose	0.83	NR	NR	[57]
2020	1D PC without coated cavity walls	Blood hemoglobin	141	0.48	NR	[58]
This work	1D PC without coated cavity walls	Red blood cells	327.7	32770	69,389.47	—

5. Conclusions

In the present piece of theoretical research work, we have explored the sensing and detection capabilities of 1D PC with defect for investigating malaria infection from preliminary to advanced stage by examining the different samples containing red blood cells A, B, C, D and E. We have used a transfer matrix formulation under normal incidence condition and MATLAB simulation software to obtain the results pertaining to the work. This study has been carried out on the five structures of different cavity layer thicknesses set as 1000 nm, 2200 nm, 3000 nm and 5000 nm to identify the structure having optimum biosensing performance. Our study shows that the biosensing performance of design maximizes corresponding to cavity thickness 5000 nm. The maximum sensitivity value obtained from this structure is 327.7 nm/RIU when the cavity is infiltrated with the RBC sample containing Cell D, which corresponds with the Schizont stage of malaria infection. Thus, our design can be very useful for identifying the person affected with different stages of malaria infection due to accuracy in the results. Additionally, the proposed work is based on minute sensing of the refractive index of different RBC samples of variation 1.408 to 1.371 corresponding to hemoglobin concentration 30.9g/dL to 15.9 g/dL, respectively. The maximum values of figure of merit and quality factor of proposed biosensing design are 32,770 RIU and 69,389.7, respectively, which is high as expected.

Author Contributions: The software handling, results preparation, investigation and initial manuscript draft preparation have been carried out by S.K.S. Conceptualization, methodology, reviewing, editing and supervision have been carried out by S.K.A. All authors have read and agreed to the published version of the manuscript.

Funding: There is no funding for the present work.

Institutional Review Board Statement: Not applicable.

Informed Consent Statement: Not applicable.

Data Availability Statement: It is not applicable to the present manuscript. The results of the present theoretical work are based on MATLAB simulations. All the relations and other relevant information have been properly cited throughout the manuscript, keeping the ease of readers of the journal. The readers can easily reproduce the results of the work with the help of theoretical details given in the manuscript, with the help of MATLAB computational software.

Conflicts of Interest: Authors do not have any conflict of interest.

References

1. Yablonovitch, E. Inhibited Spontaneous Emission in Solid-State Physics and Electronics. *Phys. Rev. Lett.* **1987**, *58*, 2059–2062. [CrossRef] [PubMed]
2. John, S. Strong localization of photons in certain disordered dielectric superlattices. *Phys. Rev. Lett.* **1987**, *58*, 2486–2489. [CrossRef] [PubMed]

3. Awasthi, S.K.; Mishra, A.; Malaviya, U.; Ojha, S. Wave propagation in a one-dimensional photonic crystal with metamaterial. *Solid State Commun.* **2009**, *149*, 1379–1383. [CrossRef]
4. Awasthi, S.K.; Malaviya, U.; Ojha, S.P. Enhancement of omnidirectional total-reflection wavelength range by using one-dimensional ternary photonic bandgap material. *J. Opt. Soc. Am. B* **2006**, *23*, 2566–2571. [CrossRef]
5. Danaie, M.; Kiani, B. Design of a label-free photonic crystal refractive index sensor for biomedical applications. *Photon-Nanostructures—Fundam. Appl.* **2018**, *31*, 89–98. [CrossRef]
6. Aly, A.H.; Awasthi, S.K.; Mohaseb, M.A.; Matar, Z.S.; Amin, A.F. MATLAB Simulation-Based Theoretical Study for Detection of a Wide Range of Pathogens Using 1D Defective Photonic Structure. *Crystals* **2022**, *12*, 220. [CrossRef]
7. Aly, A.H.; Awasthi, S.K.; Mohamed, D.; Matar, Z.S.; Al-Dossari, M.; Amin, A.F. Study on a one-dimensional defective photonic crystal suitable for organic compound sensing applications. *RSC Adv.* **2021**, *11*, 32973–32980. [CrossRef] [PubMed]
8. Aly, A.H.; Awasthi, S.K.; Mohamed, A.M.; Matar, Z.S.; Mohaseb, M.A.; Al-Dossari, M.; Tammam, M.T.; Zaky, Z.A.; Amin, A.F.; Sabra, W. Detection of Reproductive Hormones in Females by Using 1D Photonic Crystal-Based Simple Reconfigurable Biosensing Design. *Crystals* **2021**, *11*, 1533. [CrossRef]
9. Zaky, Z.A.; Moustafa, B.; Aly, A.H. Plasma cell sensor using photonic crystal cavity. *Opt. Quantum Electron.* **2021**, *53*, 591. [CrossRef]
10. Elsayed, A.M.; Ahmed, A.M.; Aly, A.H. Glucose sensor modeling based on Fano resonance excitation in titania nanotube photonic crystal coated by titanium nitride as a plasmonic material. *Appl. Opt.* **2022**, *61*, 1668. [CrossRef]
11. Mehaney, A.; Abadla, M.M.; Elsayed, H.A. 1D porous silicon photonic crystals comprising Tamm/Fano resonance as high performing optical sensors. *J. Mol. Liq.* **2021**, *322*, 114978. [CrossRef]
12. Troia, B.; Paolicelli, A.; De, F.; Passaro, V.M.N. Photonic Crystals for Optical Sensing: A Review. In *Advances in Photonic Crystals*; IntechOpen: Rijeka, Croatia, 2013. [CrossRef]
13. Zhou, X.; Zhang, L.; Armani, A.M.; Liu, J.; Duan, X.; Zhang, D.; Zhang, H.; Pang, W. An Integrated Photonic Gas Sensor Enhanced by Optimized Fano Effects in Coupled Microring Resonators With an Athermal Waveguide. *J. Light. Technol.* **2015**, *33*, 4521–4530. [CrossRef]
14. Bougriou, F.; Bouchemat, T.; Bouchemat, M.; Paraire, N. Optofluidic sensor using two-dimensional photonic crystal waveguides. *Eur. Phys. J. Appl. Phys.* **2013**, *62*, 11201. [CrossRef]
15. Pursiainen, O.L.J.; Baumberg, J.J.; Ryan, K.; Bauer, J.; Winkler, H.; Viel, B.; Ruhl, T. Compact strain-sensitive flexible photonic crystals for sensors. *Appl. Phys. Lett.* **2005**, *87*, 101902. [CrossRef]
16. Roggan, A.; Friebel, M.; Dörschel, K.; Hahn, A.; Müller, G. Optical Properties of Circulating Human Blood in the Wavelength Range 400–2500 nm. *J. Biomed. Opt.* **1999**, *4*, 36–46. [CrossRef]
17. A Report by the World Economic Forum and the Harvard School of Public Health. September 2011. Available online: www.weforum.org/EconomicsOfNCDappendix (accessed on 1 September 2019).
18. Lee, V.S.; Tarassenko, L. Absorption and multiple scattering by suspensions of aligned red blood cells. *J. Opt. Soc. Am. A* **1991**, *8*, 1135–1141. [CrossRef] [PubMed]
19. Zhernovaya, O.; Sydoruk, O.; Tuchin, V.; Douplik, A. The refractive index of human hemoglobin in the visible range. *Phys. Med. Biol.* **2011**, *56*, 4013–4021. [CrossRef]
20. Bayer, R.; Çağlayan, S.; Guenther, B. Discrimination between orientation and elongation of RBC in laminar flow by means of laser diffraction. *Proc. SPIE* **1994**, *2136*, 105–114. [CrossRef]
21. Prahl, S.A. 1999. Available online: http://omlc.ogi.edu/spectra/hemoglobin/index.html (accessed on 1 September 2019).
22. Enejder, A.M.K.; Swartling, J.; Aruna, P.; Andersson-Engels, S. Influence of cell shape and aggregate formation on the optical properties of flowing whole blood. *Appl. Opt.* **2003**, *42*, 1384–1394. [CrossRef]
23. Lindberg, L.-G.; Öberg, P. Optical properties of blood in motion. *Opt. Eng.* **1993**, *32*, 253–257. [CrossRef]
24. Sharma, P.; Sharan, P. Design of photonic crystal based ring resonator for detection of different blood constituents. *Opt. Commun.* **2015**, *348*, 19–23. [CrossRef]
25. Tangpukdee, N.; Duangdee, C.; Wilairatana, P.; Krudsood, S. Malaria Diagnosis: A Brief Review. *Korean J. Parasitol.* **2009**, *47*, 93–102. [CrossRef] [PubMed]
26. Gentilini, M. Maladies parasitaires. In *MédecineTropicale*, 5th ed.; 2ème Tirage Actualisé; Flammarion Médecine Science: Paris, France, 1995; pp. 51–122. Available online: https://www.decitre.fr/livres/medecine-tropicale-9782257143945.html (accessed on 1 September 2019).
27. WHO Report. 2019. Available online: https://www.who.int/malaria/publications/world-malaria-report-2019/en/ (accessed on 1 September 2019).
28. Agnero, M.A.; Konan, K.; Tokou, Z.G.C.S.; Kossonou, Y.T.A.; Dion, B.S.; Kaduki, K.A.; Zoueu, J.T. Malaria-Infected Red Blood Cell Analysis through Optical and Biochemical Parameters Using the Transport of Intensity Equation and the Microscope's Optical Properties. *Sensors* **2019**, *19*, 3045. [CrossRef]
29. Erdman, L.K.; Kain, K.C. Molecular diagnostic and surveillance tools for global malaria control. *Travel Med. Infect. Dis.* **2008**, *6*, 82–99. [CrossRef] [PubMed]
30. Bell, D.; Wongsrichanalai, C.; Barnwell, J.W. Ensuring quality and access for malaria diagnosis: How can it be achieved? *Nat. Rev. Genet.* **2006**, *4*, S7–S20. [CrossRef]

31. E Clendennen, T.; Long, G.W.; Baird, J.K.; Iii, T.E.C. QBC® and Giemsa-stained thick blood films: Diagnostic performance of laboratory technologists. *Trans. R. Soc. Trop. Med. Hyg.* **1995**, *89*, 183–184. [CrossRef]
32. Shaheen, S.A.; Taya, S.A. Propagation of p-polarized light in photonic crystal for sensor application. *Chin. J. Phys.* **2017**, *55*, 571–582. [CrossRef]
33. Sovizi, M.; Aliannezhadi, M. Design and simulation of high-sensitivity refractometric sensors based on defect modes in one-dimensional ternary dispersive photonic crystal. *J. Opt. Soc. Am. B* **2019**, *36*, 3450–3456. [CrossRef]
34. Taya, S. Ternary photonic crystal with left-handed material layer for refractometric application. *Opto-Electronics Rev.* **2018**, *26*, 236–241. [CrossRef]
35. Banerjee, A. Design of enhanced sensitivity gas sensors by using 1D defect ternary photonic band gap structures. *Indian J. Phys.* **2019**, *94*, 535–539. [CrossRef]
36. Xie, Q.; Chen, Y.; Li, X.; Yin, Z.; Wang, L.; Geng, Y.; Hong, X. Characteristics of D-shaped photonic crystal fiber surface plasmon resonance sensors with different side-polished lengths. *Appl. Opt.* **2017**, *56*, 1550–1555. [CrossRef]
37. Zhang, Z.; Shen, T.; Wu, H.; Feng, Y.; Wang, X. Polished photonic crystal fiber refractive index sensor based on surface plasmon resonance. *J. Opt. Soc. Am. B* **2021**, *38*, F61. [CrossRef]
38. Nie, T.; Han, Z.; Gou, Z.; Wang, C.; Tian, H. High anti-interference dual-parameter sensor using EIT-like effect photonic crystal cavity coupled system. *Appl. Opt.* **2022**, *61*, 1552–1558. [CrossRef] [PubMed]
39. Baraket, Z.; Soltani, O.; Kanzari, M. Design of magnetic field direction's sensor based on a 1D tunable magneto-photonic crystal. *Opt. Quantum Electron.* **2022**, *54*, 637. [CrossRef]
40. Parandin, F.; Heidari, F.; Aslinezhad, M.; Parandin, M.M.; Roshani, S.; Roshani, S. Design of 2D photonic crystal biosensor to detect blood components. *Opt. Quantum Electron.* **2022**, *54*, 618. [CrossRef]
41. Parandin, F.; Heidari, F.; Rahimi, Z.; Olyaee, S. Two-Dimensional photonic crystal Biosensors: A review. *Opt. Laser Technol.* **2021**, *144*, 107397. [CrossRef]
42. Liu, Y.; Salemink, H.W.M. Photonic crystal-based all-optical on-chip sensor. *Opt. Express* **2012**, *20*, 19912–19920. [CrossRef]
43. Olyaee, S.; Dehghani, A.A. Ultrasensitive Pressure Sensor Based on Point Defect Resonant Cavity in Photonic Crystal. *Sens. Lett.* **2013**, *11*, 1854–1859. [CrossRef]
44. Pacholski, C. Photonic Crystal Sensors Based on Porous Silicon. *Sensors* **2013**, *13*, 4694–4713. [CrossRef]
45. Matar, Z.S.; Al-Dossari, M.; Awasthi, S.K.; El-Gawaad, N.S.A.; Hanafy, H.; Amin, R.M.; Fathy, M.I.; Aly, A.H. Theoretical Study on Polycarbonate-Based One-Dimensional Ternary Photonic Structures from Far-Ultraviolet to Near-Infrared Regions of Electromagnetic Spectrum. *Crystals* **2022**, *12*, 642. [CrossRef]
46. Del Villar, I.; Matias, I.R.; Arregui, F.J.; Claus, R. Analysis of one-dimensional photonic band gap structures with a liquid crystal defect towards development of fiber-optic tunable wavelength filters. *Opt. Express* **2003**, *11*, 430–436. [CrossRef] [PubMed]
47. Saravanan, S.; Dubey, R. Optical and morphological studies of TiO2 nanoparticles prepared by sol–gel method. *Mater. Today: Proc.* **2021**, *47*, 1811–1814. [CrossRef]
48. Awasthi, S.K.; Panda, R.; Shiveshwari, L. Multichannel tunable filter properties of 1D magnetized ternary plasma photonic crystal in the presence of evanescent wave. *Phys. Plasmas* **2017**, *24*, 072111. [CrossRef]
49. Awasthi, S.K.; Panda, R.; Chauhan, P.K.; Shiveshwari, L. Multichannel tunable omnidirectional photonic band gaps of 1D ternary photonic crystal containing magnetized cold plasma. *Phys. Plasmas* **2018**, *25*, 052103. [CrossRef]
50. Ankita; Suthar, B.; Bhargava, A. Biosensor Application of One-Dimensional Photonic Crystal for Malaria Diagnosis. *Plasmonics* **2020**, *16*, 59–63. [CrossRef]
51. Agnero, M.A.; Konan, K.; Kossonou, A.T.; Bagui, O.K.; Zoueu, J.T. A New Method to Retrieve the Three-Dimensional Refractive Index and Specimen Size Using the Transport Intensity Equation, Taking Diffraction into Account. *Appl. Sci.* **2018**, *8*, 1649. [CrossRef]
52. Mazeron, P.; Muller, S.; Elazouzi, H. Deformation of erythrocytes under shear: A small-angle light scattering study. *Biorheology* **1997**, *34*, 99–110. [CrossRef]
53. Phillips, K.G.; Jacques, S.L.; Mccarty, O.J.T. Measurement of Single Cell Refractive Index, Dry Mass, Volume, and Density Using a Transillumination Microscope. *Phys. Rev. Lett.* **2012**, *109*, 118105. [CrossRef]
54. Tycko, D.H.; Metz, M.H.; Epstein, E.A.; Grinbaum, A. Flow-cytometric light scattering measurement of red blood cell volume and hemoglobin concentration. *Appl. Opt.* **1985**, *24*, 1355. [CrossRef]
55. El-Khozondar, H.J.; Mahalakshmi, P.; El-Khozondar, R.J.; Ramanujam, N.; Amiri, I.; Yupapin, P. Design of one dimensional refractive index sensor using ternary photonic crystal waveguide for plasma blood samples applications. *Phys. E: Low-Dimensional Syst. Nanostructures* **2019**, *111*, 29–36. [CrossRef]
56. Mohammed, N.A.; Hamed, M.M.; Khalaf, A.A.; Alsayyari, A.; El-Rabaie, S. High-sensitivity ultra-quality factor and remarkable compact blood components biomedical sensor based on nanocavity coupled photonic crystal. *Results Phys.* **2019**, *14*, 102478. [CrossRef]

57. Lidiya, A.E.; Raja, R.V.J.; Pham, V.D.; Ngo, Q.M.; Vigneswaran, D. Detecting hemoglobin content blood glucose using surface plasmon resonance in D-shaped photonic crystal fiber. *Opt. Fiber Technol.* **2019**, *50*, 132–138. [CrossRef]
58. Abadla, M.M.; Elsayed, H.A. Detection and sensing of hemoglobin using one-dimensional binary photonic crystals comprising a defect layer. *Appl. Opt.* **2020**, *59*, 418–424. [CrossRef] [PubMed]

Disclaimer/Publisher's Note: The statements, opinions and data contained in all publications are solely those of the individual author(s) and contributor(s) and not of MDPI and/or the editor(s). MDPI and/or the editor(s) disclaim responsibility for any injury to people or property resulting from any ideas, methods, instructions or products referred to in the content.

Article

Modified Electrode with ZnO Nanostructures Obtained from Silk Fibroin for Amoxicillin Detection

Cristina Dumitriu, Alexandra Constantinescu, Alina Dumitru and Cristian Pîrvu *

Faculty of Chemical Engineering and Biotechnologies, Politehnica University of Bucharest, 1-7 Polizu Street, 011061 Bucharest, Romania
* Correspondence: cristian.pirvu@upb.ro; Tel.: +40-214-023-930; Fax: +40-213-111-796

Abstract: Antibiotics are a novel class of contaminants that represent a substantial risk to human health, making their detection an important task. In this study, ZnO nanostructures were prepared starting from Bombyx mori silk fibroin and $Zn(NO_3)_2$, using thermal treatment. The resulting ZnO structures were characterized using SEM, FT-IR, and XRD. They had a fibrous morphology with a wurtzite crystalline structure, with nanometric dimensions. FT-IR and XRD confirmed silk fibroin's disappearance after thermal treatment. To prepare modified electrodes for amoxicillin (AMX) antibiotic detection, ZnO nanostructures were mixed with Nafion polymer and drop-casted on an electrode's surface. Parameters such as drying time and concentration appeared to be important for electrochemical detection. Differential pulse voltammetry (DPV) was sensitive for AMX detection. The measurements revealed that the novel electrode based on ZnO nanostructures embedded in Nafion polymer has potential to be used for AMX electrochemical detection.

Keywords: nanostructures; silk fibroin; electrochemical sensor

1. Introduction

One of the most frequently prescribed antibiotics for infectious diseases that can affect humans (such as Anthrax, Lyme disease, pneumonia, gastroenteritis, otitis, vaginal infections, urinary infections, and oral infections) is amoxicillin (AMX) [1,2]. AMX, 6-(p-hydroxy-α-amino phenyl acetoamido) penicillanic acid [3], is also a common antibiotic used in animal husbandry, and because of this, its residues may be found in animal products, other agricultural products, and the agricultural environment [4]. Such contaminants in processed beef pose a serious threat to human health and can harm kidney and liver functions [4]. Beta-lactam antibiotics such as AMX have a stable chemical structure, high levels of toxicity, and slow rates of biodegradation [2]. They can therefore readily be discharged into the environment through wastewater from food and livestock production, human excretion, hospital wastewater discharge, and poor wastewater treatment [5]. Considering these, the monitoring of antibiotics such as AMX is very important.

Multiple analytical methods have been used to detect AMX, primarily surface plasmon resonance, chromatography, capillary electrophoresis, liquid chromatography coupled to tandem mass spectrometry (LC–MS-MS), spectrofluorometric and microbiological methods, enzymatic quantification, fluorescence, and spectrophotometry [2,3,6]. There is a need for simpler, less expensive, quicker, more sensitive, and more selective amoxicillin determination methods, because most current methods need time-consuming and expensive sample pre-treatment procedures [2]. Due to their benefits, including low cost, easy operation, quick measurement, and strong sensitivity and selectivity, electrochemical techniques have been used extensively in the study of antibiotics in recent decades. Electrochemical methods can also be used to quickly test for pollution on-site [5].

In the electrochemical approach, the reduction–oxidation reactions of analytes appear on the surface of the working electrode, which is a very important part of the analytical measurement. Therefore, in an effort to improve the analyte signal, scientists have attempted to modify the electrode surface using cutting-edge techniques [5]. Most investigations on the electrochemical detection of amoxicillin in the literature are based on metallic nanoparticles combined with CNTs and graphene to produce various geometries and stacking arrangements [6].

According to the literature [7], the chemical stability of the oxides and surface OH bonds enables the covalent grafting of probe molecules in sensor applications. An example of an oxide used for AMX detection is TiO_2 [7]. Another oxide, ZnO, has generated a great deal of interest because of its admirable qualities, including inexpensive cost, great abundance, excellent catalytic activity against biological and chemical species, and antifouling characteristics [8–10]. ZnO nanoparticles also offer unique physical characteristics, including high electron mobility, a tunable band position, high chemical and thermal stability, and non-toxicity [11]. ZnO is a metal oxide semiconductor with a straight and wide band gap. Its hexagonal wurtzite crystal structure makes a variety of micro- and nanostructured materials easy to produce, including tubes, combs, and rods [8].

Various techniques can be used to create zinc oxide nanoparticles, one of them being microfluidic reactor-based synthesis [12–14]. Chemical processes including hydrothermal, sol–gel, and microemulsion, as well as physical methods such as ball milling, physical vapor deposition, and laser ablation, were also used for the synthesis of ZnO nanoparticles [15]. Biological (green) synthesis techniques are also included among the conventional methods. Various efforts have been made to create ZnO NPs from various green sources, including bacteria, fungi, algae, plants, and others [16].

Recently, ZnO nanostructures have been prepared starting from silk fibroin from silkworm *Bombyx Mori* cocoons [17]. Fibroin is a highly adaptable biopolymer that has enabled the development of a diverse range of materials whose properties and architectures may be tailored to meet specific application requirements [18]. Additionally, the B. mori silkworm is practical for industrial-scale breeding, making it simple to obtain silkworms and their silk, and exhibits strong biocompatibility and biodegradability [19]. It contains in situ active sites for the anchoring of a zinc nitrate—ZnO—precursor [17]. The mechanical stability, flexible biomimetic form, and biocompatibility of silk fibroin fibers were carried over into the as-prepared nanostructures [17].

This paper proposes a new electrochemical device based on ZnO nanostructures derived from a natural compound named silk fibroin. These nanostructures were incorporated into a Nafion polymer matrix. Some characteristics, such as drying time and concentration, were discovered to be crucial in the modification of a glassy carbon (GC) electrode. This type of modified electrode was proven to be useful for AMX electrochemical detection.

2. Materials and Methods

2.1. Preparation of ZnO

First, *Bombix Mori* cocoons from a local farmer were used; the larvae were extracted from the cocoons, and they were cut into small pieces. In a 2 L Berzelius beaker, ultrapure water (obtained with a Millipore Direct-Q UV3 water filtration system) was brought to a boil (100 °C) on a thermostatic hot plate (Stuart, model UC152) and sodium carbonate (anhydrous Na_2CO_3, 99.9%, Sigma Aldrich, Poznań, Poland) was added to 0.02 M. Boiling time was 30 min. In the last stage, the fibroin was rinsed with ultrapure water under magnetic stirring. The operation was repeated 2–3 times to remove the sodium carbonate and sericin dissolved in this solution. The silk fibroin (SF) that was produced in this way was left to dry at room temperature for 24 h.

Dried silk fibroin was then mixed with a solution of 1.462 g of $Zn(NO_3)_2 \cdot 6H_2O$ (99% metal basis, Alpha Aesar, Kandel, Germany) in 40 mL of a 1:1 distilled water : ethanol mixture at room temperature for 12 h. The $Zn(NO_3)_2$-soaked fibroin was subjected to thermal treatment for 1 h at 200 °C, followed by 6 h at 600 °C, in a muffle furnace (LEF-1035, Daihan Labtech, Namyangju-city, Korea). This step led to the burning of fibroin and the crystallization of ZnO.

2.2. Preparation of Modified Glassy Carbon Electrodes

Glassy carbon (GC) electrodes (3 mm, Metrohm, Bucharest, Romania) were mirror-polished on polishing cloth (PRESI) with 1 μm diamond paste and 0.3 μm alumina slurry, followed by ultrasonic treatment in ultrapure water for 10 min.

Next, 25 μL Nafion solution (20 % in lower aliphatic alcohol and water from Sigma Aldrich, St Louis, Missouri, USA) was diluted to 1 mL with 0.1 M phosphate buffer solution (PBS) with pH 7. PBS was prepared from 0.2 M stock aqueous solutions of K_2HPO_4 and KH_2PO_4. The purity of both was 98% and they were acquired from Sigma Aldrich. After cleaning the GC electrode, 3 μL of this solution was dripped onto the surface and it was left to dry. This electrode will be known as GC/Nafion.

In the meantime, 10 mg of ZnO nanostructure was added to a diluted Nafion solution in PBS, and the resulting mixture was ultrasonicated for 1 h. Then, 6 μL or 9 μL (in 3 μL portions) of the ZnO nanoparticles and Nafion solution was drop-casted onto the surface of the cleaned GC electrode with a micropipette. The electrode prepared with 6 μL and a drying time of 1 h was named GC/Nafion/ZnO 6 (1 h), the one prepared with 6 μL and a drying time of 24 h was named GC/Nafion/ZnO 6, and the last one prepared with 9 μL and a drying time of 24 h was named GC/Nafion/ZnO 9.

2.3. ZnO Structures and Modified Electrode Characterization

The modified electrode and the resulting ZnO structures were characterized using a Thermo Fisher Scientific Quanta 650 FEG Scanning Electron Microscope (SEM, Hillsborough, CA, USA), which features ESEM technology and high-resolution scanning, equipped with a Bruker energy-dispersive X-ray system (EDX) for chemical analysis.

A Perkin Elmer Spectrum 100 ATR FT-IR (Perkin Elmer, Waltham, MA, USA) was used to record the Fourier transform infrared spectrum (FTIR). Four sequential scans were used to record between 4000 and 600 cm^{-1}. The corresponding program was used to process spectra acquired at a resolution of 4 cm^{-1}, performing background correction and smoothing.

The crystalline phases and their structures were identified using a Shimadzu X-ray diffraction (XRD) 6000 diffractometer with Ni-filtered Cu K radiation (λ = 0.154 nm), 2θ, varying between 10° and 70°.

2.4. Amoxicillin Detection

Two electrochemical methods were used for amoxicillin detection: cyclic voltammetry (CV) and DPV. Experiments were carried out in a single-compartment electrochemical cell, in a solution of PBS Ph = 7. Three electrodes were connected to a potentiostat/galvanostat (Autolab 302 N, Metrohm, Barendrecht, Nederland): a GC working electrode, an AG/AgCl 3 M KCl reference electrode (Metrohm, Bucharest, Romania), and a Pt rod (Metrohm, Bucharest, Romania). For CV measurements, the scan rate was 25 mV/s between 0.3 and 1.3 V with 0.02 V steps. DPV was recorded between 0.7 and 1.2 V with a 50 mV modulation amplitude and 20 mV/s scan rate. In addition, 0.1 M PBS pH 7 prepared from 0.2 M stock aqueous solutions of K_2HPO_4 and KH_2PO_4 (Sigma Aldrich, St Louis, Missouri, USA) with or without amoxicillin was employed as the electrolyte.

3. Results

3.1. Modified Electrode Characterization

Prepared ZnO structures and modified GC electrodes were subjected to SEM analysis and the obtained images are presented in Figure 1. In Figure 1a,b, it can be observed that the ZnO structures had a fibrous morphology with diameters between 1 and 2 μm.

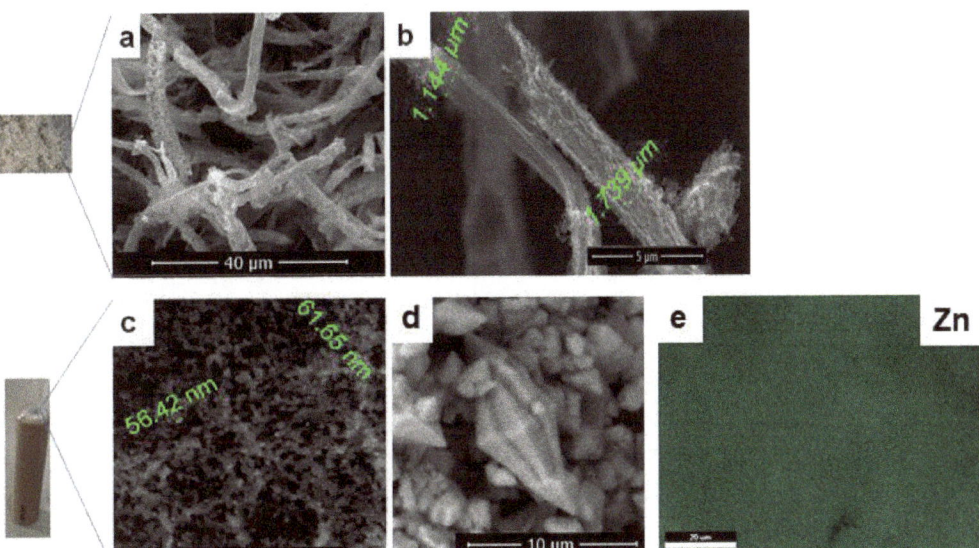

Figure 1. SEM images of obtained ZnO nanostructures (**a,b**) and modified GC/Nafion ZnO 6 electrode (**c,d**). EDS mapping of Zn from GC/Nafion ZnO 6 electrode (**e**).

ZnO structures were subjected to 1 h ultrasonic treatment before depositing on GC. In Figure 1c,d, there are corresponding top-view SEM images of the GC/Nafion ZnO 6 electrode. It is visible that ZnO was embedded in the Nafion matrix as nanoparticles with uniform distribution.

According to the elemental EDS mapping (Figure 1e), the polymer matrix contributed to the homogeneous dispersion of the ZnO nanoparticles without agglomeration. The uniform Zn distribution appeared on the entire surface of the electrode, as can be seen in Figure 1e, in a percentage of 15%, which helped to ensure sensor reproducibility. The Nafion matrix, with its antifouling qualities, was successfully used for nanoparticle dispersion, which led to a greater specific surface area at the electrode surface. Moreover, 51% oxygen was found, a part of the percentage being a component of Nafion and the other part associated with ZnO.

Figure 2a shows the FT-IR spectra recorded for zinc oxide structures deposited on fibroin templates compared to the spectrum recorded for silk fibroin. Thus, in the spectrum of fibroin, the peaks specific to the β-sheet structure can be observed: 1511 cm^{-1} and 1621 cm^{-1}. In the spectra of the zinc oxide nanostructures, the absorption peak at 3500 cm^{-1}, characteristic of the stretching vibration of the -OH group, can be observed. The absorption peaks at 2300 cm^{-1} and 2400 cm^{-1} are attributed to the CO_2 group, and the absorption peak at 1416 cm^{-1} is attributed to the C-C stretching vibration. The absorption band formed at 995 cm^{-1} is attributed to ZnO. These bands are consistent with data found in the literature [11,20].

In Figure 2b, the spectra obtained for zinc oxide structures obtained on the fibroin template and for the glassy carbon-modified electrode GC/Nafion/ZnO 6 are presented. In the GC/Nafion/ZnO 6 spectrum, we can observe -OH stretching vibration at 3336 cm^{-1} corresponding to adsorbed water. The peak recorded at 1005 cm^{-1} is due to S-O symmetric stretching, the band obtained at 1148 cm^{-1} is due to symmetric stretching of C-F, and the band obtained at 1215 cm^{-1} is due to the asymmetric C-F stretching. The peak obtained at 1005 cm^{-1} is attributed to the S-O group [21]. The peak at 988 cm^{-1} corresponding to ZnO [11] is covered by the much larger peak specific to Nafion. The band obtained at 637 cm^{-1} is attributed to the stretching of the C-S group from the Nafion structure [21].

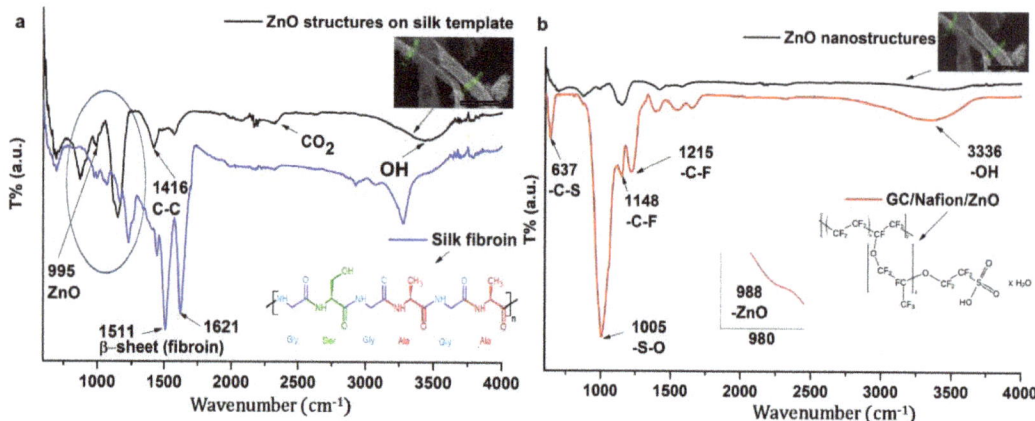

Figure 2. FT-IR spectra: (**a**) comparison between SF and ZnO structures on SF template and (**b**) comparison between ZnO nanostructures and GC/Nafion/ZnO 6.

The quality of the crystals and the orientation of the synthesized ZnO nanoparticles were studied using X-ray diffraction (XRD), and the patterns are represented in Figure 3. The sharp and narrow peaks demonstrate the sample's high crystallinity. By comparison with the data from JCPDS card No. 89-7102, all the XRD peaks are very well correlated with the hexagonal phase (wurtzite structure), with no evidence of a secondary phase. The first three peaks, defined in Table 1, were utilized to calculate particle sizes using the Scherrer equation:

$$D = \frac{k\lambda}{\beta \cos \theta} \quad (1)$$

where D is the crystallite size, k = 0.9 (Scherrer constant); λ is the light wavelength utilized for diffraction, which is equal to 1.54 Å; β is the full width at half maximum (FWHM) of the diffraction peak, and θ is the reflection angle (Bragg's angle).

The peaks of the ZnO nanostructures match the hexagonal wurtzite structure of zinc oxide. The Miller indices of the ZnO hexagonal phase's typical peaks, presented in Figure 3, are (100), (002), (101), (102), (110), (103), and (112). The average size of crystallites was 37 nm.

3.2. Selecting Proper Voltametric Method for AMX Detection

The first tested method was cyclic voltammetry. CV curves were obtained in phosphate buffer solution and phosphate buffer with a high concentration of amoxicillin (100 µM). They are presented in Figure 4.

The unmodified GC electrode (Figure 4a) and GC/Nafion/ZnO 6 (Figure 4c) gave a signal in the presence of amoxicillin, while the GC electrode on which only the Nafion polymer was deposited (Figure 4b) did not show a signal in the presence of amoxicillin. The current values recorded for GC/Nafion are much lower compared to those recorded for GC.

Further, differential pulse voltammetry was used. The DPV curves corresponding to the GC and GC/Nafion/ZnO 6 (1 h) in phosphate buffer solution and phosphate buffer with amoxicillin 100 µM are shown in Figure 5. It is observed that a signal can be obtained in the presence of amoxicillin with the unmodified electrode. With the help of the modified electrode, a much better signal is obtained, the peak current being of the order of 8.2×10^{-8} A, compared to 4.4×10^{-8} A in the case of GC.

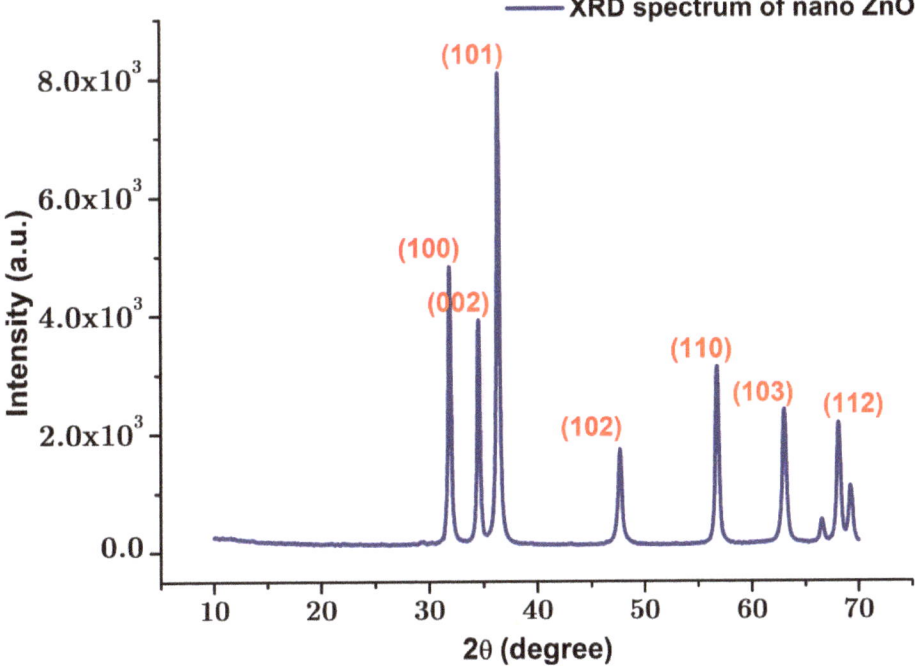

Figure 3. The X-ray diffraction (XRD) pattern of ZnO nanotructures prepared on silk fibroin template.

Table 1. The first three peaks obtained from XRD spectrum.

No.	2-Theta [θ]	d [Å]	FWHM [θ]	Crystallite Size [nm]	Phase Name
1	31.4815	2.83945	0.1528	54	ZnO (100)
2	34.5518	2.59384	0.2720	30	ZnO (002)
3	36.3804	2.46755	0.2991	27	ZnO (101)

The drying time was increased from 1 h to 24 h and the obtained DPV curves are presented in Figure 6b. Consecutive scans were again performed in 100 μM AMX solution. It is observed that the signal, with the peak at 0.9 V, suffers a slight drop but is more stable.

3.3. Improving the Electrode Signal for AMX Detection

In order to improve the signal, an electrode with a higher concentration of Nafion + ZnO, namely GC/Nafion/ZnO 9, was prepared. The signal obtained in DPV with GC/Nafion/ZnO 6 was compared to the one obtained with GC/Nafion/ZnO 9, with the results being presented in Figure 7. The drying time was 24 h for both electrodes. The obtained peak height was greatly improved, from 6.8×10^{-8} A to 1.46×10^{-7} A.

To check the electrode stability, a series of three consecutive scans were performed in the 100 μM AMX solution in PBS with the GC/Nafion/ZnO 6 electrode (1 h). It was observed (Figure 6a) that the signal gradually decreased, with a very large difference between scans 1 and 3.

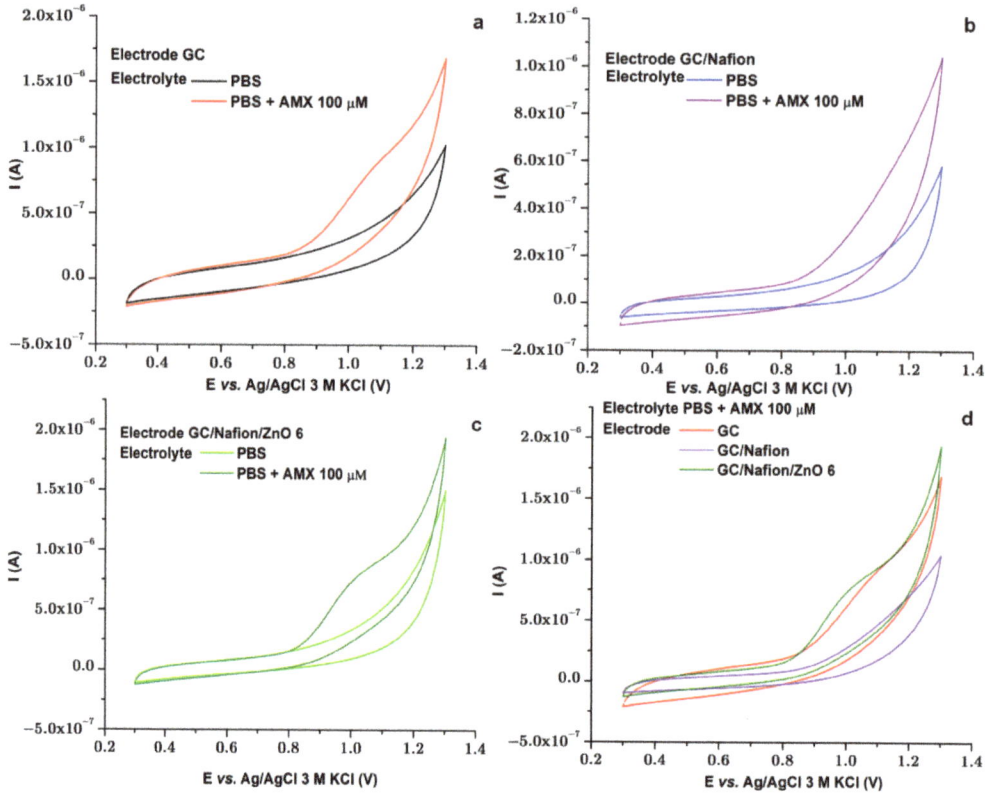

Figure 4. Cyclic voltammetry curves recorded in PBS and PBS + AMX 100 μM corresponding to (**a**) GC electrode, (**b**) GC/Nafion electrode, (**c**) GC/Nafion/ZnO 6, (**d**) GC/Nafion, and GC/Nafion/ZnO 6 compared with an uncoated GC electrode.

The stability of the signal was also checked for this electrode. It is observed from Figure 8a that the signal is stable for three successive scans (scan 1–3). The electrode was tested again after one week (Figure 8a, scans 4–6) and the signal decreased by 8%. It was also checked whether this last method of electrode modification could be reproduced. Thus, two similar GC/Nafion_ZnO 9 electrodes were prepared. Scans were made in freshly prepared 80 μM amoxicillin solution, with the results being presented in Figure 8b. It is observed that the obtained signal is similar.

3.4. Calibration Curve

To obtain the calibration curve, AMX solutions in PBS with concentrations between 5 and 110 μM were prepared. An increase in the peak height corresponding to the increase in the concentration of amoxicillin is observed in Figure 9b. The AMX redox reaction (Figure 9a) on the GC/Nafion/ZnO 9 electrode leads to a calibration curve with good linearity, r^2 being 0.998, as can be observed in the Figure 9b inset.

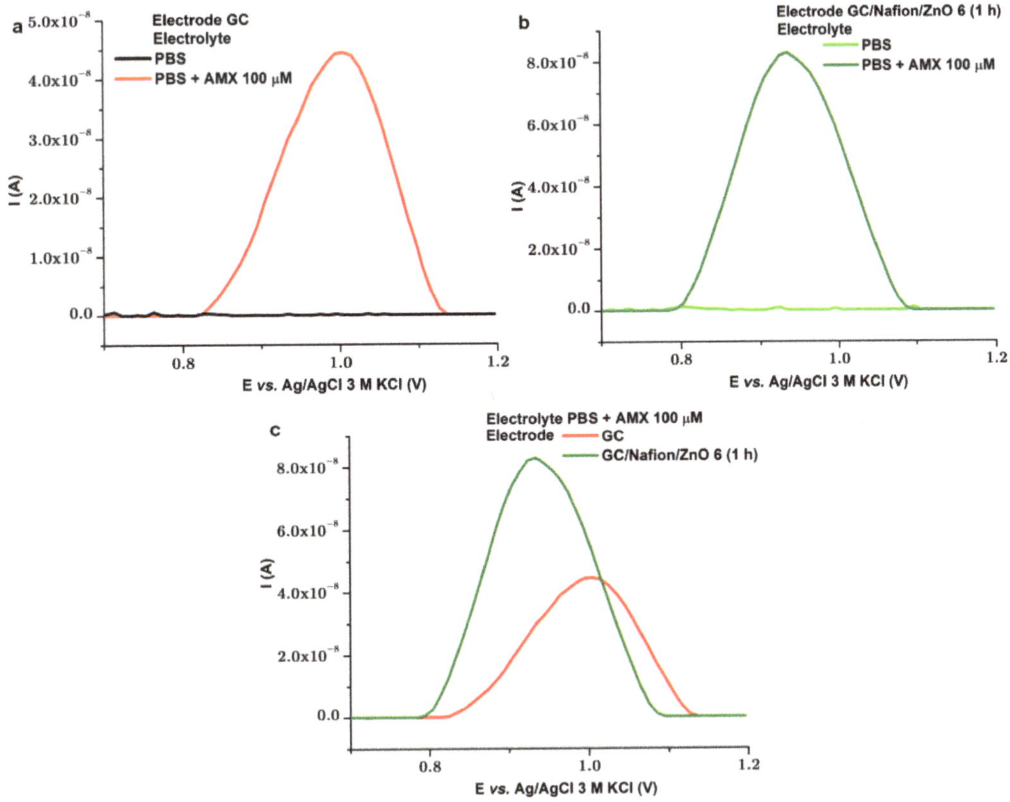

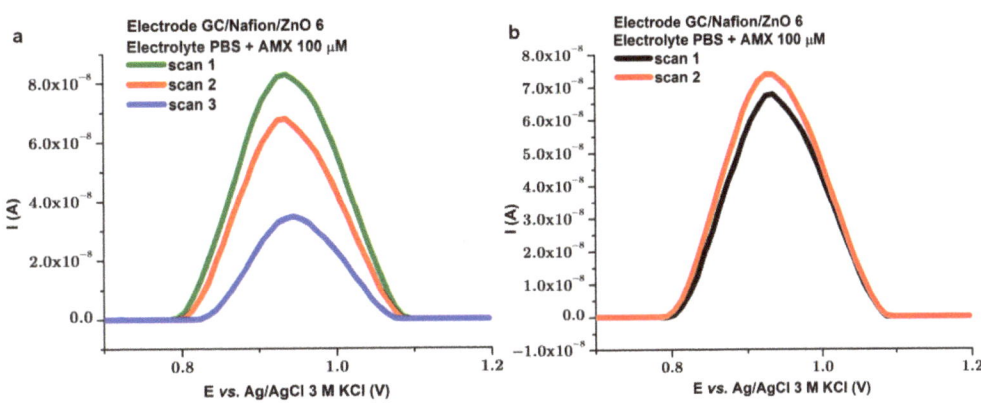

Figure 5. DPV curves corresponding to (**a**) GC in PBS and PBS with AMX 100 μM, (**b**) GC/Nafion/ZnO 6 (1 h) in PBS and PBS with AMX 100 μM and (**c**) comparation between GC and GC/Nafion/ZnO 6 (1 h) in PBS with AMX 100 μM.

Figure 6. Differential pulse voltammetry curves in PBS with AMX 100 μM consecutive scans using GC/Nafion ZnO 6 electrode with (**a**) 1 h drying time, (**b**) 24 h drying time.

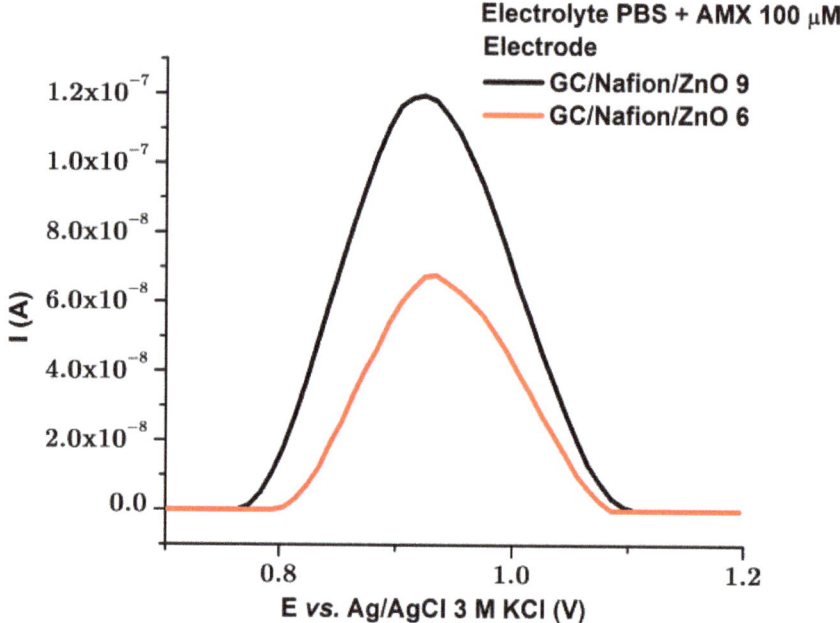

Figure 7. DPV curves in PBS solution + AMX 100 μM corresponding to GC/Nafion/ZnO 9 and GC/Nafion/ZnO 6.

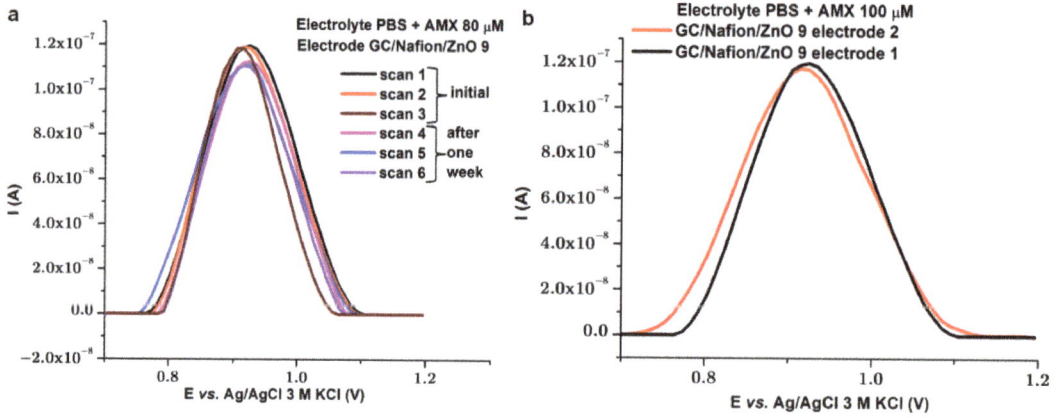

Figure 8. DPV curves in PBS + AMX 80 μM recorded with GC/Nafion/ZnO 9: (**a**) consecutive scans with the same electrode, (**b**) scans recorded with two freshly prepared electrodes.

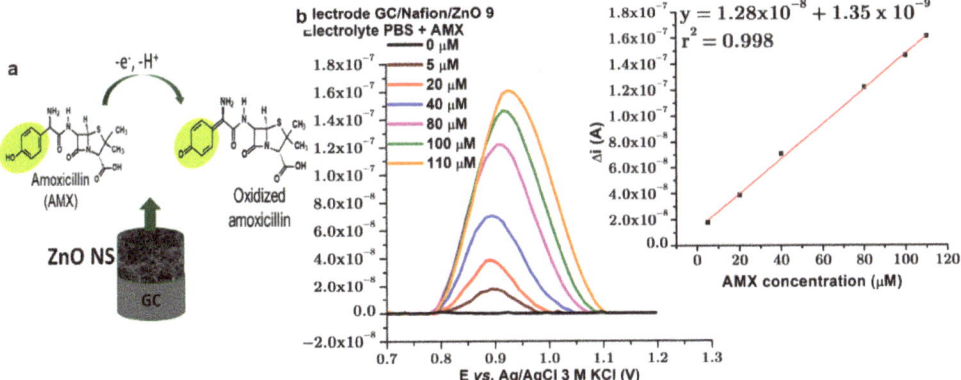

Figure 9. (**a**) Redox reaction of AMX; (**b**) DPV curves in PBS + AMX 6 different concentrations; and calibration curve using GC/Nafion/ZnO 9 electrode.

4. Discussion

4.1. ZnO Structures and Electrode Modification

To create ZnO nanostructures, the template method was used in different studies [22]. In this method, nanostructures are grown through the pores or channels of a template using a nucleation and growth process, using a template with a preset size and shape. Hard and soft templates are examples of the two different types of templates used for ZnO synthesis. A soft template is made up of flexible polymers and biomolecular and single molecular-based templates, whereas a hard template often consists of CNT and porous alumina [22]. In this study, flexible biopolymer silk fibroin was used. The ZnO-prepared structures have a fibrous morphology (Figure 1) and larger specific surface area, as can be seen from the SEM images. ZnO structures prepared in a similar manner were reported to have a wurtzite crystallite structure [17]. The wurtzite's tetrahedral atom coordination leads to a non-centrosymmetric crystal structure with polar Zn^{2+} and O_2 surfaces, which gives rise to piezoelectric qualities that make it possible to use these nanostructures in mechanical actuators and sensing technology [23].

In the scientific literature, two types of conformations specific to silk fibroin are specified: β-sheets—with a crystalline, organized, hydrophobic domain—and a random coil/α-helix domain specific to amides I, II, and III. For the random coil/α-helix conformation, the corresponding peaks are found in the ranges of 1638–1660 cm^{-1}, 1536–1545 cm^{-1}, and 1235 cm^{-1} for amides I, II, and III, respectively. In the case of the crystal conformation, the β-sheet, the corresponding peaks are found in the ranges of 1616–1637 cm^{-1}, 1513–1525 cm^{-1}, and 1265 cm^{-1} for amides I, II, and III, respectively [20]. From the obtained spectra shown in Figure 2a, our prepared silk fibroin has a β-sheet configuration.

For ZnO structures, the silk fibroin template was removed after thermal treatment at 600 °C and the $Zn(NO_3)_2$ was decomposed into ZnO. As can be seen from Figure 2a, in the FT-IR spectra, peaks specific to fibroin are not present in the spectra corresponding to ZnO structures.

These ZnO structures were successfully embedded in the Nafion polymer matrix, as can be seen from Figure 2b. Nafion is a well-known ionic polymer with benefits of good ionic conductivity, cation selectivity, chemical inertness, and thermal stability. It is made up of sulfonate groups and a stable hydrophobic polytetrafluoroethylene backbone. It adheres well to the majority of electrode surfaces and can prevent them from fouling or degrading [24]. Nafion polymer was selected because it helps in enhancing the stability of the sensor; water can pass through Nafion quite easily, and it can withstand chemical attacks [25,26]. It helps in increasing ion exchange during the oxidation reduction process with its ion-exchanging abilities.

From the XRD pattern (Figure 3), the strongest line in the ZnO sample under examination falls along the (1 0 1) plane. Additionally, the diffraction peaks are narrow and intense, this being a sign of a well-formed crystalline structure, in the nanoscale range, confirmed by Scherrer equation calculations. Pure ZnO nanoparticles having a hexagonal wurtzite phase were created, because no diffraction peaks matching other ZnO phases or organic compounds were found. This could be proof of the successful burning of silk fibroin by thermal treatment.

4.2. Amoxicillin Electrochemical Detection

For the sensitive detection of organic compounds, including medicines and related substances, in pharmaceutical samples and biological fluids, electrochemical techniques have proven to be excellent methods [27]. Among all electrochemical techniques utilized for AMX detection, there are several voltametric approaches, such as linear voltammetry, CV, square wave voltammetry (SWV), and DPV [28]. For example, CV was used to demonstrate the catalytic oxidation of AMX using a chemically modified nickel-based (Ni(II)-curcumin) modified carbon paste electrode by Reza Ojani and coworkers [27]. According to their experimental findings, the amoxicillin catalytic oxidation current at this electrode can be utilized to assess the amount of AMX in aqueous solution, allowing for the achievement of a detection limit and linear dynamic range that are suitable [27].

Adding Nafion to the GC surface leads to a current decrease, and ZnO nanostructures' incorporation into Nafion leads to a slight improvement, but this is not sufficient. In our case, for the GC/Nafion/ZnO 6 complex electrode, the highest signal is found in the presence of AMX 100 µM, as can be seen in Figure 4. With our proposed modified electrode based on Nafion/ZnO, it can be concluded that CV is not sensitive enough to be used as an electrochemical method for the detection of amoxicillin.

Another tested method was DPV. In the first experiment, presented in Figure 5, we used GC/Nafion/ZnO 6 (1 h) to decrease the electrode preparation time. It seems that DPV is suitable for AMX detection with the proposed modified electrode, leading to an almost two-fold improvement in the signal.

When consecutive scans were performed with GC/Nafion/ZnO 6 (1 h), the signal was not stable (Figure 6a); one possible explanation could be that the Nafion film was not sufficiently dried. It is possible that 1 h is not sufficient time for the PBS solvent used for Nafion solution preparation to be evaporated, leading to ZnO nanostructures' detachment from the electrode surface between scans.

When the drying time of the electrode is increased to 24 h (Figure 6b), the signal is more stable, so, in all further experiments, we used a 24 h drying time for electrode preparation.

According to the results presented in Figure 7, another important parameter for this modified electrode is the concentration. Upon increasing the concentration from 6 µL of solution to 9 µL, the peak height was doubled.

The GC/Nafion/ZnO 9 electrode's stability during three consecutive scans is presented in Figure 8. The electrode prepared by this method gives a stable signal in the presence of AMX. After one week, a slight decrease in peak height is observed, but there is not a large difference during three consecutive scans. Reproducibility was tested (Figure 8) and it could be concluded that the modification method can be reproduced.

With the proposed modified electrode, a calibration curve can be obtained (Figure 9). With the Excel program, the LOD and LOQ were calculated—respectively, the limit of detection and the limit of quantification—according to the linear regression method described in the literature [29]. The LOD was 0.02 µM and LOQ 0.05 µM.

Comparing the created sensor's sensitivity to some of the existing reported voltametric sensors, as presented in Table 2, GC/Nafion/ZnO was more sensitive. The linear response range is comparable to other electrochemical sensors' linear ranges. Another benefit of the created sensor is the ease of its fabrication and the use of nanocomposite ZnO based on fibroin, which does not affect the environment, which is significant considering the principles of green chemistry.

Table 2. Analytical characteristics evaluated for amoxicillin electrochemical sensors.

WE [1]	Transduction Method	Linear Range (μM)	LOD (μM)	Reference
Treated GPE [2]	SWV	1–80	0.2 μM	[28]
TiO$_2$/CMK/AuNPs/Nafion/GCE [3]	CV	0.5–2.5 2.5–133.0	0.3	[30]
ZnO NRs/gold/glass electrode [4]	CV	5.0–2.5	1.9	[31]
MWCNT/GCE [5]	CV	0.6–8	0.2	[32]
Poly-4-vinylpyridine/CPE [6]	CV	-	8	[33]
[VO(Salen)]/CPE [7]	DPV	18.3–35.5	16.6	[34]
FeCr$_2$O$_4$/MWCNTs/GCE [8]	DPV	0.1–10.0 10.0–70.0	0.05	[35]
CILE [9]	CV	5.0–400	0.8	[36]
ZnO/Nafion/GC	DPV	5–110	0.02	This work

[1] Working electrode. [2] Graphite pencil mechanically polished. [3] Titanium dioxide/mesoporous carbon CMK-3-type/gold nanoparticles/Nafion on glassy carbon electrode. [4] Zinc oxide nanorod/gold/glass electrode. [5] Multiwalled carbon nanotube modified glassy carbon electrode. [6] Carbon paste electrode modified with poly-4-vinylpyridine. [7] Carbon paste electrode modified with [N,N-ethylenebis(salicylideneaminato)]oxovanadium. [8] Nanoparticle-decorated multiwall carbon nanotubes on glassy carbon electrode. [9] Carbon ionic liquid electrode.

5. Conclusions

A new electrochemical sensor based on ZnO nanostructures grown in a natural silk fibroin matrix was obtained. The ZnO nanostructures, with high porosity and a fibrillar shape, were deposited on the glassy carbon electrode surface by incorporation into a Nafion polymer matrix.

Glassy carbon surface modification was highlighted by scanning electron microscopy, energy-dispersive X-ray spectroscopy, X-ray diffraction, and Fourier-transform infrared spectroscopy analysis. The ZnO structures prepared on the silk fibroin template had a fibrous morphology and their diameters ranged from 1 to 2 μm. X-ray diffraction revealed a hexagonal wurtzite phase. The size of crystallites was around 37 nm. X-ray diffraction and Fourier-transform infrared spectroscopy showed that silk fibroin was successfully burned by thermal treatment. ZnO was then distributed uniformly as nanoparticles within the Nafion matrix on the glassy carbon electrode's surface.

The new GC/Nafion/ZnO electrode, with good stability, was successfully used for AMX electrochemical detection, with an LOD of 0.02 μM, which was better than or comparable to values reported in other papers. The built-in standard calibration has a high degree of correlation.

Measurements in solutions containing different antibiotics will be performed to test the AMX selectivity.

Author Contributions: Conceptualization, C.D. and C.P.; methodology, C.D. and C.P; software, C.D., A.C., and A.D.; validation, C.D., A.C., A.D., and C.P.; formal analysis, C.D., A.C., and A.D.; investigation, A.D. and C.D.; resources, C.D. and C.P.; data curation, C.D.; writing—original draft preparation, C.D.; writing—review and editing, C.D., A.C., and C.P.; supervision, C.P.; project administration, C.D. All authors have read and agreed to the published version of the manuscript.

Funding: This research received no external funding.

Informed Consent Statement: Not applicable.

Data Availability Statement: Not applicable here.

Conflicts of Interest: The authors declare no conflict of interest.

References

1. Wong, A.; Santos, A.M.; Cincotto, F.H.; Moraes, F.C.; Fatibello-Filho, O.; Sotomayor, M.D.P.T. A new electrochemical platform based on low-cost nanomaterials for sensitive detection of the amoxicillin antibiotic in different matrices. *Talanta* **2020**, *206*, 120252. [CrossRef] [PubMed]
2. Essousi, H.; Barhoumi, H.; Karastogianni, S.; Girousi, S.T. An Electrochemical Sensor Based on Reduced Graphene Oxide, Gold Nanoparticles and Molecular Imprinted Over-oxidized Polypyrrole for Amoxicillin Determination. *Electroanalysis* **2020**, *32*, 1546–1558. [CrossRef]
3. Norouzi, B.; Mirkazemi, T. Electrochemical sensor for amoxicillin using Cu/poly (o-toluidine) (sodium dodecyl sulfate) modified carbon paste electrode. *Russ. J. Electrochem.* **2016**, *52*, 37–45. [CrossRef]
4. Li, S.; Ma, X.; Pang, C.; Li, H.; Liu, C.; Xu, Z.; Luo, J.; Yang, Y. Novel molecularly imprinted amoxicillin sensor based on a dual recognition and dual detection strategy. *Anal. Chim. Acta* **2020**, *1127*, 69–78. [CrossRef] [PubMed]
5. Pham, T.H.Y.; Mai, T.T.; Nguyen, H.A.; Chu, T.T.H.; Vu, T.T.H.; Le, Q.H. Voltammetric Determination of Amoxicillin Using a Reduced Graphite Oxide Nanosheet Electrode. *J. Anal. Methods Chem.* **2021**, *2021*, 8823452. [CrossRef]
6. Hrioua, A.; Loudiki, A.; Farahi, A.; Bakasse, M.; Lahrich, S.; Saqrane, S.; El Mhammedi, M.A. Recent advances in electrochemical sensors for amoxicillin detection in biological and environmental samples. *Bioelectrochemistry* **2021**, *137*, 107687. [CrossRef]
7. Song, J.; Huang, M.; Jiang, N.; Zheng, S.; Mu, T.; Meng, L.; Liu, Y.; Liu, J.; Chen, G. Ultrasensitive detection of amoxicillin by TiO_2-g-C_3N_4@AuNPs impedimetric aptasensor: Fabrication, optimization, and mechanism. *J. Hazard. Mater.* **2020**, *391*, 122024. [CrossRef]
8. Shetti, N.P.; Bukkitgar, S.D.; Reddy, K.R.; Reddy, C.V.; Aminabhavi, T.M. ZnO-based nanostructured electrodes for electrochemical sensors and biosensors in biomedical applications. *Biosens. Bioelectron.* **2019**, *141*, 111417. [CrossRef]
9. Lefatshe, K.; Muiva, C.M.; Kebaabetswe, L.P. Extraction of nanocellulose and in-situ casting of ZnO/cellulose nanocomposite with enhanced photocatalytic and antibacterial activity. *Carbohydr. Polym.* **2017**, *164*, 301–308. [CrossRef]
10. Ong, C.B.; Ng, L.Y.; Mohammad, A.W. A review of ZnO nanoparticles as solar photocatalysts: Synthesis, mechanisms and applications. *Renew. Sustain. Energy Rev.* **2018**, *81*, 536–551. [CrossRef]
11. Hoseinpour, V.; Souri, M.; Ghaemi, N.; Shakeri, A. Optimization of green synthesis of ZnO nanoparticles by *Dittrichia graveolens* (L.) aqueous extract. *Health Biotechnol. Biopharma* **2017**, *1*, 39–49. [CrossRef]
12. Jin, S.E.; Jin, H.E. Synthesis, Characterization, and Three-Dimensional Structure Generation of Zinc Oxide-Based Nanomedicine for Biomedical Applications. *Pharmaceutics* **2019**, *11*, 575. [CrossRef] [PubMed]
13. Gutul, T.; Rusu, E.; Condur, N.; Ursaki, V.; Goncearenco, E.; Vlazan, P. Preparation of poly(N-vinylpyrrolidone)-stabilized ZnO colloid nanoparticles. *Beilstein J. Nanotechnol.* **2014**, *5*, 402–406. [CrossRef] [PubMed]
14. Pan, Y.; Zuo, J.; Hou, Z.; Huang, Y.; Huang, C. Preparation of Electrochemical Sensor Based on Zinc Oxide Nanoparticles for Simultaneous Determination of AA, DA, and UA. *Front. Chem.* **2020**, *8*, 592538. [CrossRef] [PubMed]
15. Weldegebrieal, G.K. Synthesis method, antibacterial and photocatalytic activity of ZnO nanoparticles for azo dyes in wastewater treatment: A review. *Inorg. Chem. Commun.* **2020**, *120*, 108140. [CrossRef]
16. Agarwal, H.; Venkat Kumar, S.; Rajeshkumar, S. A review on green synthesis of zinc oxide nanoparticles—An eco-friendly approach. *Resour. Effic. Technol.* **2017**, *3*, 406–413. [CrossRef]
17. Chen, L.; Xu, X.; Cui, F.; Qiu, Q.; Chen, X.; Xu, J. Au nanoparticles-ZnO composite nanotubes using natural silk fibroin fiber as template for electrochemical non-enzymatic sensing of hydrogen peroxide. *Anal. Biochem.* **2018**, *554*, 1–8. [CrossRef]
18. Bucciarelli, A.; Motta, A. Use of Bombyx mori silk fibroin in tissue engineering: From cocoons to medical devices, challenges, and future perspectives. *Biomater. Adv.* **2022**, *139*, 212982. [CrossRef]
19. Yao, X.; Zou, S.; Fan, S.; Niu, Q.; Zhang, Y. Bioinspired silk fibroin materials: From silk building blocks extraction and reconstruction to advanced biomedical applications. *Mater. Today Bio* **2022**, *16*, 100381. [CrossRef]
20. Park, H.J.; Lee, J.S.; Lee, O.J.; Sheikh, F.A.; Moon, B.M.; Ju, H.W.; Kim, J.-H.; Kim, D.-K.; Park, C.H. Fabrication of microporous three-dimensional scaffolds from silk fibroin for tissue engineering. *Macromol. Res.* **2014**, *22*, 592–599. [CrossRef]
21. Sigwadi, R.; Dhlamini, M.S.; Mokrani, T.; Ņemavhola, F.; Nonjola, P.F.; Msomi, P.F. The proton conductivity and mechanical properties of Nafion®/ZrP nanocomposite membrane. *Heliyon* **2019**, *5*, e02240. [CrossRef] [PubMed]
22. Bhati, V.S.; Hojamberdiev, M.; Kumar, M. Enhanced sensing performance of ZnO nanostructures-based gas sensors: A review. *Energy Rep.* **2020**, *6*, 46–62. [CrossRef]
23. Sha, R.; Basak, A.; Maity, P.C.; Badhulika, S. ZnO nano-structured based devices for chemical and optical sensing applications. *Sens. Actuators Rep.* **2022**, *4*, 100098. [CrossRef]
24. Chen, Z.; Patel, R.; Berry, J.; Keyes, C.; Satterfield, C.; Simmons, C.; Neeson, A.; Cao, X.; Wu, Q. Development of Screen-Printable Nafion Dispersion for Electrochemical Sensor. *Appl. Sci.* **2022**, *12*, 6533. [CrossRef]
25. Marie, M.; Mandal, S.; Manasreh, O. An enzymatic glucose detection sensor using ZnO nanostructure. *MRS Adv.* **2016**, *1*, 847–853. [CrossRef]
26. Mandal, S.; Marie, M.; Manasreh, O. Fabrication of an Electrochemical Sensor for Glucose Detection using ZnO Nanorods. *MRS Adv.* **2016**, *1*, 861–867. [CrossRef]
27. Ojani, R.; Raoof, J.-B.; Zamani, S. A novel voltammetric sensor for amoxicillin based on nickel–curcumin complex modified carbon paste electrode. *Bioelectrochemistry* **2012**, *85*, 44–49. [CrossRef] [PubMed]

28. Yen, P.T.H.; Anh, N.H.; Ha, V.T.T.; Hung, L.Q.; Phong, P.H.; Hien, C.T.T. Electrochemical properties of amoxicillin on an economical, simple graphite pencil electrode and the ability of the electrode in amoxicillin detection. *Vietnam. J. Chem.* **2020**, *58*, 201–205. [CrossRef]
29. Miller, J.N.; Miller, J.C. *Statistics and Chemometrics for Analytical Chemistry*; Prentice Hall/Pearson: Hoboken, NJ, USA, 2010.
30. Pollap, A.; Knihnicki, P.; Kuśtrowski, P.; Kozak, J.; Gołda-Cępa, M.; Kotarba, A.; Kochana, J. Sensitive voltammetric amoxicillin sensor based on tio2 sol modified by cmk-3-type mesoporous carbon and gold ganoparticles. *Electroanalysis* **2018**, *30*, 2386–2396. [CrossRef]
31. Hatamie, A.; Echresh, A.; Zargar, B.; Nur, O.; Willander, M. Fabrication and characterization of highly-ordered Zinc Oxide nanorods on gold/glass electrode, and its application as a voltammetric sensor. *Electrochim. Acta* **2015**, *174*, 1261–1267. [CrossRef]
32. Rezaei, B.; Damiri, S. Electrochemistry and adsorptive stripping voltammetric determination of amoxicillin on a multiwalled carbon nanotubes modified glassy carbon electrode. *Electroanal. Int. J. Devoted Fundam. Pract. Asp. Electroanal.* **2009**, *21*, 1577–1586. [CrossRef]
33. Biryol, I.; Uslu, B.; Küçükyavuz, Z. Voltammetric determination of amoxicillin using a carbon paste electrode modified with poly (4-vinyl pyridine). *STP Pharma Sci.* **1998**, *8*, 383–386.
34. Bergamini, M.F.; Teixeira, M.F.; Dockal, E.R.; Bocchi, N.; Cavalheiro, É.T. Evaluation of different voltammetric techniques in the determination of amoxicillin using a carbon paste electrode modified with [N, N′-ethylenebis (salicylideneaminato)] oxovanadium (IV). *J. Electrochem. Soc.* **2006**, *153*, E94. [CrossRef]
35. Ensafi, A.A.; Allafchian, A.R.; Rezaei, B. Multiwall carbon nanotubes decorated with FeCr2O4, a new selective electrochemical sensor for amoxicillin determination. *J. Nanoparticle Res.* **2012**, *14*, 1244. [CrossRef]
36. Absalan, G.; Akhond, M.; Ershadifar, H. Highly sensitive determination and selective immobilization of amoxicillin using carbon ionic liquid electrode. *J. Solid State Electrochem.* **2015**, *19*, 2491–2499. [CrossRef]

Article

Impact of Mo-Doping on the Structural, Optical, and Electrocatalytic Degradation of ZnO Nanoparticles: Novel Approach

Vanga Ganesh [1,*], Mai S. A. Hussien [2,3], Ummar Pasha Shaik [4], Ramesh Ade [5], Mervat I. Mohammed [2], Thekrayat H. AlAbdulaal [1], Heba Y. Zahran [1,2,6], Ibrahim S. Yahia [1,2,6,*] and Mohamed Sh. Abdel-wahab [7]

1. Laboratory of Nano-Smart Materials for Science and Technology (LNSMST), Department of Physics, Faculty of Science, King Khalid University, Abha P.O. Box 9004, Saudi Arabia
2. Nanoscience Laboratory for Environmental and Bio-medical Applications (NLEBA) (Metallurgical Lab.1.), Department of Physics, Faculty of Education, Ain Shams University, Roxy, Cairo 11757, Egypt
3. Department of Chemistry, Faculty of Education, Ain Shams University, Roxy, Cairo 11757, Egypt
4. Department of Physics, Anurag Engineering College (An UGC Autonomous Institution), Ananthagiri (V&M), Suryapet(D), Suryapet 508206, Telangana, India
5. Department of Physics, Koneru Lakshmaiah Education Foundation, R V S Nagar, Aziz Nagar (PO), Moinabad Road, Hyderabad 500075, Telangana, India
6. Research Center for Advanced Materials Science (RCAMS), King Khalid University, Abha P.O. Box 9004, Saudi Arabia
7. Materials Science and Nanotechnology Department, Faculty of Postgraduate Studies for Advanced Sciences, Beni-Suef University, Beni-Suef 62511, Egypt
* Correspondence: vangaganesh@gmail.com (V.G.); dr_isyahia@kku.edu.sa (I.S.Y.)

Abstract: Pure and Molybdenum (Mo)-doped zinc oxide (ZnO) nanoparticles were prepared by a cost-effective combustion synthesis route. XRD results revealed the decrement in crystallite size of ZnO with an increase in Mo-doping concentration. Optical bandgap (E_g) values were determined using optical reflectance spectra of these films measured in the range of 190–800 nm. The E_g values decreased with increasing the Mo-doping concentration. The dielectric properties of these samples were studied to determine the dielectric constant values. Raman spectra of these samples were recorded to know the structure. These sample absorption spectra were recorded for electrocatalytic applications. All the prepared samples were subjected to electrocatalytic degradation of Rhodamine B. The 0.01 wt% Mo doped ZnO showed 100% in 7 min electrocatalytic degradation.

Keywords: Mo-doped ZnO; nanostructured metal oxides; XRD/SEM; optical properties; electrocatalysis

1. Introduction

It is a known fact that all the world's human beings strongly depend on natural resources, for example, soil, water, air, etc. Unfortunately, in recent years, contamination of these resources has been very high from textile, leather, chemical laboratories, and paper industries, which were the primary sources of water pollution. They release unprocessed wastewater directly into the water bodies [1–4]. Using this unsafe water, many health problems will occur to human beings [1,5]. Currently, wastewater treatment and recycling are getting significant attention. Therefore, great attention has to be paid to investigating cost-effective and eco-friendly techniques for water purification. Photocatalysis is a technique used for water purification that is simple and cheap. The importance of photocatalysis for water electrolysis and hydrogen fuel cells has been discussed by many researchers [6].

ZnO is one of the promising II–VI semiconducting materials, which displays a large optical band gap (3.37 eV) and exciton binding energy (60 meV), high linear refractive index, and electrical properties. These properties make ZnO suitable for electronics, photonic devices, UV laser diodes, piezoelectric transducers, light-emitting diodes sensors, solar cells, and optoelectronic devices [7–11]. Semiconducting nanostructures with different morphologies such as nanoparticles, nanowires, nanorods, nanobelts, nanoprisms, nanodots, and nanostructured thin films help understand the physical and chemical properties of nanoscale systems. The behavior of electron, photon, and phonon of nanoscale systems was studied and reported by the earlier workers [12]. Different methods to fabricate ZnO nanoparticles include sol-gel, ultrasonic, chemical vapor deposition, microemulsion, organometallic precursor, solvothermal, microemulsion, and spray pyrolysis electrodeposition, sonochemical, microwave-assisted, and hydrothermal [13–21].

ZnO acts as a photocatalyst upon the photoirradiation by transferring electrons from the valence band to the conduction band in the shorter wavelength region [13,22]. Due to the similarities in the bandgap of ZnO and TiO_2, many researchers tried to replace expensive TiO_2 photocatalyst with ZnO photocatalyst. Since ZnO is a low-cost and non-toxic material, it has received significant interest. Furthermore, during UV- light irradiation, ZnO exhibits remarkable electrocatalytic ability for degrading organic pollutants in aqueous suspension by generating powerful oxidizing agents such as hydroxyl radicals and superoxide radical anions, which decompose the organic pollutants [9], but they learned that ZnO is less active under visible light than TiO_2, and several attempts were made to improve ZnO properties.

The electrocatalytic activity of various ZnO crystal surfaces was investigated. They discovered that the surface atomic structures of the material had a significant impact on its photostability and electrocatalytic activity. When the surface is polar and has high surface energy, photolysis of ZnO occurs quickly [19]. Recently, reports enhanced electrocatalytic properties by adding dopants such as transition metals to ZnO lattice [13,23,24]. It is also reported that Mg-doped ZnO nanoparticles showed an adequate capacity to remove contaminating organic molecules, specifically MO (Ceramics International 47 (2021) 15668–15681). Recently, by introducing metal dopants, enhancement in electronic, magnetic, and catalytic properties of biphenylene network has been reported [25–28]. In the Mo-doped ZnO incorporation, Zn-O-M instead of Zn-O-Zn occurs, leading to new energy levels between the valence and conduction bands by lowering bandgap values. Mo-doped ZnO can be used as a photocatalyst to degrade bacteria and water purification [13,29]. Mo has an ionic radius similar to Zn^{2+}, which is a good match. As a result, it produces stable mixed metal oxides with the composition ZnO/MoO_3. Recently some workers also investigated the photocatalysis studies of Mo, Mn, Co, and Cu-doped ZnO thin films using different dye solutions and concluded that the pH of the dye solution plays a crucial role in the dye degrading mechanism [30].

In the current work, we demonstrated the synthesis of pure and Mo-doped ZnO nanoparticles by a cost-effective combustion synthesis method. The main objective of this study is pure, and Mo-doped ZnO samples were examined for water purification by using them as photocatalysts.

2. Experimental Details
2.1. Synthesis of Mo-Doped ZnO Nanostructures

The present study synthesized nanostructures of pure and Mo-doped ZnO using a low-cost combustion technique. As starting material, we prepared a homogeneous solution by adding 5gm of zinc nitrate (Zn $(NO_3)_2.6H_2O$) with 5 gm of citric acid in a crucible. Various concentrations of Ammonium pentamolybdate (0 wt%, 0.001 wt%, 0.01 wt%, 0.1 wt%, 0.5 wt%, and 1 wt%) with 30 mL of distilled water were added to the homogeneous solution. The solution was continuously stirred at 170 °C on a hot plate for 2 h. Finally, these solutions were heated in a furnace at 550 °C for 2 h; the final product is well-grounded using mortar and pestle. In the current study, six different samples were prepared by varying the concentration of ammonium pentamolybdate as 0 wt% (sample named ZNCM-1), 0.001 wt% (sample named ZNCM-2), 0.01 wt% (sample named ZNCM-3) 0.1 wt% (sample named as ZNCM-4), 0.5 wt% (sample named as ZNCM-5), and 1 wt% (sample named as ZNCM-6).

2.2. Devices and Measurements

The structural properties of the present samples were examined by using a Shimadzu LabX-XRD-6000 X-ray diffractometer, with CuKa = 1.54 Å radiation in the angle range from 5° to 80°.

The structural morphology of the present samples was investigated by scanning electron microscopy techniques (SEM-Jeol. JSM-6360 type) operated at 20 kV.

Optical reflectance (190–800 nm) and the absorption spectra (200–1600 nm) were recorded using a 3600 UV/Vis/NIR spectrophotometer (Shimadzu, Japan).

The present sample's Raman spectra were recorded using an FT-Raman spectrometer (Thermo Fisher Scientific, Waltham, MA, USA). In addition, Fourier transforms infrared (FT-IR) spectra (400 to 4000 cm^{-1}) of the prepared samples were recorded by XDR FT-IR Spectrometer, THERMO SCIENTIFIC.

The dielectric properties were recorded using Keithley 4200-SCS in the frequency range of 1 kHz and 10 MHz.

A Rhodamine B dye (50 mg/L) was used as a typical organic pollutant in the experiment of electrocatalytic degradation. The electrocatalytic degradation processes and electrocatalysis single-cell reactor (EC) reactor were used, with 0.01 gm of catalyst provided and two graphite electrodes functioning as a working electrode and a counter electrode. The electrodes were separated by 5 cm under biasing DC voltage of 10 volts.

3. Results and Discussion

3.1. X-ray Diffraction Studies

The X-ray diffraction patterns of pure (ZNCM-1) and Mo-doped (ZNCM-2, ZNCM-3, ZNCM-4, ZNCM-5, and ZNCM-6) ZnO nanoparticles samples are demonstrated in Figure 1. The diffraction peaks of the ZNCM-1 sample matched with hexagonal ZnO structure. These peaks were indexed according to COD Card entry 9004179. We did not find any significant Mo or Zn residues from the XRD patterns of ZNCM-2 and ZNCM-3 samples. It indicates that Mo replaces Zn in the hexagonal ZnO lattice. Similar results were obtained for Mo and Al-doped ZnO and Mo-doped In_2O_3 films [31–39]. Further, the intensity of all the ZnO peaks decreased, whereas the intensity of Mo-based Mo5O14 peaks increased with an increase in dopant concentration observed in the XRD pattern of ZNCM-4, ZNCM-5, and ZNCM-6 samples. The peaks corresponding to Mo5O14 are indexed according to COD Card entry 1537518. K. Ravichandran et al. [1] reported similar results. The unit cell parameters of hexagonal ZnO calculated from the XRD pattern were a = 3.2 ± 0.2 Å and c = 5.2 ± 0.2 Å, indicating independence of dopant concentration of the samples. Scherer's formula was used to calculate crystallite size (D), dislocation density (δ), and dislocation density (δ), using the following relationships [21,22]:

$$D = 0.9\lambda/\beta\cos\theta \quad (1)$$

$$\delta = 1/D^2 \quad (2)$$

$$\varepsilon = \beta\cos\theta/4 \quad (3)$$

Here, θ is the diffraction angle, λ is the X-ray wavelength in the nm unit, and $\hat{A}$ is the full width at half maximum (FWHM). All the calculated structural parameters of the present samples are depicted in Table 1. The average D value ZnO peaks [34] was 23 ± 2 nm for the samples ZNCM-1 and ZNCM-2. The D of ZNCM-3, ZNCM-4 and ZNCM-5 are 18 ± 2 nm. However, ZNCM-6 had a D of 14 ± 2 nm. The lower D-value for ZNCM-6 can be attributed to residues in the XRD pattern and lower the intensity of the ZnO peaks. It is clear that with increasing dopant concentration D-value of ZnO decreases. At the same time, ε and δ values increase with increasing Mo-dopants in the host ZnO material. The value of δ determines a defect density in a crystal, which arises due to misregistration of the lattice in one segment of another crystal section. Khalid Umar et al. (2015) observed similar results on Mo and Mn-doped ZnO and their electrocatalytic activity studies, and the possible reason for decreasing crystallite size discussed as the addition of dopant may hinder the growth of ZnO particles [35].

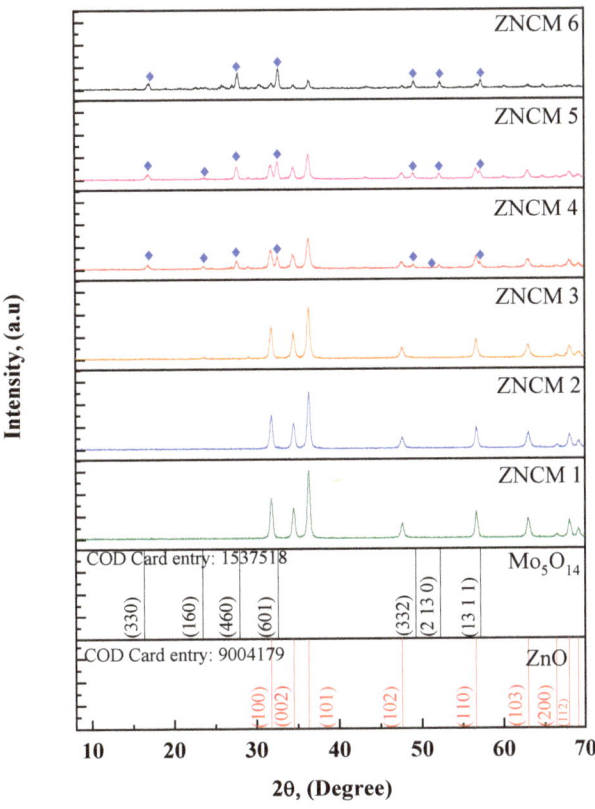

Figure 1. XRD patterns of pure and Mo-doped ZnO nanoparticle samples.

Table 1. The structural parameters of the pure and Mo-doped ZnO nanoparticles.

Sample		2θ	FWHM	d-Spacing	a, (nm)	c, (nm)	D, (nm)	Å×10⁻³	δ (×10⁻¹⁵)
ZNCM-1 (0 wt% Mo)	ZnO	36.316 (101)	0.3738	2.47177	3.2445	5.1979	23.37	5	1.831
ZNCM-2 (0.001 wt% Mo)	ZnO	36.338 (101)	0.3789	2.47035	3.2425	5.1956	23.06	5	1.881
ZNCM-3 (0.01 wt% Mo)	ZnO	36.262 (101)	0.4800	2.47531	3.2494	5.2038	18.2	6.4	3.019
ZNCM-4 (0.1 wt% Mo)	ZnO	36.303 (101)	0.4800	2.47263	3.24523	5.2017	18.2	6.4	3.019
	Mo$_5$O$_{14}$	32.484 (601)	0.5200	2.75408	23.0115	3.9571	16.63	7.8	3.616
ZNCM-5 (0.5 wt% Mo)	ZnO	31.773 (100)	0.4399	2.81406	3.2494	5.2038	19.62	6.7	2.597
	Mo$_5$O$_{14}$	32.570 (601)	0.4144	2.7470	22.8809	3.9334	20.87	6.2	2.296
ZNCM-6 (1.0 wt% Mo)	ZnO	31.945 (100)	0.5836	2.79931	3.2323	5.1795	14.8	8.9	4.565
	Mo$_5$O$_{14}$	32.708 (601)	0.3600	2.7357	23.0176	3.9108	24.03	5.4	1.732

3.2. SEM Studies

The microstructure of prepared nano samples is displayed in Figure 2. All the microstructures exhibit highly agglomerated and soapy porous structures. The reason for this kind of structure is released out of the reaction mixture during combustion. Upon increasing dopant concentration, the size of the nanoparticles is reduced from 177, 168, 165, 154, to 135 nm for the samples ZNCM-1, ZNCM-2, ZNCM-3, ZNCM-4, to ZNCM-5, respectively. The same results were observed in the earlier reports [35–38]. There is a remarkable change in the microstructure of 1 wt% of Mo-doped ZnO sample (ZNCM-6) instead of having nanoparticles. This could be due to the occupation of Mo- in the additional interstitial site of ZnO [35–40]. Petronela Pascariu et al. [41] reported platelet structured microparticles of diameters between 4 and 5 μm for molybdenum oxide. In the present study, the microstructure of ZNCM-6 exhibits a nanoflakes structure. The observed nanoflakes are typical of the order of 300 nm few μm. It was observed that the specific surface area of Mo-doped ZnO samples increases slightly with increasing Mo concentration, which may be due to a decrease in D-values. Due to the large surface area, these samples are suitable for the electrocatalytic activity of ZnO by adsorption of a dye [39,41]

Figure 2. SEM images (**a**) ZNCM-1, (**b**) ZNCM-2, (**c**) ZNCM-3, (**d**) ZNCM-4, (**e**) ZNCM-5, and (**f**) ZNCM-6.

3.3. Diffuse Reflectance Analysis of Mo-Doped ZnO Nanostructures

The Optical Diffuse Reflectance (ODR) spectroscopy is a specialized optical technique for determining the electronic structure of nanostructured materials. Reflection from the loaded samples generated by diffuse illumination is well studied because of its non-destructive nature and ability to generate a mirror-like reflection from loaded samples. To determine the optical bandgap and absorption coefficient of semiconducting material, the ODR technique is used. ODR of undoped and Mo-doped ZnO nanoparticle samples as a function of light wavelength in the range of 200–800 nm are shown in Figure 3. To summarize, the optical bandgaps are produced by the absorption of light through the nanomaterial; the optical energy band gap was calculated using the Kubelka–Munk model [42], which is as follows:

$$F(R) = \frac{(1-R^2)}{2R} \quad (4)$$

where $F(R)$ is the Kubelka–Munk function, and R is the absolute reflectance. The absorption coefficient (α) is calculated by the following equation [21,22]:

$$\alpha = \frac{absorbance}{t} = \frac{F(R)}{t} \quad (5)$$

where t is the height of the sample holder, which is equal to 2 mm, and the optical bandgap (E_g) is calculated from the following equation:

$$\alpha h\vartheta = \left(\frac{(F(R))h\vartheta}{t}\right)^n = A\left(h\vartheta - E_g\right)^n \quad (6)$$

where h is the photon energy, ϑ is the photon frequency, h is Planck's constant, and A is the band tailing factor, has values ranging from 1×10^5 to 1×10^6 cm$^{-1}\cdot$eV^{-1}. The optical transition (n) values equal 1/2 for the direct bandgap, and the plots (optical band) are shown in Figure 4. The reflectance spectra of all samples were found to be identical in the wavelength range of 200–370 nm. The optical bandgap of the samples was revealed by a sharp increase in the ODR spectra between 370 and 410 nm. In the visible region of 410–800 nm, all ZnO nanoparticles have a high reflectance. ZNCM-1, ZNCM-2, ZNCM-3, ZNCM-4, ZNCM-5, and ZNCM-6 have estimated band gap values of 3.261, 3.259, 3.252, 3.256, 3.246, and 3.235 eV, respectively. E.g., values decrease as Mo-doping concentration increases, which could be due to new energy states between the valence and conduction bands. C. Aydn et al. [43] obtained comparable results for sol-gel prepared ZnO: Fe nanoparticles. ZnO nanoparticles have high reflectance values in the visible region that decrease as dopant concentration increases, indicating that these samples are suitable for electrocatalytic applications under visible light irradiation. Similar decreasing band gap values were observed by Peyman Gholami et al. (2019) [44] on biochar-supported ZnO nanorods in the degradation of Gemifloxacin.

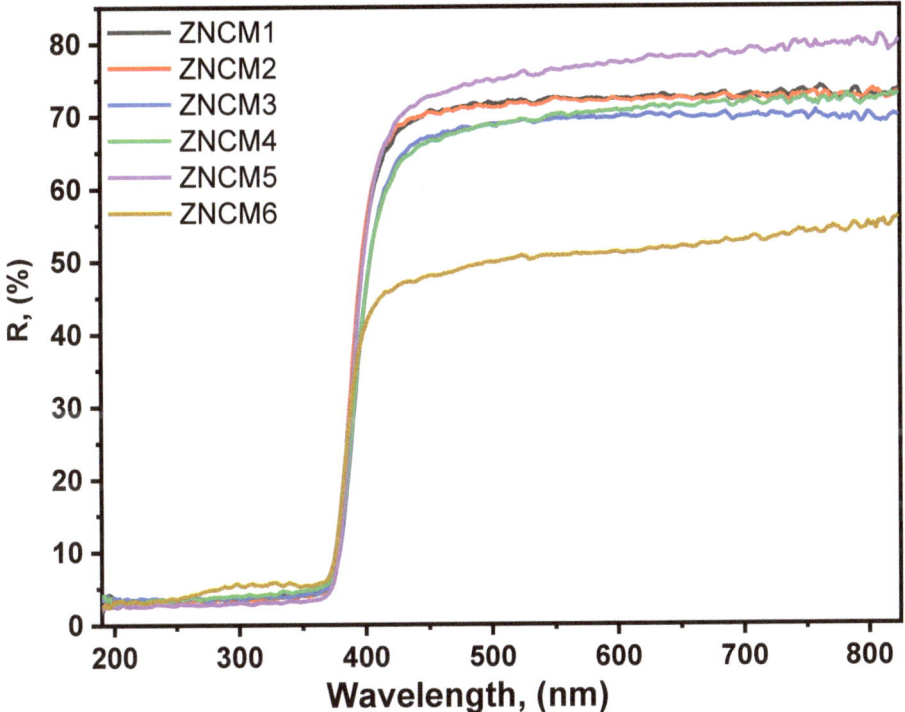

Figure 3. ODR spectra of pure and Mo-doped ZnO nano-samples.

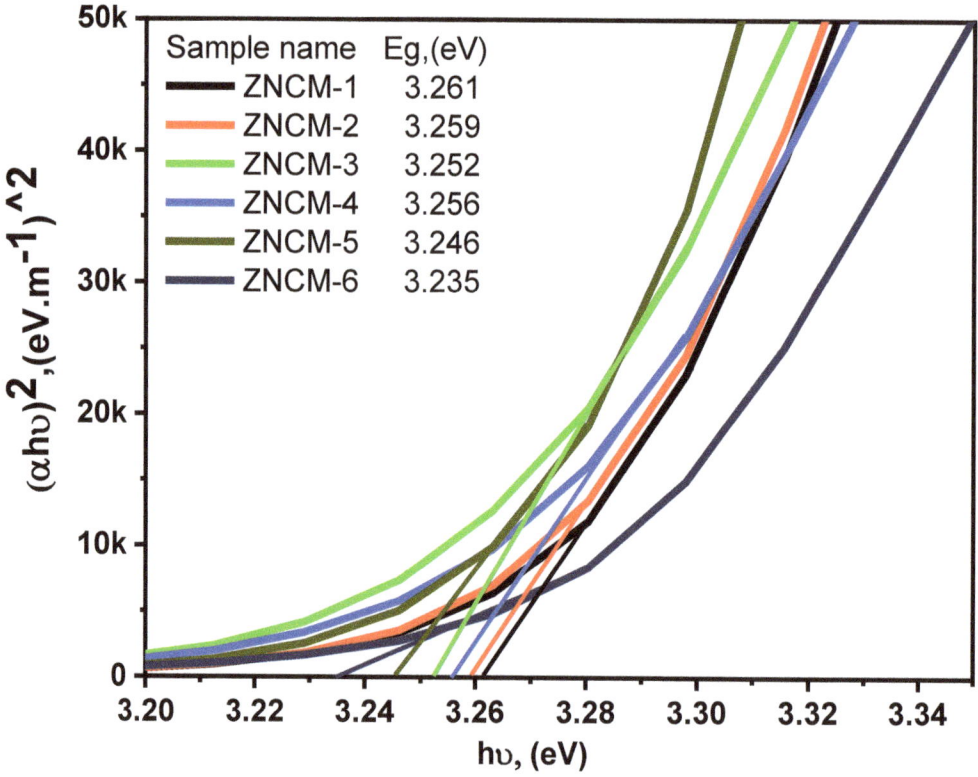

Figure 4. Tauc's plots of pure and Mo-doped ZnO nano samples.

3.4. Raman Spectroscopy of Mo-Doped ZnO Nanostructures

Figure 5 depicts the Raman spectra of pure and Mo-doped ZnO samples. Wurtzite ZnO is an associate of the C6V (P63mc) space group, and its possible vibrational modes [6,37] include U opt = $A_1 + 2B_1 + E_1 + 2E_2$. A_1, E_1, and E_2 are Raman active, while B_1 is Raman prohibited. The Raman peaks are observed at 95 cm^{-1}, 328, 431, 808, 843, and 903 cm^{-1}, corresponding to E_2 low, E_{2H}–E_{2L}, E_2 high, E_2 low + A_1 (TO), E_2 high + A_1 (TO), and A_1 mode, respectively. These findings corroborated previous findings [7,45–49]. A_1 and E_1 are polar modes that are both infrared and Raman active, whereas E_2 is non-polar and only Raman active. E_2 high is the peak at 431 cm^{-1}. Wurtzite ZnO's characteristic peak indicates good crystallinity [50]. The intensity of the E_2 high mode decreases as dopant concentration increases, whereas the intensity of the A_1 mode increases.

3.5. FT-IR Spectroscopy of Mo-Doped ZnO Nanostructures

The types of bonds and functional groups present in pure and Mo-doped ZnO samples were investigated using FTIR spectra. Figure 6 depicts the FTIR spectra of pure and doped ZnO nanoparticle samples. Zn–O stretching vibration is responsible for the peak at around 440 cm^{-1}. Earlier researchers [51,52] reported characteristic peaks for MoO_3 at 553, 876, 995, 1630, and 3445 cm^{-1}. In the current study, we found one such peak at 809 cm^{-1}, and the intensity of this peak increases as dopant concentration increases. Peaks at 1462 and 1675 cm^{-1} have previously been observed for ZnO nanoparticles and are attributed to asymmetrical and symmetrical stretching of the carboxylate group [53]. The stretching and bending vibrations of surface hydroxyl groups on ZnO were assigned to the peak at around 3280 cm^{-1}. The intensity of this peak increases as dopant concentration increases. These results agree well with the XRD results. The surface area of Mo-doped ZnO samples increases, resulting in more hydroxyl groups on the samples' surfaces. The Hydroxyl group is important in dye decomposition because it transfers photogenerated holes (h+) to an OH radical [54].

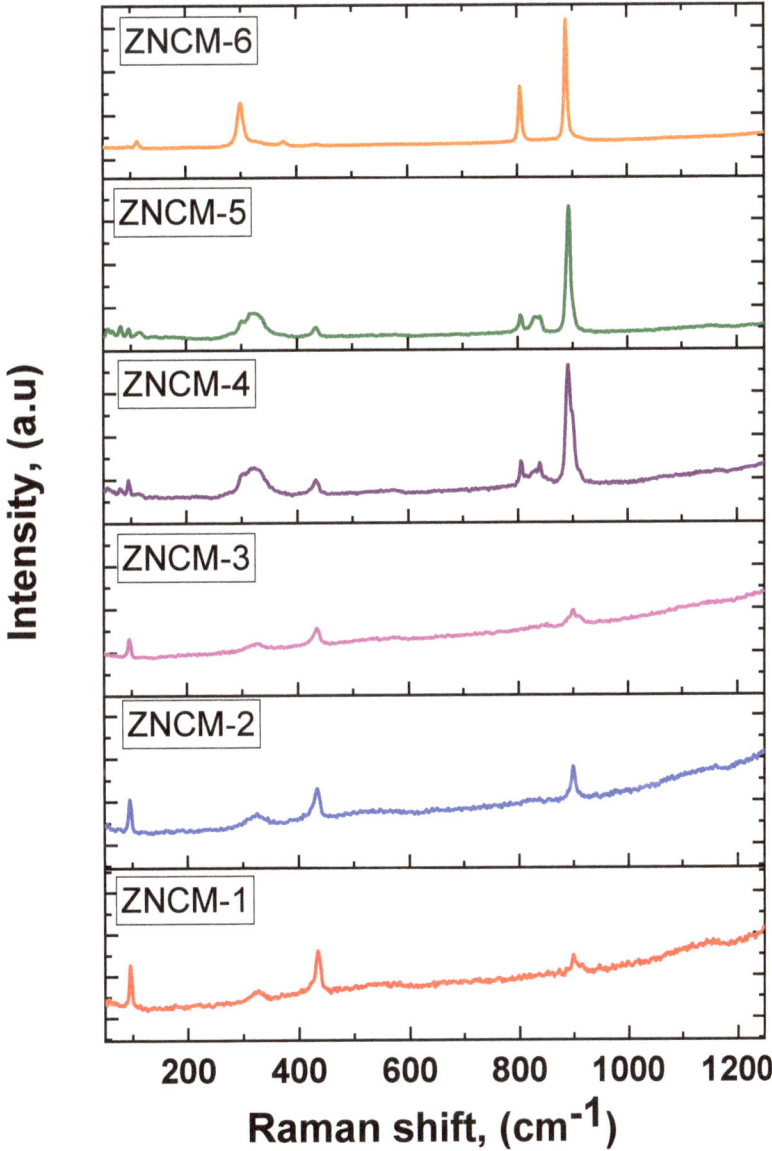

Figure 5. Raman spectra of pure and Mo-doped ZnO nano samples.

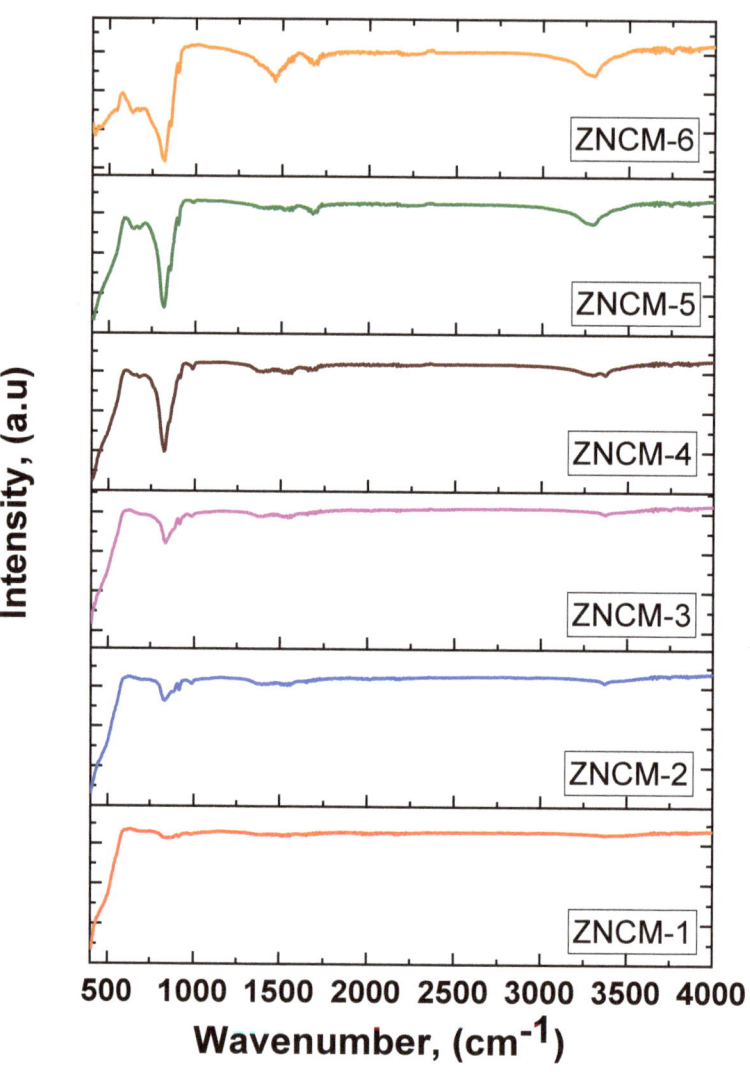

Figure 6. FTIR spectra of pure and Mo-doped ZnO nanoparticle samples.

3.6. Dielectric Properties of Mo-Doped ZnO Nanostructures

The dielectric response of the ZnO nano is typically high compared to bulk materials. A surface with a large volume can produce micro-porosities, dangling bonds, and vacancy clusters. All these defects can alter the space charge distribution present in the sample. The electric field's positive and negative poles attract negative and positive space charge distributions in interfaces. The dipole moments are created when trapped at the defect site because the volume percentage of nano-size sample interfaces is more significant than bulk materials. Because of these essential properties of nanostructured material, the dielectric constant (ε_1), dielectric loss (tanδ), and total AC electrical conductivity ($\sigma_{AC.Total}$) for pure and Mo-doped ZnO nanomaterials in the range of 1 MHz–10 MHz are illustrated in Figure 7a–c, respectively, using the following equations [47,48]:

$$\varepsilon_1 = \frac{C \times l}{\varepsilon_0 \times A} \tag{7}$$

$$\varepsilon_2 = \tan \delta \times \varepsilon_1, \tag{8}$$

$$\sigma_{AC.Total} = \frac{l}{Z \times A}, \tag{9}$$

where the real part of the dielectric constant is ε_1, and the imaginary component is ε_2. A represents the electrode area, $\tan \delta$ represents the loss tangent, C, l, and Z represents the sample capacitance, thickness, and impedance.

$$\sigma_{AC.\,Total} = \sigma_{DC} + B\omega^s, \text{ where } \sigma_{AC} = B\omega^s \tag{10}$$

The σ_{DC} denotes direct current conductivity, B is a constant, ω is the angular frequency, and s is a frequency exponent. The dielectric constant of the present samples decreased with increasing frequency and became saturated at higher frequencies (Figure 7a). The decreasing dielectric constant is explained as a high frequency, and the dipoles do not change in response to changes in the field variations. The decrease in dielectric constant values is caused by interfacial polarization/grain boundaries [40,41]. At lower frequencies, the dielectric constants of ZNCM-1, ZNCM-2, ZNCM-3, ZNCM-4, ZNCM-5, and ZNCM-6 are in the 40–48 range. The dielectric constant is frequency-dependent throughout the frequency range. It rises dramatically as the Mo content increases and stabilizes at high frequencies without significant change, allowing current samples to be used for various microwave device applications [46,55]. The frequency dependence of the dielectric loss with various concentrations of Mo is shown in Figure 7b. The dielectric loss also follows the same trend as the dielectric constant. It decreases with doping concentration at lower frequencies, indicating the dispersion phenomenon. It is worth mentioning that the dielectric constant and loss are purely dependent on Mo-doping concentration, suggesting that present samples found huge applications in microelectronic, sensors, and memory device applications. The electrical conductivity (σ_{AC}) gives a different Mo concentration with frequency variation, as shown in Figure 7c. From the figure, the conductivity is increasing with increasing frequency. The increase in conductivity suggests the clear influence of Mo doping in ZnO nanoparticles.

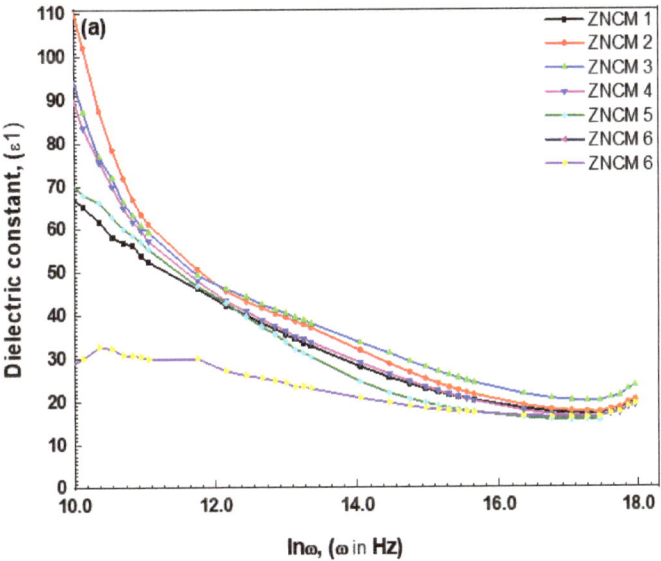

Figure 7. *Cont.*

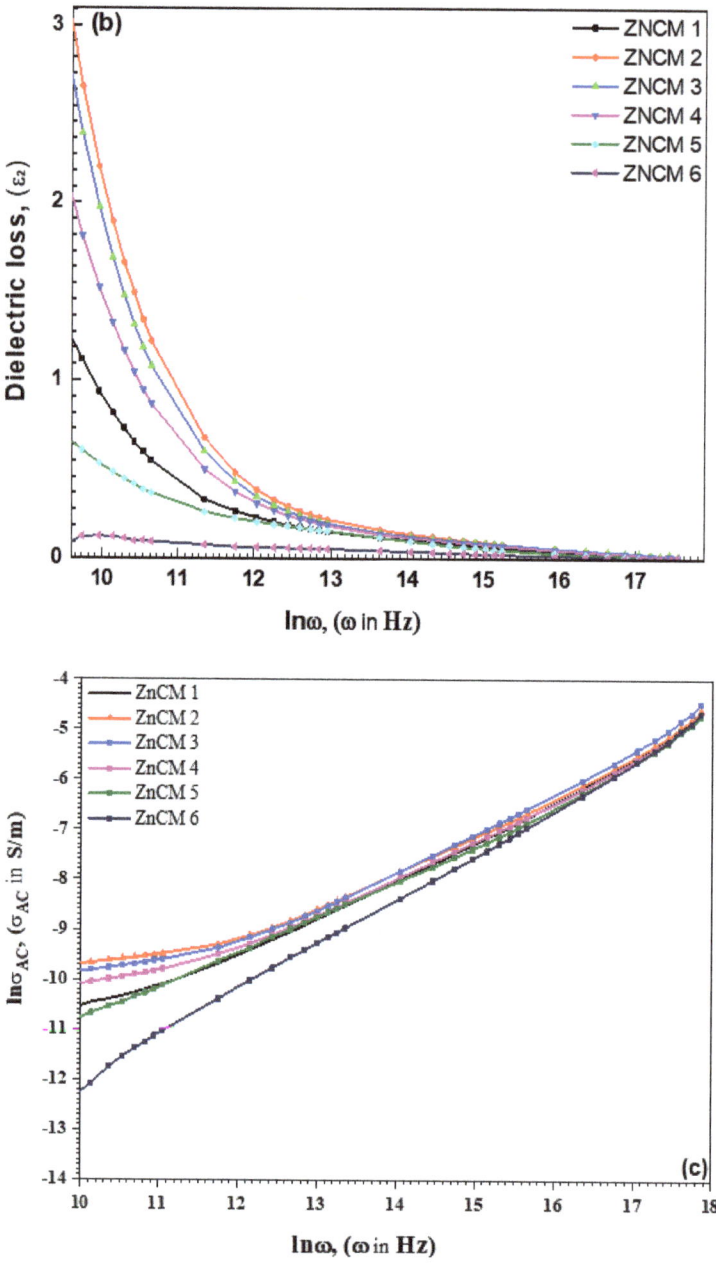

Figure 7. (**a**) Dielectric constant, (**b**) dielectric loss, and (**c**) electrical conductivity of pure and Mo-doped ZnO nanoparticle samples.

3.7. Kinetic Study of Electrocatalysis of RhB of Mo-Doped ZnO Nanostructures

The electrocatalysis (EC) degradation of RhB dye was thoroughly investigated as a model organic species for the ZNCM electrode. To encourage charge carrier transfer via the external circuit, a voltage of 10 V was applied between the photoanode and cathode in the single-cell reactor. The data collected and recorded throughout the degradation show a decrease in organic species absorbance with reaction time, indicating decreased organic species concentration. In the presence of seven different ZNCM samples, EC degraded RhB. As shown in Figure 8, EC degradation of RhB was consistent with previously reported pseudo-first-order kinetics [53,55,56].

$$ln(A/A_o) = -kt \qquad (11)$$

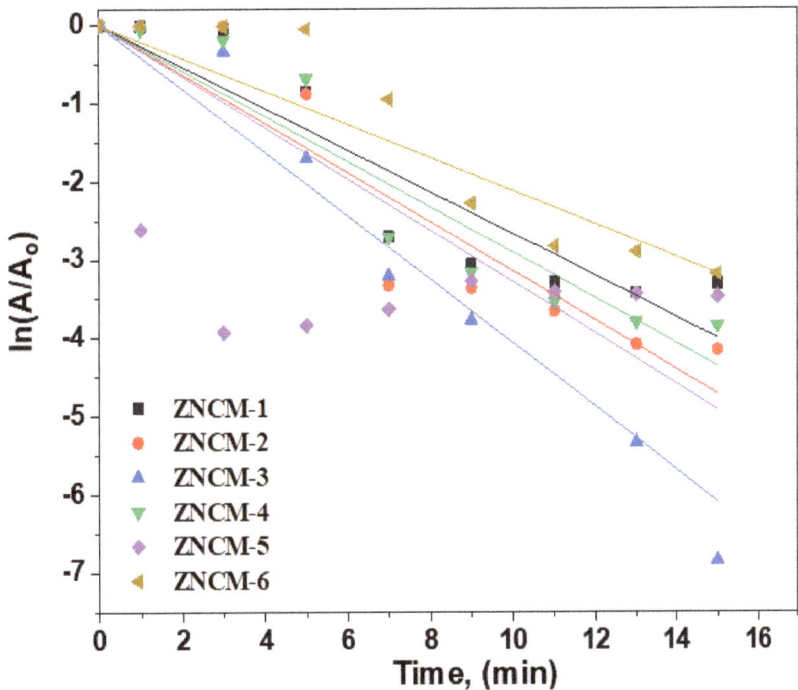

Figure 8. Kinetic study of EC degradation of RhB in the presence of ZNCM nanoparticles.

The degradation efficiencies were computed using the relationship shown below [57]:

$$\% \text{ of degradation} = (A_o - A_t/A_o) \times 100\% \qquad (12)$$

The initial absorbance is A_o. A_t is the absorbance at various time intervals is A_t, K is the rate constant value, and the reaction time is t. In the electrocatalytic reaction, a drop in the height of the absorbance peak at 525 nm wavelength was recorded with an increase in reaction time, providing a degradation efficiency of about 97 percent within 7 min, as shown in Figure 9. In the conduction band, excited electrons interact with electron acceptors to produce reactive species. As a result, the accumulation of electrons at the conduction band's bottom may slow charge carrier recombination even more. As a result, charge carrier separation is an excellent method for improving the electrocatalytic degradation of RhB [58]. RhB deterioration results demonstrate rapid ZNCM-3 over 7 min, as seen in Figure 9.

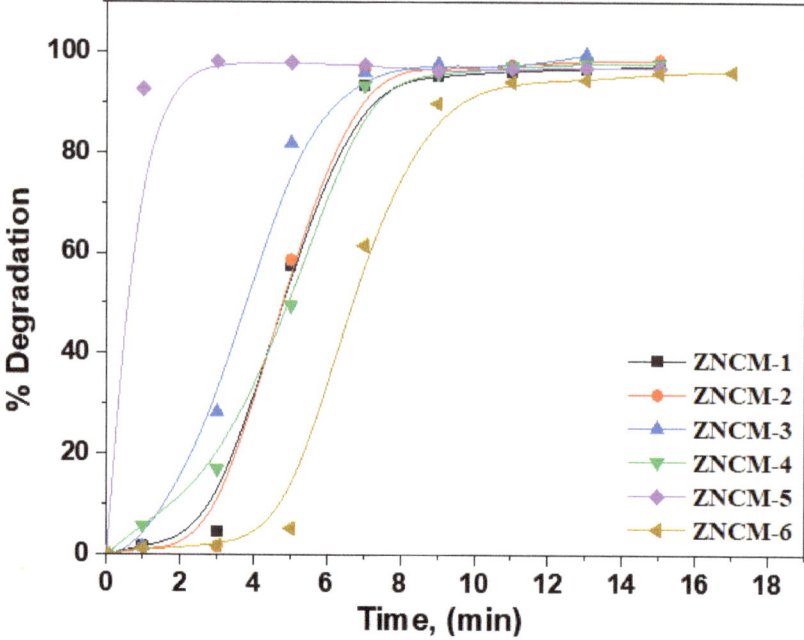

Figure 9. The % of EC degradation of RhB in the presence of ZNCM nanoparticles.

3.7.1. Mechanism of EC of RhB in the Presence of ZNCM

A significant benefit of EC is the influence of electrical energy, which results in better organic pollutant elimination efficiency [51]. Electrons are stimulated to the conduction band by electric charge, whereas empty holes reside in the valence band. When photogenerated holes interact with water molecules, they form powerful oxidants capable of breaking down refractory organic structures. Hydroxyl radicals are the second most powerful oxidants after fluorine due to their high standard reduction potential (Eo OH/H_2O) of approximately 2.8 V [59,60]. Unfortunately, photogenerated holes have a short lifetime due to their rapid recombination with electrons, a disadvantage in photocatalysis. However, in EC, an applied bias potential helps push photogenerated electrons away from the anode surface, extending the lifetime of generated holes [61]. Because current creates more holes, more hydroxyl radicals can be generated inside the solution, causing organic molecules to mineralize faster. Other reactive oxygen species are produced inside the reaction system, and the holes and hydroxyl radicals attack organic compounds in EC, which contributes to the mineralization of organic molecules. Superoxide radicals and hydroperoxyl radicals are weaker oxidizing entities [62].

According to previous work, creating a gradient at the ZNCM surface efficiently isolates generated charge carriers, according to previous work [63]. Electrons from the valence band (VB) are stimulated to the conduction band after an electric path more significant than the Eg on the ZNCM, leaving holes in the VB. During the degradation process, electrons pass through the external circuit to counter electro, assisting in forming highly reactive superoxide anion radicals (O_2). The valance band holes oxidize water, producing OH radicals interacting with organic contaminants to create a less harmful response. In the case of ZNCM photocatalyst, the primary degradation species are O_2 and generated holes (depending on the bonding of the catalyst with the pollutant). The combined impact of electrocatalytic oxidation processes influences ZNCM degradation performance.

As previously mentioned, the following were the probable processes of RhB [64] photodegradation utilizing ZNCM in the presence of EC as mentioned in Scheme 1 and the following equations:

$$O_2 + e^- \rightarrow {}^-O_2 \quad (13)$$

$$^-O_2 + 2H^+ + 2e^- \rightarrow H_2O_2 \quad (14)$$

$$O_2 + e^- \rightarrow {}^-O_2 \quad (15)$$

$$^-O_2 + 2H^+ + 2e^- \rightarrow H_2O_2 \quad (16)$$

$$H_2O_2 + e^- \rightarrow OH + {}^-OH \quad (17)$$

$$(OH/{}^-O_2) + RhB \rightarrow products \quad (18)$$

$$(h^+/H_2O_2) + RhB \rightarrow CO_2 + H_2O \quad (19)$$

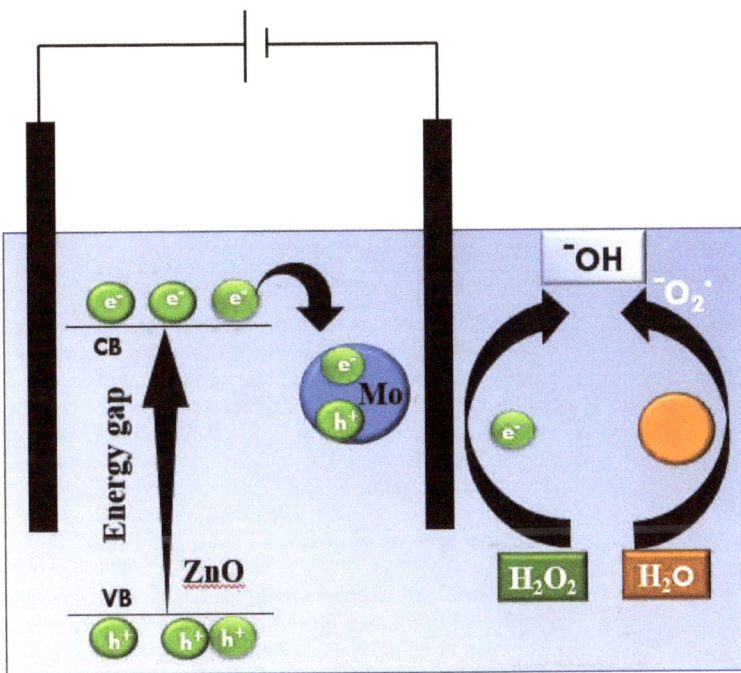

Scheme 1. EC degradation mechanism/cell.

3.7.2. Comparison of PEC of RhB in the Presence of ZNCM with Previous Work

Table 2 compares electrocatalytic degradation of RhB in the presence of ZNCM to other prior study samples, highlighting their contributions to the synthesis technique, electrocatalytic conditions, and RhB concentration under consideration. It was discovered that ZNCM-3 has the highest electron degradation of RhB, with a 0.4 min^{-1} rate of degradation and 100% degradation in 7 min. Nabil et al. [65] found that CoFe$_2$O$_4$ thin films electrochemically degraded an aqueous solution of rhodamine B (RhB) at a rate of 99 percent within the first three minutes of reaction time. The trapping studies revealed that the dominant active species were hydroxyl radicals, which resulted in the rapid elimination of RhB at an initial concentration of 10 mg/L. Ali et al. [66] investigated the electrochemical oxidation of Rhodamine B dye (RhB) on DSA and SnO$_2$ electrodes as active and non-active electrode models. They discovered that by using the DSA electrode in a NaCl 0.05 mol L1 + Na$_2$SO$_4$ 0.1 mol L1 solution as a supporting electrolyte, they could achieve 100% color removal after 90 min of electrolysis. In various operating conditions, DSA outperformed SnO2 and was shown to be more cost-effective and efficient. Zhao et al. [67] concluded that using $ZnWO_4$ films as an anode resulted in a significant synergetic effect in rhodamine B (RhB) degradation via simultaneous electro-oxidation and photocatalysis. They discovered that RhB degradation follows the pseudo-first kinetic equation at bias potentials of 1.0, 1.5, and 2.0 V. The respective rate constants are 0.022, 0.026, and 0.138 hr^{-1}. Furthermore, as the bias increases, the rate of RhB degradation slows. There is almost no deterioration of RhB after the first hour at 3.5 V.

Table 2. The calculated data of rate constants of electrocatalytic degradation of Rhodamine B in the presence of **ZNCM** prepared samples with the previous work.

Samples	Synthesis Method	Dye Concentration	K, (min^{-1})	% Removal	Refs.
ZNCM-1 (0 wt% Mo)	Combustion method	Rhodamine B (50 mgL^{-1})	0.2671	97% after 15 min	Present work
ZNCM-2 (0.001 wt% Mo)	Combustion method	Rhodamine B (50 mgL^{-1})	0.29064	98% after 15 min	Present work
ZNCM-3 (0.01 wt% Mo)	Combustion method	Rhodamine B (50 mgL^{-1})	0.40605	100 % after 7 min	Present work
ZNCM-4 (0.1 wt% Mo)	Combustion method	Rhodamine B (50 mgL^{-1})	0.31455	98 % after 15 min	Present work
ZNCM-5 (0.5 wt% Mo)	Combustion method	Rhodamine B (50 mgL^{-1})	0.32801	97% after 15 min	Present work
ZNCM-6 (1.0 wt% Mo)	Combustion method	Rhodamine B (50 mgL^{-1})	0.21218	96% after 15 min	Present work
$CoFe_2O_4$	Electrodeposition	Rhodamine B (10 mgL^{-1})		99% after 3 min	[65]
Ti/RuO_2-IrO_2	DSA (De Nora Company)	Rhodamine B		100% after 90 min	[66]
$ZnWO_4$	Dip-coating	Rhodamine B (5 mgL^{-1})	0.0023	after 60 min	[67]
$Ca_{10}(PO_4)_6(OH)_2$	Electrodeposition	Rhodamine B (5 mgL^{-1})	0.0399	98 % after 105 min	[68]
$Zn_3(PO_4)_2 \cdot 4H_2O$	Electrodeposition	Rhodamine B (30 mgL^{-1})		90 % after 10 min	[69]
TiO_2-NTs/Ce-PbO_2	Electrodeposition	Rhodamine B (30 mgL^{-1})		90 % after 10 min	[70]
$BiVO_4/TiO_2$	Sol-gel and hydrothermal	Rhodamine B (10 mgL^{-1})		93.9 % after 5 h	[71]

Furthermore, Ahmed et al. [68] electrodeposited hydroxyapatite $Ca_{10}(PO_4)_6(OH)_2$ on stainless-steel substrates in aqueous solutions of calcium nitrate tetrahydrate and dihydrogen phosphate using the chronopotentiometry mode. They discovered that at time t = 105 min, the ratio Ct/Co decreases by 17% for a current density of 5 mA/cm^2 and by 98% for a current density of 30 mA/cm^2. A pseudo-first-order kinetics rule guided the degrading mechanism. Chennai et al. [69] discovered that $Zn_3(PO_4)2H_2O$ films (hydrated zinc phosphate, abbreviated h-ZP) are electrodeposited on different substrates and used as active anodes to electrode grade (RhB). It should be noted that after 12 min, the h-ZP/FTO electrodes have 90% RhB degradation. Qin Li et al. [70] developed cerium-doped lead dioxide (TiO2-NTs/Ce-PbO2) that degrades RhB 90 percent in 10 min. Wang et al. demonstrated improved electrocatalytic rhodamine B dye degradation [71]. The heterojunction was created using the sol-gel and hydrothermal methods. After a 5 h reaction time with an applied potential of 4.0 V, the degradation efficiency for eliminating rhodamine B dye was 93.9%.

4. Conclusions

In conclusion, pure and Mo-doped ZnO nanoparticles were prepared via the combustion route. With increasing Mo-dopant concentration, D-value of ZnO decreased. The microstructure of these samples exhibits a highly agglomerated and soapy porous structure. The ZnO nanoparticles exhibit high reflectance values in the visible region. The reflectance values of these samples decrease with the increase in dopant concentration. ZNCM-1, ZNCM-2, ZNCM-3, ZNCM-4, ZNCM-5, and ZNCM-6 have estimated band gap values of 3.261, 3.259, 3.252, 3.256, 3.246, and 3.235 eV, respectively. Eg values decrease as Mo-doping concentration increases. At lower frequencies, the dielectric constants of ZNCM-1, ZNCM-2, ZNCM-3, ZNCM-4, ZNCM-5, and ZNCM-6 are in the 40–48 range. In the case of ZNCM photocatalyst, the main degradation species are O_2 and generated holes (depending on the bonding of the catalyst with the pollutant). The combined impact of electrocatalytic oxidation processes influences ZNCM degradation performance. The dielectric constant is frequency-dependent throughout the frequency range. ZNCM-3 was a promising nanoparticle in EC of rhodamine b with 0.4 min^{-1} exhibited 100% degradation in 7 min.

Author Contributions: Conceptualization, V.G., M.S.A.H. and I.S.Y.; methodology, U.P.S., R.A. and M.I.M.; formal analysis, T.H.A., H.Y.Z. and M.S.A. investigation, M.S.A. and I.S.Y.; writing—original draft preparation, V.G., M.S.A.H. and M.S.A.; writing—review and editing, I.S.Y. and H.Y.Z.; visualization, T.H.A., R.A. and M.I.M.; funding acquisition, I.S.Y. and V.G. All authors have read and agreed to the published version of the manuscript.

Funding: The authors extend their appreciation to the Deputyship for Research & Innovation, Ministry of Education in Saudi Arabia, for funding this research work, project number IFP-KKU-2020/6.

Institutional Review Board Statement: Not applicable.

Informed Consent Statement: Not applicable.

Data Availability Statement: Data are contained within the manuscript.

Conflicts of Interest: The authors declare no conflict of interest.

References

1. Ravichandran, K.; Dhanraj, C.; Ibrahim, M.M.; Kavitha, P. Enhancing the electrocatalytic efficiency of ZnO thin films by adding Mo and rGO. *Mater. Today Proc.* **2022**, *48*, 234–244. [CrossRef]
2. Saravanan, R.; Mansoob Khan, M.; Gupta, V.K.; Mosquera, E.; Gracia, F.; Narayanan, V.; Stephen, A. ZnO/Ag/CdO nanocomposite for visible light-induced electrocatalytic degradation of industrial textile effluents. *J. Colloid Interface Sci.* **2015**, *452*, 126–133. [CrossRef] [PubMed]
3. Thejaswini, T.V.L.; Prabhakaran, D.; Maheswari, M.A. Soft synthesis of potassium co-doped Al–ZnO nanocomposites: A comprehensive study on their visible light driven electrocatalytic activity on dye degradation. *J. Mater. Sci.* **2016**, *51*, 8187–8208. [CrossRef]
4. Bizarro, M. High electrocatalytic activity of ZnO and ZnO: Al nanostructured films deposited by spray pyrolysis. *Appl. Catal. B Environ.* **2010**, *97*, 198–203. [CrossRef]
5. Bizarro, M.; Martínez-Padilla, E. Visible light responsive electrocatalytic ZnO: Al films decorated with Ag nanoparticles. *Thin Solid Films.* **2014**, *553*, 179–183. [CrossRef]
6. Luo, Y.; Wu, Y.; Huang, C.; Menon, C.; Feng, S.-P.; Chu1, P.K. Plasma modified and tailored defective electrocatalysts for water electrolysis and hydrogen fuel cells. *EcoMat* **2022**, *4*, e12197. [CrossRef]
7. Shaik, U.P.; Kumar, P.A.; Krishna, M.G.; Rao, S.V. Morphological manipulation of the nonlinear optical response of ZnO thin films grown by thermal evaporation. *Mater. Res. Express* **2014**, *1*, 046201. [CrossRef]
8. Yan, R.; Gargas, D.; Yang, P. Nanowire photonics. *Nat. Photon.* **2009**, *3*, 569. [CrossRef]
9. Özgür, Ü.; Alivov, Y.I.; Liu, C.; Teke, A.; Reshchikov, M.A.; Doğan, S.; Avrutin, V.; Cho, S.J.; Morkoç, H.J. A comprehensive review of ZnO materials and devices. *Appl. Phys.* **2005**, *98*, 041301. [CrossRef]
10. Panda, D.; Tseng, T.-Y. One-dimensional ZnO nanostructures: Fabrication, optoelectronic properties, and device applications. *J. Mater. Sci.* **2013**, *48*, 6849. [CrossRef]
11. Dhara, S.; Giri, P.K. ZnO nanowire heterostructures: Intriguing photophysics and emerging applications. *Rev. Nanosci. Nanotech.* **2013**, *2*, 1. [CrossRef]
12. Shaik, U.P.; Krishna, G.M. Single-step formation of indium and tin-doped ZnO nanowires by thermal oxidation of indium–zinc and tin–zinc metal films: Growth and optical properties. *Ceram. Int.* **2014**, *40*, 13611–13620. [CrossRef]
13. Jose, A.; Devi, K.R.S.; Pinheiro, S.; Narayana, S.L. Electrochemical synthesis, photodegradation and antibacterial properties of PEG capped zinc oxide nanoparticles. *J. Photochem. Photobiol. B Biol.* **2018**, *187*, 25–34. [CrossRef]
14. Fan, L.; Song, H.; Li, T.; Yu, L.; Liu, Z.; Pan, G.; Lei, Y.; Bai, X.; Wang, T.; Zheng, Z.; et al. Hydrothermal synthesis and photoluminescent properties of ZnO nanorods. *J. Lumin.* **2007**, *122*, 819–821. [CrossRef]
15. Yiamsawas, D.; Boonpavanitchakul, K.; Kangwansupamonkon, W. Preparation of ZnO Nanostructures by Solvothermal Method. *J. Microsc. Soc.* **2009**, *23*, 75–78.
16. Singhal, M.; Cbbabra, V.; Kang, P.; Shah, D.O. Pergamon PI1 SOO2s-5408(96)00175-4 synthesis of ZnO nanoparticles for varistor application using zn-substituted aerosol ot microemulsion. *Mater. Res. Bull.* **1997**, *32*, 239–247. [CrossRef]
17. Rataboul, F.; Nayral, C.; Casanove, M.-J.; Maisonnat, A.; Chaudret, B. Synthesis and characterization of monodisperse zinc and zinc oxide nanoparticles from the organometallic. *J. Organomet. Chem.* **2002**, *643*, 307–312. [CrossRef]
18. Moghaddam, A.B.; Nazari, T.; Badraghi, J.; Kazemzad, M. Synthesis of ZnO nanoparticles and electrodeposition of polypyrrole/ZnO nanocomposite film. *Int. J. Electrochem. Sci.* **2009**, *4*, 247–257.
19. Hu, X.L.; Zhu, Y.J.; Wang, S.W. Sonochemical and microwave-assisted synthesis of linked single-crystalline ZnO rods. *Mater. Chem. Phys.* **2004**, *88*, 421–426. [CrossRef]

20. Wu, J.J.; Wu, J.J.; Liu, S.C.; Liu, S.C. Low-temperature growth of well-aligned ZnO nanorods by chemical vapor deposition. *Synth. Met.* **1999**, *102*, 1091. [CrossRef]
21. Okuyama, K.; Lenggoro, W.W. Preparation of nanoparticles via spray route. *Chem. Eng. Sci.* **2003**, *58*, 537–547. [CrossRef]
22. Talam, S.; Karumuri, S.R.; Gunnam, N. Synthesis, Characterization, and Spectroscopic Properties of ZnO Nanoparticles. *Int. Sch. Res. Not.* **2012**, *2012*, 1–6. [CrossRef]
23. Ananda, S.S. Synthesis and Characterization of Mo-Doped ZnO Nanoparticles by Electrochemical Method: Photodegradation Kinetics of Methyl Violet Dye and Study of Antibacterial Activities of Mo-Doped ZnO Nanoparticles. *Int. J. Nanomater. Biostruct.* **2015**, *5*, 7–14.
24. Tachikawa, S.; Noguchi, A.; Tsuge, T.; Hara, M.; Odawara, O.; Wada, H. Optical properties of ZnO nanoparticles capped with polymers. *Materials* **2011**, *4*, 1132–1143. [CrossRef]
25. Vargas, M.A.; Rivera-Muñoz, E.M.; Diosa, J.E.; Mosquera, E.E.; Rodríguez-Páez, J.E. Nanoparticles of ZnO and Mg-doped ZnO: Synthesis, characterization and efficient removal of methyl orange (MO) from aqueous solution. *Ceram. Int.* **2021**, *47*, 15668–15681. [CrossRef]
26. Ren, K.; Shu, H.; Huo, W.; Cui, Z.; Xu, Y. Tuning electronic, magnetic and catalytic behaviors of biphenylene network by atomic doping. *Nanotechnology* **2022**, *33*, 345701. [CrossRef]
27. Mohammadi, R.; Alamgholiloo, H.; Gholipour, B.; Rostamnia, S.; Khaksar, S.; Farajzadeh, M.; Shokouhimehr, M. Visible-light-driven photocatalytic activity of ZnO/g-C3N4 heterojunction for the green synthesis of biologically interest small molecules of thiazolidinones. *J. Photochem. Photobiol. A Chem.* **2020**, *402*, 112786. [CrossRef]
28. Doustkhah, E.; Esmat, M.; Fukata, N.; Ide, Y.; Hanaor, D.A.; Assadi, M.H. MOF-derived nanocrystalline ZnO with controlled orientation and photocatalytic activity. *Chemosphere* **2022**, *303*, 134932. [CrossRef]
29. Sushma, C.; Girish Kumar, S. Advancements in the zinc oxide nanomaterials for efficient photocatalysis. *Chem. Pap.* **2017**, *71*, 2023–2042. [CrossRef]
30. Jyothi, N.S.; Ravichandran, K. Optimum pH for effective dye degradation: Mo, Mn, Co and Cu doped ZnO photocatalysts in thin film form. *Ceram. Int.* **2020**, *46*, 23289–23292. [CrossRef]
31. Xiu, X.; Pang, Z.; Lv, M.; Dai, Y.; Ye, L.; Han, S. Transparent conducting molybdenum-doped zinc oxide films deposited by RF magnetron sputtering. *Appl. Surf. Sci.* **2007**, *253*, 3345–3348. [CrossRef]
32. Zhang, D.H.; Yang, T.L.; Ma, J.; Wang, Q.P.; Gao, R.W.; Ma, H.L. Preparation of transparent conducting ZnO: Al films on polymer substrates by rf magnetron sputtering. *Appl. Surf. Sci.* **2000**, *158*, 43. [CrossRef]
33. Meng, Y.; Yang, X.L.; Chen, H.X.; Shen, J.; Jiang, Y.M.; Zhang, Z.J.; Hua, Z.Y. A new transparent conductive thin film In2O3: Mo. *Thin Solid Film.* **2001**, *394*, 219. [CrossRef]
34. Cullity, B.D. *Elements of X-ray Diffraction*; Addison Wesley: Reading, MA, USA, 1956.
35. Umar, K.; Aris, A.; Parveen, T.; Jaafar, J.; Majid, Z.A.; Reddy, A.V.; Talib, J. Synthesis, characterization of Mo and Mn doped ZnO and their photocatalytic activity for the decolorization of two different chromophoric dyes. *Appl. Catal. A Gen.* **2015**, *505*, 507–514. [CrossRef]
36. Prashanth, G.K.; Prashanth, P.A.; Bora, U.; Gadewar, M.; Nagabhushana, B.M.; Ananda, S.; Krishnaiah, G.M.; Sathyananda, H.M. In vitro antibacterial and cytotoxicity studies of ZnO nanopowders prepared by combustion assisted facile green synthesis. *Karbala Int. J. Mod. Sci.* **2015**, *1*, 67–77.
37. Nehru, L.C.; Swaminathan, V.; Sanjeeviraja, C. Rapid synthesis of nanocrystalline ZnO by a microwave-assisted combustion method. *Powder Technol.* **2012**, *226*, 29–33. [CrossRef]
38. Naik, E.I.; Naik, H.S.B.; Swamy, B.E.K.; Viswanath, R.; Suresh Gowda, I.K.; Prabhakara, M.C.; Chetankumar, K. Influence of Cu doping on ZnO nanoparticles for improved structural, optical, electrochemical properties and their applications in efficient detection of latent fingerprints. *Chem. Data Collect.* **2021**, *33*, 100671. [CrossRef]
39. Swapna, R.; Santhosh Kumar, M.C. Growth and characterization of molybdenum doped ZnO thin films by spray pyrolysis. *J. Phys. Chem. Solids* **2013**, *74*, 418–425. [CrossRef]
40. Kaleemulla, S.; Rao, N.M.; Joshi, M.G.; Reddy, A.S.; Uthanna, S.; Reddy, P.S. Electrical and optical properties of In2O3: Mo thin films prepared at various Mo-doping levels. *J. Alloys. Compd.* **2010**, *504*, 351. [CrossRef]
41. Pascariu, P.; Homocianu, M.; Olaru, N.; Airinei, A.; Ionescu, O. New Electrospun ZnO:MoO3 Nanostructures: Preparation, Characterization and Electrocatalytic Performance. *Nanomaterials* **2020**, *10*, 1476. [CrossRef]
42. Rao, T.P.; Santhoshkumar, M.C.; Safarulla, A.; Ganesan, V.; Barman, S.R.; Sanjeeviraja, C. Physical properties of ZnO thin films deposited at various substrate temperatures using spray pyrolysis. *Phys. B Condensed Matter* **2010**, *405*, 2226.
43. Aydın, C.; Abd El-Sadek, M.S.; Zheng, K.; Yahia, I.S.; Yakuphanoglu, F. Synthesis, diffused reflectance and electrical properties of nanocrystalline Fe-doped ZnO via sol-gel calcination technique. *Opt. Laser Technol.* **2013**, *48*, 447–452. [CrossRef]
44. Gholami, P.; Dinpazhoh, L.; Khataee, A.; Orooji, Y. Sonocatalytic activity of biochar-supported ZnO nanorods in degradation of gemifloxacin: Synergy study, effect of parameters and phytotoxicity evaluation. *Ultrason. Sonochem.* **2019**, *55*, 44–56. [CrossRef] [PubMed]
45. Rajalakshmi, M.; Arora, A.K.; Bendre, B.S.; Mahamuni, S.J. Optical Phonon Confinement in Zinc oxide nanoparticles. *Appl. Phys.* **2000**, *87*, 2445. [CrossRef]

46. Pavithra, N.S.; Lingaraju, K.; Raghu, G.K.; Nagaraju, G. Citrus maxima (Pomelo) juice mediated eco-friendly synthesis of ZnO nanoparticles: Applications to electrocatalytic, electrochemical sensor and antibacterial activities. *Spectrochim. Acta Part A Mol. Biomol. Spectrosc.* **2017**, *185*, 11–19. [CrossRef]
47. Sánchez Zeferino, R.; Barboza Flores, M.; Pal, U. Photoluminescence and Raman Scattering in Ag-doped ZnO Nanoparticles. *J. Appl. Phys.* **2011**, *109*, 014308. [CrossRef]
48. Sharma, A.; Singh, B.P.; Dhar, S.; Gondorf, A.; Spasova, M. Effect of surface groups on the luminescence property of ZnO nanoparticles synthesized by sol-gel route. *Surf. Sci.* **2012**, *606*, L13–L17. [CrossRef]
49. Muravitskaya, A.; Rumyantseva, A.; Kostcheev, S.; Dzhagan, V.; Stroyuk, O.; Adam, P.-M. Enhanced Raman scattering of ZnO nanocrystals in the vicinity of gold and silver nanostructured surfaces. *Opt. Express* **2016**, *24*, A168–A173. [CrossRef]
50. Yu, C.; Yang, K.; Shu, Q.; Yu, J.C.; Cao, F.; Li, X.; Zhou, X. Preparation, characterization, and electrocatalytic performance of Mo-doped ZnO photocatalysts. *Sci. China Chem.* **2012**, *55*, 9. [CrossRef]
51. Chen, Y.P.; Lu, C.; Xu, L.; Ma, Y.; Hou, W.H.; Zhu, J.J. Single-crystallineorthorhombic molybdenum oxide nanobelts: Synthesis and electrocatalytic properties. *CrystEngComm* **2010**, *12*, 3740–3747. [CrossRef]
52. Xiong, G.; Pal, U.; Serrano, J.G.; User, K.B.; Williams, K.B. Photoluminescence and FTIR study of ZnO nanoparticles: The impurity and defect perspective. *Phys. Status Solidi* **2006**, *3*, 3577. [CrossRef]
53. Hufschmidt, D.; Liu, L.; Seizer, V.; Bahnemann, D. Electrocatalytic water treatment: Fundamental knowledge required for its practical application. *Water Sci. Technol.* **2004**, *49*, 135–140. [CrossRef] [PubMed]
54. Ravichandran, K.; Anbazhagan, A.; Dineshbabu, N.; Ravidhas, C. Influence of Mo doping on transparent conducting properties of ZnO films prepared by a simplified spray technique. *J. Mater. Sci. Mater. Electron.* **2015**, *26*, 7649–7654. [CrossRef]
55. Hussien, M.S.A. Facile Synthesis of Nanostructured Mn-doped Ag3PO4 for Visible Photodegradation of Emerging Pharmaceutical Contaminants: Streptomycin Photodegradation. *J. Inorg. Organomet. Polym. Mater.* **2021**, *31*, 945–959. [CrossRef]
56. AnjuChanu, L.; Joychandra Singh, W.; Jugeshwar Singh, K.; Nomita Devi, K. Effect of operational parameters on the electrocatalytic degradation of Methylene blue dye solution using Manganese doped ZnO nanoparticles. *Results Phys.* **2019**, *12*, 1230–1237. [CrossRef]
57. Shkir, M.; Yahia, I.S.; Al-Qahtani, A.M.A. Bulk monocrystal growth, optical, dielectric, third-order nonlinear, thermal, and mechanical studies on HCl added l-alanine: An organic NLO material. *Mater. Chem. Phys.* **2016**, *184*, 12–22. [CrossRef]
58. Jeseentharani, V.; Dayalan, A.; Nagaraja, K.S. Co-precipitation synthesis, humidity sensing and photoluminescence properties of nanocrystalline Co^{2+} substituted zinc (II) molybdate ($Zn_{1-x} Co_x MoO_4$; x = 0, 0.3, 0.5, 0.7, 1). *Solid-State Sci.* **2017**, *67*, 46–58. [CrossRef]
59. Srinivas, K.; Venugopal Reddy, P. Synthesis, Structural, and Magnetic Properties of Nanocrystalline $Ti_{0.95}Co_{0.05}O_2$-Diluted Magnetic Semiconductors. *J. Supercond. Nov. Magn.* **2014**, *27*, 2521–2538. [CrossRef]
60. Shejwal, N.N.; Anis, M.; Hussaini, S.S.; Shirsat, M.D. Investigation on structural, UV- visible, SHG efficiency, dielectric, mechanical and thermal behavior of L-cystine doped zinc thiourea sulphate crystal for NLO device applications. *Int. J. Mod. Phys. B* **2016**, *30*, 1650159. [CrossRef]
61. Štengl, V.; Bakardjieva, S. Molybdenum-Doped Anatase and Its Extraordinary Electrocatalytic Activity in the Degradation of Orange II in the UV and vis Regions. *J. Phys. Chem. C* **2010**, *1145*, 19308–19317. [CrossRef]
62. Al Farsi, B.; Souier, T.M.; Al Marzouqi, F.; Al Maashani, M.; Bououdina, M.; Widatallah, H.M.; Al-Abri, M. Structural and optical properties of visible active electrocatalytic Al-doped ZnO nanostructured thin films prepared by dip coating. *Opt. Mater.* **2021**, *113*, 110868. [CrossRef]
63. Sagadevan, S.; Vennila, S.; Lett, J.A.; Marlinda, A.R.; Hamizi, N.A.B.; Johan, M.R. Tailoring the structural, morphological, optical, thermal, and dielectric characteristics of ZnO nanoparticles using starch as a capping agent. *Results Phys.* **2019**, *15*, 102543. [CrossRef]
64. Putri, N.A.; Fauzia, V.; Iwan, S.; Roza, L.; Ali, A.; Umar, A.A.; Budi, S. Mn-doping-induced electrocatalytic activity enhancement of ZnO nanorods prepared on glass substrates. *Appl. Surf. Sci.* **2018**, *439*, 285–297. [CrossRef]
65. Labchir, N.; Amaterz, E.; Hannover, A.; Ait hssi, A.; Vincent, D.; Ihlal, A.; Sajieddine, M. Highly efficient nanostructured $CoFe_2O_4$ thin-film electrodes for electrochemical degradation of rhodamine B. *Water Environ. Res.* **2020**, *92*, 759–765. [CrossRef] [PubMed]
66. Baddour, A.; Bessegato, G.G.; Rguiti, M.M.; El Ibrahimi, B.; Bazzi, L.; Hilali, M.; Zanoni, M.V.B. Electrochemical decolorization of Rhodamine B dye: Influence of anode material, chloride concentration and current density. *J. Environ. Chem. Eng.* **2018**, *6*, 2041–2047. [CrossRef]
67. Xu, Z.; Zhu, Y. Synergetic degradation of rhodamine B at a porous ZnWO4 film electrode by combined electro-oxidation and photocatalysis. *Environ. Sci. Technol.* **2006**, *40*, 3367–3372.
68. Nareejun, W.; Poncho, W. Novel electrocatalytic/solar cell improvement for organic dye degradation based on simple dip-coating WO3/BiVO4 photoanode electrode. *Sol. Energy Mater. Sol. Cells* **2020**, *212*, 110556. [CrossRef]
69. Pedanekar, R.S.; Madake, S.B.; Narewadikar, N.A.; Mohite, S.V.; Patil, A.R.; Kumbhar, S.M.; Rajpure, K.Y. Electrocatalytic degradation of Rhodamine B by spray deposited Bi2WO6 photoelectrode under solar radiation. *Mater. Res. Bull.* **2022**, *147*, 111639. [CrossRef]

70. Chennai, A.; Naciri, Y.; Taoufyq, A.; Bakiz, B.; Bazzi, L.; Guinneton, F.; Villain, S.; Gavarri, J.R.; Benlhachemi, A. Electrodeposited zinc phosphate hydrate electrodes for electrocatalytic applications. *J. Appl. Electrochem.* **2019**, *42*, 163–177. [CrossRef]
71. Wang, Y.; Lu, N.; Luo, M.; Fan, L.; Zhao, K.; Qu, J.; Guan, J.; Yuan, X. Enhancement mechanism of fiddlehead-shaped TiO_2-$BiVO_4$ type II heterojunction in SPEC towards RhB degradation and detoxification. *Appl. Surf. Sci.* **2019**, *463*, 234–243. [CrossRef]

MDPI AG
Grosspeteranlage 5
4052 Basel
Switzerland
Tel.: +41 61 683 77 34

Crystals Editorial Office
E-mail: crystals@mdpi.com
www.mdpi.com/journal/crystals

Disclaimer/Publisher's Note: The statements, opinions and data contained in all publications are solely those of the individual author(s) and contributor(s) and not of MDPI and/or the editor(s). MDPI and/or the editor(s) disclaim responsibility for any injury to people or property resulting from any ideas, methods, instructions or products referred to in the content.